Topics in Fluorescence Spectroscopy

Volume 9
Advanced Concepts in
Fluorescence Sensing
Part A: Small Molecule Sensing

Topics in Fluorescence Spectroscopy

Edited by JOSEPH R. LAKOWICZ and CHRIS D. GEDDES

Volume 1: Techniques
Volume 2: Principles
Volume 3: Biochemical Applications
Volume 4: Probe Design and Chemical Sensing
Volume 5: Nonlinear and Two-Photon-Induced Fluorescence
Volume 6: Protein Fluorescence
Volume 7: DNA Technology
Volume 8: Radiative Decay Engineering
Volume 9: Advanced Concepts in Fluorescence Sensing
    Part A: Small Molecule Sensing
Volume 10: Advanced Concepts in Fluorescence Sensing
    Part B: Macromolecular Sensing

Topics in Fluorescence Spectroscopy

Volume 9
Advanced Concepts in
Fluorescence Sensing
Part A: Small Molecule Sensing

Edited by

CHRIS D. GEDDES
The Institute of Fluorescence
Medical Biotechnology Center
University of Maryland Biotechnology Institute
Baltimore, Maryland

and

JOSEPH R. LAKOWICZ
Center for Fluorescence Spectroscopy and
Department of Biochemistry and Molecular Biology
University of Maryland School of Medicine
Baltimore, Maryland

The Library of Congress cataloged the first volume of this title as follows:

Topics in fluorescence spectroscopy/edited by Chris D. Geddes and Joseph R. Lakowicz.
p. cm.
Includes bibliographical references and index.
Contents: v. 1. Techniques
1. Fluorescence spectroscopy. I. Geddes, Chris D. II. Lakowicz, Joseph R.

QD96.F56T66 1991
543'.0858—dc20

91-32671
CIP

ISSN: 1574-1036
ISBN 0-387-23334-2   Printed on acid-free paper

©2005 Springer Science+Business Media, Inc.
All rights reserved. This work may not be translated or copied in whole or in part without the written permission of the publisher (Springer Science+Business Media, Inc., 233 Spring Street, New York, NY 10013, USA), except for brief excerpts in connection with reviews or scholarly analysis. Use in connection with any form of information storage and retrieval, electronic adaptation, computer software, or by similar or dissimilar methodology now known or hereafter developed is forbidden.
The use in this publication of trade names, trademarks, service marks and similar terms, even if they are not identified as such, is not to be taken as an expression of opinion as to whether or not they are subject to proprietary rights.

Printed in the United States of America

9 8 7 6 5 4 3 2 1   SPIN 11319535

springeronline.com

CONTRIBUTORS

Joanne M. Bedlek-Anslow, DuPont Nylon, South Carolina.

Paul D. Beer. Inorganic Chemistry Laboratory, Department of Chemistry, University of Oxford, South Parks Road, Oxford, OX1 3QR, UK.

Bruce F. Carroll. Department of Mechanical and Aerospace Engineering, University of Florida, Gainesville, FL, 32611.

David Curiel. Inorganic Chemistry Laboratory, Department of Chemistry, University of Oxford, South Parks Road, Oxford, OX1 3QR, UK.

Luisa Stella Dolci. Università degli studi di Bologna, Dipartimento di Chimica "G. Ciamician", Via Selmi 2, I 40126, Bologna, Italy.

Ute Resch-Genger. Federal Institute for Materials Research and Testing, Richard-Willstätter-Str. 11, D- 12489 Berlin, Germany.

Elizabeth J. Hayes. Inorganic Chemistry Laboratory, Department of Chemistry, University of Oxford, South Parks Road, Oxford, OX1 3QR, UK.

Gunther Hennrich. Dpto. de Química Orgánica, L 101, Facultad de Ciencias, Universidad Autónoma de Madrid, Cantoblanco 28049-Madrid, Spain.

Stephanie Hodgen. Department of Pure and Applied Chemistry, University of Strathclyde, Glasgow, G1 1XL, UK

Scott A. Hilderbrand. Department of Chemistry, Massachusetts Institute of Technology, 77 Massachusetts Avenue, Cambridge 02139.

John P. Hubner. Department of Mechanical and Aerospace Engineering, University of Florida, Gainesville, FL, 32611

Muhammet E. Kose. Department of Chemistry, University of Florida, Gainesville, FL, 32611.

Kanji Kubo. Kyushu University, Fukuoka, Japan 816-8580.

Mi Hee Lim. Department of Chemistry, Massachusetts Institute of Technology, 77 Massachusetts Avenue, Cambridge, 02139.

Stephen J. Lippard. Department of Chemistry, Massachusetts Institute of Technology, 77 Massachusetts Avenue, Cambridge, MA 02139.

Andrew Mills. Department of Pure and Applied Chemistry, University of Strathclyde, Glasgow, G1 1XL, UK

Marco Montalti. Università degli studi di Bologna, Dipartimento di Chimica "G. Ciamician", Via Selmi 2, I 40126, Bologna, Italy.

Luca Prodi. Università degli studi di Bologna, Dipartimento di Chimica "G. Ciamician", Via Selmi 2, I 40126, Bologna, Italy.

Kirk S. Schanze. Department of Chemistry, University of Florida, Gainesville, FL, 32611.

Jan W. Verhoeven. Laboratory of Organic Chemistry, University of Amsterdam, Nieuwe Achtergracht 129, 1018WS Amsterdam, The Netherlands.

Nelsi Zaccheroni. Università degli studi di Bologna, Dipartimento di Chimica "G. Ciamician", Via Selmi 2, I 40126, Bologna, Italy.

PREFACE

Over the last decade fluorescence has become the dominant tool in biotechnology and medical imaging. These exciting advances have been underpinned by the advances in time-resolved techniques and instrumentation, probe design, chemical / biochemical sensing, coupled with our furthered knowledge in biology.

Ten years ago Volume 4 of the Topics in Fluorescence Spectroscopy series outlined the emerging trends in time resolved fluorescence in analytical and clinical chemistry. These emerging applications of fluorescence were the result of continued advances in both laser and computer technology and a drive to develop red/near-infrared fluorophores. Based on the advancements in these technologies, it was envisaged that small portable devices would find future common place in a doctor's office or for home health care.

Today, these past emerging trends in fluorescence sensing are now widely used as either standard practices in clinical assessment or commercialized health care products. Miniature lasers in the form of laser diodes and even light emitting diodes are widely used in applications of time-resolved fluorescence. Computer clock-speed is now not considered a hurdle in data analysis. Even our choice of fluorophores has changed dramatically in the last decade, the traditional fluorophore finding continued competition by fluorescent proteins and semi-conductor quantum dots, to name but just a few.

This volume "Advanced Concepts in Fluorescence Sensing: Small Molecule Sensing" aims to summarize the current state of the art in fluorescence sensing. For this reason we have invited chapters, encompassing a board range of fluorescence sensing techniques. Chapters in this volume deal with small molecule sensors, such as for anions, cations and CO_2.

While many of the changes in recent fluorescence have been well received, its continued growth in the world has created a challenge in trying to archive and document its use. Subsequently Chris D. Geddes has now become co-series editor of the Topics in Fluorescence Spectroscopy series. We have also recently launched the Reviews in Fluorescence series, which co-edited also by Dr's Geddes and Lakowicz and published annually, is meant to directly compliment the Topics in Fluorescence Spectroscopy series, with small chapters summarizing the yearly progress in fluorescence.

Finally we would like to thank all the authors for their excellent contributions, Mary Rosenfeld for administrative support and Kadir Aslan for help in typesetting the volume.

Chris D. Geddes
Joseph R. Lakowicz
Baltimore, Maryland, US.
August 2004

CONTENTS

1. PROBES AND SENSORS FOR CATIONS 1

Luca Prodi, Marco Montalti, Nelsi Zaccheroni, and Luisa Stella Dolci

    1.1. INTRODUCTION 1
        1.1.1. The Need of Chemical Sensors 1
        1.1.2. The Power of Luminescence Spectroscopy 2
        1.1.3. Sensory Devices and Chemosensors: the Role of Chemists 2
        1.1.4. Chemosensors and Probes for Metal Ions 5
    1.2. CHEMOSENSORS WITH ACYCLIC RECEPTOR 6
    1.3. CHEMOSENSORS WITH CYCLIC RECEPTOR 21
    1.4. SENSORS WITH DENDRIMER-, PEPTIDE- AND PROTEIN-BASED RECEPTOR 42
    1.5. SENSORS BASED ON SILICA NANOPARTICLES 47
    1.6. CONCLUSIONS 51
        1.6.1 Acknowledgments 51
    1.7. REFERENCES 52

2. FLUORESCENT ANION COMPLEXATION AGENTS 59

David Curiel, Elizabeth J. Hayes and Paul D. Beer

    2.1. INTRODUCTION 59
    2.2. ORGANIC ANION SENSORS 60
        2.2.1. Sensors with Naphthalene as Signalling Unit 60
        2.2.2. Sensors with Anthracene as Signalling Unit 66
        2.2.3. Sensors with Pyrene as Signalling Unit 79
        2.2.4. Sensors with Other Hydrocarbonated Signalling Unit 82
        2.2.5. Sensors with Condensed or Conjugated Heteroaromatic Rings as Signalling Unit 82
        2.2.6. Sensors with Other Fluorescent Probes as Signalling Unit 89
    2.3. INORGANIC SENSORS FOR ANIONS 94

2.3.1. Ruthenium (II) - and Rhenium (I)-Bipyridyl-based Receptors 94
2.3.2 Sensors containing other fluorophores 105
2.3.3 Lanthanide-based receptors 108
2.4. REFERENCES 113

3. FLUORESCENT CARBON DIOXIDE INDICATORS 119

Andrew Mills and Stephanie Hodgen

3.1. INTRODUCTION 119
3.2. THE TWO TYPES OF CARBON DIOXIDE OPTICAL SENSOR
     SYSTEMS 121
     3.2.1 Wet Optical Sensors for Carbon Dioxide 121
     3.2.2 Dry Optical Sensors for Carbon Dioxide 137
3.3. APPLICATIONS AND PRACTICAL SYSTEMS 155
3.4. LIST OF TERMS AND SYMBOLS 158
3.5. REFERENCES 159

4. FLUORESCENCE-BASED NITRIC OXIDE DETECTION 163

Scott A. Hilderbrand, Mi Hee Lim, and Stephen J. Lippard*

4.1. INTRODUCTION 163
4.2. CURRENT NON-FLUORESCENT DETECTION METHODS 164
4.3. EARLY FLUOROMETRIC IMAGING OF NO 166
4.4. THE DIAMINOFLUORESCEIN PLATFORM 167
4.5. OTHER o-DIAMINE NO SENSORS 169
4.6. BIOLOGICAL NO DETECTION WITH o-DIAMINE-BASED
     SENSORS 169
4.7. OTHER ORGANIC SENSORS 171
4.8. TRANSITION METAL SYSTEMS AND FIBER-OPTIC
     NO DETECTION 173
4.9. IRON COMPLEXES 174
4.10. COBALT COMPLEXES 176
      4.10.1. Cobalt SATI Sensors 176
      4.10.2. Cobalt FATI Sensors 180
4.11. DIRHODIUM TETRACARBOXYLATES AS REVERSIBLE NO
      SENSORS 181
4.12. SUMMARY 183
4.13. REFERENCES 184

5. FLUORESCENT REDOX-SWITCHABLE DEVICES 189

Ute Resch-Genger and Gunther Hennrich

5.1. INTRODUCTION 189
5.2. REDOX-ACTIVATED FLUORESCENCE SIGNAL GENERATION 190
  5.2.1. Redox-Active Control Units 192
  5.2.2. Composite Fluorescent Redox Switches 194
  5.2.3. Fluorescence Signalling of Redox-Active Analytes 196
5.3. FLUORESCENT REDOX-ACTIVE AND ANALYTE-RESPONSIVE DEVICES 199
  5.3.1. Fluorosensors with Redox Control of Analyte Recognition and Fluorescence 202
5.4. CONCLUSION AND OUTLOOK 211
5.5. ACKNOWLEDGEMENT 211
5.6. REFERENCES 211

6. PET SENSORS 219

Kanji Kubo

6.1. INTRODUCTION 219
6.2. PRINCIPLE OF PET SENSORS 220
  6.2.1. "OFF-ON" Switches 220
  6.2.2. "ON-OFF" Switches 222
  6.2.3. "OFF-ON-OFF" Switches 224
6.3. DESIGN OF PET SENSORS 225
  6.3.1. Design of the Fluorophore Unit 226
  6.3.2. Design of the Spacer Unit 226
  6.3.3. Design of the Receptor Unit 226
6.4. CLASSIFICATION OF PET SENSORS 227
  6.4.1. Coronand (Crown)-Based PET Sensors 227
  6.4.2. Chelator- and Podand-Based PET Sensors 231
  6.4.3. Cryptand-Based PET Sensors 236
  6.4.4. Cavitand-Based PET Sensors 238
  6.4.5. Polymer-Supported PET Sensors 238
  6.4.6. PET Sensors Involving Excimer Formation 239
  6.4.7. PET Sensors Involving Energy Transfer 240
6.5. PET SENSORS AS LOGIC GATES 241
6.6. CONCLUSIONS 243
6.7. REFERENCES 243

7. SIGMA-COUPLED CHARGE-TRANSFER PROBES OF THE FLUOROPROBE AND FLUOROTROPE TYPE 247

Jan W. Verhoeven

    7.1. INTRODUCTION 249
        7.1.1. Maximizing the Solvatochromism of CT Probes 250
    7.2. SIGMA-COUPLED CHARGE TRANSFER PROBES; D-σ-A 253
        7.2.1. Fluoroprobe (FP) an Extremely Solvatochromic Fluorescent D-σ-A Probe 253
        7.2.2. Quantum Yields and Fluorescence Lifetimes of FP and Related Probes as a Function of Solvent 258
        7.2.3. Improving the Absorption Characteristics of Fluoroprobe, from FP (1) to FT (3) 262
        7.2.4. Fluorogenic Derivatives of FP and FT which Allow Covalent Attachment of the Probes; from FP and FT to MFP (4) and MFT (5) .. 267
    7.3. APPLICATIONS OF D-σ-A PROBES 270
        7.3.1. Probing Solvation and its Dynamics as a Function of Temperature in Glass Forming Solvents 270
        7.3.2. Probing of Mobility Changes and Phase Transitions in (Semi)Crystalline Ma-terials 274
        7.3.3. Probing of Polymerization Processes 276
        7.3.4. Penetration of Solvents and Vapors in Polymers 278
        7.3.5. Thin Film Electroluminescence, OLEDs and PLEDs 279
    7.4. CONCLUDING REMARKS 280
    7.5. ACKNOWLEDGEMENTS 280
    7.6. REFERENCES 280

8. A DUAL LUMINOPHORE PRESSURE SENSITIVE PAINT: ELIMINATING THE TEMPERATURE INTERFERENCE IN THE MEASUREMENT OF OXYGEN PARTIAL PRESSURE 285

Muhammet E. Kose, Joanne M. Bedlek-Anslow, James P. Hubner, Bruce F. Carroll, Kirk S. Schanze

    8.1. BACKGROUND 285
    8.2. TEMPERATURE EFFECTS 288
    8.3. DUAL-LUMINOPHORE PRESSURE SENSITIVE PAINT 289
    8.4. SCANNING ELECTRON MICROSCOPY: STRUCTURE OF THE DOCI LOADED MICROSPHERES 292
    8.5. FLUORESCENCE MICROSCOPY OF DUAL LUMINOPHORE PRESSURE SENSITIVE PAINT 294
    8.6. DEMONSTRATION OF THE DUAL LUMINOPHORE PRESSURE SENSITIVE PAINT 296
    8.7. SUMMARY AND RELATIONSHIP TO THE OTHER WORK IN THE FIELD 300

CONTENTS

    8.8. REFERENCES 301

COLOR INSERTS 303

INDEX 307

PROBES AND SENSORS FOR CATIONS

Luca Prodi,* Marco Montalti, Nelsi Zaccheroni, and Luisa Stella Dolci

1.1. INTRODUCTION

1.1.1. The Need of Chemical Sensors

Sensors of any kind are nowadays substantially ubiquitous with the aim of improving the quality of our lives in any technologically advanced application. Following a definition given by IUPAC, a sensor is a system that, stimulated by any form of energy, reacts changing its own state and thus one or more of its characteristics. Among the different kinds of sensors, chemical sensors, i.e., those sensors that transform chemical information (ranging from the concentration of a specific sample component to total composition analysis) into an analytically useful signal, are of particular importance.[1] They have, in fact, already found a wide application in many fields, such as environmental monitoring, process control, food and beverage analysis, medical diagnosis, and, lately, toxic gases and explosives detection. It is evident that all these fields are of great importance from a social and an economical point of view. The development of chemical sensors seems thus predestined to revolutionise the potentialities of chemical analysis. Up to now the convenience of characterising from a chemical point of view an environment was strongly conditioned by many practical factors, mostly related to time and cost, and in many cases these two variables, in the final balance, could make this kind of analysis unsuitable or even useless.[2] Classical methodologies require collection, transportation, eventual pretreating of the sample, and, in many cases, expensive instrumentation manageable only by trained personnel. Chemical sensory devices have been conceived to bypass these restrictions and cover a large field of applications where conventional strategies result to be, even when feasible, inadequate. Chemical sensors, however, are valuable not only since they are cheap and user-friendly analytical tools; they indeed offer more than this: if properly designed they allow monitoring analyte concentrations in real-time and real-space.[3-5]

---

* Università degli studi di Bologna, Dipartimento di Chimica "G. Ciamician", Via Selmi 2, I 40126, Bologna, Italy; email: luca.prodi@unibo.it

1.1.2. The Power of Luminescence Spectroscopy

Among the different chemical sensors, fluorescence-based ones present many advantages: fluorescence measurements are usually very sensitive (single molecule detection is possible), low cost, easily performed, and versatile, offering subnanometer spatial resolution with submicron visualisation and sub millisecond temporal resolution. Furthermore, many opportunities exist for modulating the photophysical properties of a luminophore, such as the introduction of proton-, energy- and electron-transfer processes, the presence of heavy-atom effects, changes of electronic density, and the destabilisation of a non emissive $n\pi^*$ excited state. The versatility of fluorescence-based chemosensors originates also from the wide number of parameters that can be tuned in order to optimize the convenient signal. Even very complex analytical problems can be indeed overwhelmed by controlling the excitation and emission wavelengths, the time window of signal collection, and the polarization of the excitation beam or of the emitted light. In most cases luminescence intensity changes represent the most direct detectable response to target recognition; more recently, however, also other properties such excited state lifetime and fluorescence anisotropy have been preferred as detectable parameters, since they are less affected by the environmental and experimental conditions.[5]

1.1.3. Sensory Devices and Chemosensors: the Role of Chemists

It is evident from what discussed above that the design of efficient chemical sensory devices requires a multidisciplinary approach: it is in fact essential that chemists, biologists, physicists and engineers collaborate in order to obtain the desired system. The main role of chemists in this multidisciplinary team consists on the design and development of the interface among the matrix to be analyzed and the device itself. Chemists are therefore responsible of the part of the device that interacts at the molecular level with the analyte, that is responsible for the affinity and selectivity of the whole device, and that, in many cases, is also responsible of the signal transduction process, determining, as a consequence, the sensitivity of the system.

A very fruitful approach recently followed by chemists for the design of new efficient chemical sensors is based on the principles of supramolecular chemistry.[3-9] This approach implies indeed to start the design of a sensory device from the molecular level, where the space resolution achievable is the lowest possible. In addition to this, the knowledge reached so far in the field of supramolecular chemistry and molecular recognition can suggest the right choice to obtain the desired affinity and selectivity towards the target analytes, and an efficient signal transduction mechanism. The molecular or, more often, supramolecular entities that carry out these functions are conventionally referred to as chemosensors, mostly to distinguish them from chemical sensors, that usually are intended as macroscopic devices.[10-13] It is noteworthy, however, that also chemosensors and probes respond to the definition of sensor given by IUPAC. In fact, chemosensors designed following the supramolecular approach are chemical species that are able to bind selectively and reversibly the analyte of interest with a concomitant change in one or more properties of the system, such as redox potentials, absorption or fluorescence properties.

Because of the two different processes occurring during analyte detection, i.e., molecular recognition and signal transduction, luminescent chemosensors can usually be schematised as the assembly of three possible different components (Figure 1.1): a receptor (responsible for the selective analyte binding), an active unity (whose properties should change upon complexation) and, eventually, a spacer that controls the geometry of the system and tunes the electronic interaction between the two former moieties.

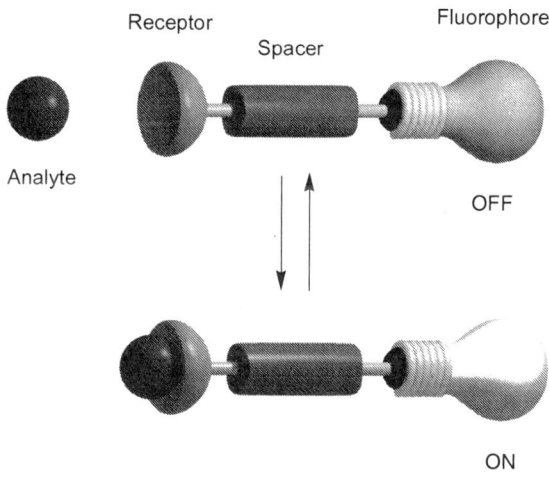

Figure 1.1. Schematic representation of a luminescent chemosensor whose signal transduction mechanism is a photoinduced electron transfer process.

This is the most common approach, that includes also the large family of PET (photoinduced-electron-transfer) based systems. In this case, the fluorophore is electronically coupled to a quencher unit constituting together the electron donor-acceptor pair involved in the PET process through which the luminescent excited state thermally deactivates. The donor, which can be in the simplest case an aminic nitrogen, is usually an integrating part of the receptor and plays an active role in coordinating the analyte, strongly decreasing its electron donating properties upon complexation. As a result the interaction with the analyte can destabilise the charge separated state to such an extent that the PET process cannot compete anymore with the radiative deactivation of the fluorophore. It is worth noticing that this mechanism operates by modifying the electronic properties of the donor moiety but usually does not alter the geometry of the donor acceptor pair.

Another possible arrangement includes all those systems where alterations of the fluorescence properties arise from direct interaction between the fluorophoric moiety and the target species.

In a third representative model of chemosensor, the reconfiguration of the system imposed by complexation plays, on the contrary, a fundamental role. In this case, the structure presents two active units involved in excimer formation or in energy or electron transfer processes between them, (Figure 1.2). Complexation of the analyte, tuning the relative distance of these two moieties, modulates the efficiency of the intervening inter-component process.

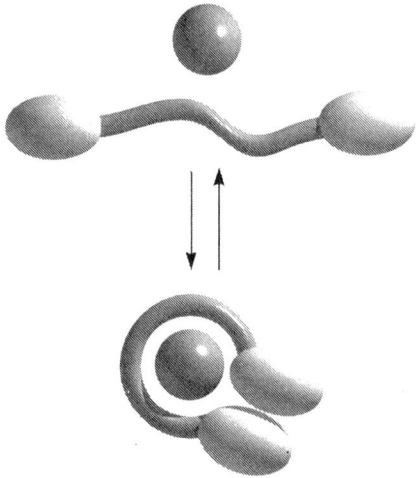

Figure 1.2. Schematic representation of a luminescent chemosensor whose signal transduction mechanism is based on conformational changes of its structure.

As already discussed, luminescent chemosensors can and do already find use in many disciplines; consequently this area of research is attracting attention in the scientific community, especially among chemists, biologists, physicists and material scientists.[3-9] In biochemistry, clinical and medical sciences, and cell biology, freely mobile sensor molecules are employed extensively in microscopy allowing to draw maps of the concentration of a given target analyte in a sample in real time.[14]

For many other applications in analytical chemistry and environmental sciences, however, only the immobilization of luminescent chemosensors to yield insoluble sensitive materials and the consequent development of a sensory device can allow continuous measurement of analyte concentrations. The characteristics of the holding up material and the methods of immobilisation are in this context crucial for the efficiency of the sensory devices, and they are again a special task for chemists. The support, in general, must be transparent to the wavelengths of the light absorbed and emitted by the chemosensor and, on the other hand, the active unit must be immobilised at such densities that a facile observation of changes in absorption and/or emission is possible. Solutions are many. One possibility is to immobilise effective chemosensors directly on glass surfaces (Pyrex or quartz).[15,16] This procedure leads however to the formation of a single layer of active units and the output signal could be too weak in the case of luminophores having low emission quantum yield. In order to obtain a larger density of chromophoric units and a stronger signal, the immobilization can be performed grafting on the glass surface a polymer branched with chemosensor molecules. Choosing polymers of different molecular weight can control the amount of chemosensor immobilized on the glass, as well as the thickness of the film. In general, immobilization does not result in large changes in affinity or selectivity of the receptors.[17] The development of chemical sensors with optical transduction can take advantage of the development of optical fibres.[11,12,18] This technology has already extended the possibility of performing remote real-time measurements with optical sensors by monitoring changes in the absorption or luminescence spectra of fluorescent compounds immobilised on a polymer matrix on the tip of a fibre. It is clear indeed that light is a very versatile signal, and micro sized optical fibres can allow analysis at almost any location.

Very recently, we also proposed the modification of silica nanoparticles as a possible way for immobilizing luminescent chemosensors.[19,20] The results obtained in this direction will be discussed at the end of this chapter.

As outlined so far, however, the development of new chemosensors and probes is the first (and, for some applications, the only one) essential step for the design of efficient sensors, and the main goal of chemists working in this field. For this reason, this chapter will mainly deal with the results and perspectives in tailoring chemosensors.

1.1.4. Chemosensors and Probes for Metal Ions

Among the different target analytes, a special interest is devoted to develop chemosensors for metal ions. Detecting cations is, in fact, of great interest for many applications. For example, sodium, potassium, magnesium, and calcium, are involved in biological processes such as transmission of nerve impulses, muscle contraction, and regulation of cell activity. In medicine, it is also important to control the serum levels of lithium in patients under treatment for manic depression, and potassium in the case of high blood pressure.

As far as the transition metals are concerned, they can represent an environmental concern when present in uncontrolled amounts, while some of them, such as zinc, copper, and cobalt are present in biological systems in trace amounts as essential elements.[21]

As we discussed in the previous section, the final goal of a luminescent chemosensor is to convert a chemical signal, represented by the concentration of the target analyte into a luminescence related signal (change in the intensity, spectral distribution, temporal decay etc.), which can be quantitatively interpreted.

This condition implies the ability of the chemosensor to recognise the analyte through a specific interaction, leading to the formation of a complex whose fluorescence properties can be easily distinguished from those of the free ligand. As a consequence, a chemosensor must be able to perform two main different functions: the metal ion binding and the signalling of this event. The simplest approach to this problem consists in using fluorescent moieties that are at the same time metal ion receptors. Classical fluorogenic ligands or fluorescent indicators (such as for example hydroxyquinoline derivatives) belong to this class of compounds. The field of applicability and the characteristics of these species are mostly well established.[22] A major limitation of this molecular approach to cation sensing resides in its lack of versatility. The recognition and the transduction mechanism cannot be modulated independently since adjustments in both these directions require modification of the same moiety. On the other hand these properties are often difficult to be tuned and it could become impossible to induce in the ligand peculiar characteristics such as a high selectivity. The need to operate individually on the recognition and the signalling process suggested, as discussed above, to confine them to separated parts of the whole structure following a supramolecular tactic, as depicted in figures 1.1-1.2. In this way a huge library of photophysically inactive well characterised receptors become available to be integrated into luminescent sensor structures. Historically this process was of course conditioned by the pioneering work of Pedersen, Cram and Lehn who started a systematic research on the synthesis of abiotic metal ions receptors. From then onwards, hundreds of possible chemosensors for metal ions were listed.[3-9,21,23-35]

This chapter will not be an exhaustive review of the systems published so far; rather, it will try do describe, using mainly examples from our laboratory, the possible approaches to the development of luminescent chemosensor for metal ions.

As a method, we will gather the various species in different classes, according to the receptor moiety present in the chemosensor, since it is usually this part that confers the required selectivity to the whole system. We have also inserted a paragraph dealing with sensors containing dendrimer-, peptide-, or protein-based receptors, and one, as discussed above, concerning nanoparticles, since we believe that these paragraphs allow a better comprehension of the state-of-the-art in the field of sensors and probes for metal ions.

1.2. CHEMOSENSORS WITH ACYCLIC RECEPTOR

As outlined above, in principle chemosensors for metal ions can be designed in a very simple and direct way by interconnecting a proper luminophore with an anchoring function, typically an amino group, though an alkyl chain.

For example, Ramachandaram et al.[36,37] and Mitchell et al.[38] have exploited this possibility for transition metal ion detection. The fluorescence of these systems is quenched by the occurrence of a photoinduced electron transfer (PET) process between the lone pair of the nitrogen and the appended chromophore. A significant enhancement of the fluorescence of **1a**, **1b**,[36] **2a**, and **2b**[37] (> 30 times for **1a**, >100 for **2a**) can be observed in acetonitrile solutions upon complexation of several transition metal ions. It is interesting noticing that some species that are usually reported to quench luminescence, such as Cr^{3+}, Fe^{3+}, Co^{2+}, Ni^{2+} and Cu^{2+} in this particular case induce an increase of the luminescence intensity. Such an atypical effect, according to the authors, is the result of the involvement of the nitrogen lone pair electrons in the coordination of the metal, which makes the PET quenching process energetically inaccessible. This effect prevails on the quenching effect due to the metal ion. The proposed mechanism is supported by the observation that the greatest enhancement effect has been obtained with the Zn^{2+} ions, which, having a d^{10} electronic configuration, do not usually introduce new electron- or energy- transfer processes for the deactivation of the excited state.

PROBES AND SENSORS FOR CATIONS

The absence of a specific receptor in the above-described systems is a major drawback of their simple design and is responsible for the lack of marked selectivity experienced by these molecular systems.

This limitation has been partially overwhelmed in the case of compounds **3a** and **3b**. Their fluorescence in 2-propanol solution (10 μM) is affected[38] by sub-ppm levels of Cu^{2+}, and only by much higher concentrations (10-100 ppm) of Ni^{2+} and Mn^{2+} while it is totally unaltered by the presence of other metal ions such as Zn^{2+}, Al^{3+}, and Ca^{2+}. A three-fold luminescence enhancement upon metal complexation has been observed in the case of ligand **3b**. It is to note that da Silva et al.[39] have described a very similar compound as a PET based pH sensor.

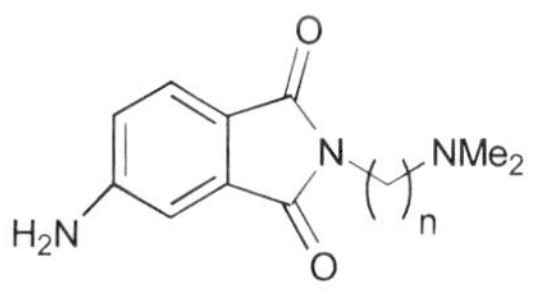

1a n = 2
1b n = 3

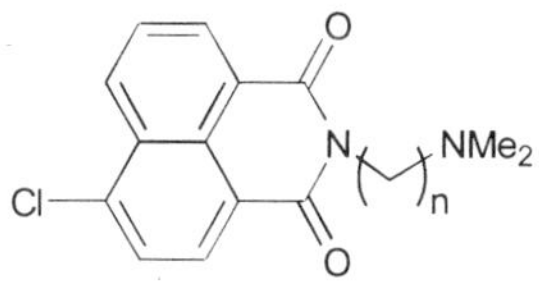

2a; n = 2
2b; n = 3

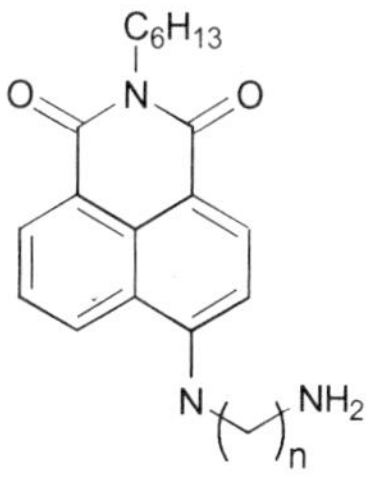

3a; n = 2
3b; n = 3

Carboxylate can also be employed as coordinative groups for metal ions recognition. **4** is a very simple and commercially available chemical species; nevertheless we proposed it as fluorescent chemosensor for metal ions[40] because of the unique ability of pyrene to give strongly luminescent dimers and exciplexes and of the co-ordinating abilities of the carboxylate group. Addition of earth metal ions to acetonitrile solutions of **4** causes in fact strong changes in the absorption and fluorescence spectra (Figures 1.3-1.4).

The results obtained were interpreted assuming that a complex with 1:2 (metal:ligand) stoichiometry was formed, in which the two pyrene units, lying very close to each other, could interact even in the ground state forming a pyrene dimer. Using this methodology, a detection limit of 4 ppb was obtained in acetonitrile for Ca^{2+} ions. Complexation with other ions, such as Sr^{2+}, Ba^{2+}, and Zn^{2+} induced very similar photophysical changes, although with lower association constants. A quenching of the fluorescence was instead observed with Cu^{2+} ions.

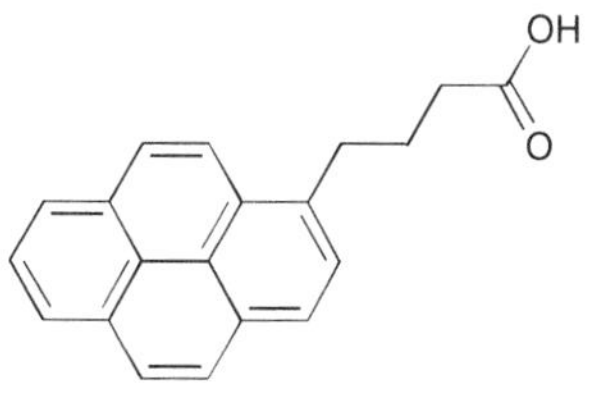

4

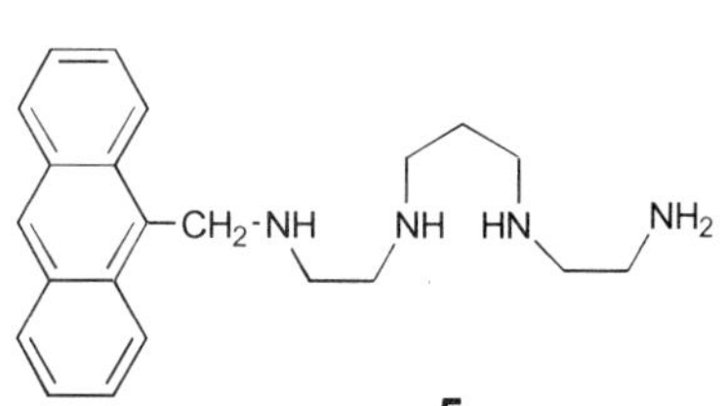

5

A more specific receptor unit is however required for working in aqueous medium, a typical condition for practical application, because of the strong solvation energy experienced by ionic species in the presence of water molecules.

With this aim in mind, Fabbrizzi et al.[41] synthesized compound **5**, having an anthracene chromophore and a polyamine chain as receptor unit. In aqueous media the metal binding properties of a ligand containing basic centres is strongly affected by the pH conditions. For this reason these authors have first examined the effect of pH on the fluorescence of **5** in acetonitrile/water solutions. In particular, they observed a fluorescence decrease at pH values above 4 where the deprotonation of the polyaminic chain starts enabling the free nitrogen lone pairs to take part to PET process towards the anthracene fragment causing its quenching. The pH dependence of the fluorescence intensity is affected by the presence of Cu^{2+}, Ni^{2+}, and Zn^{2+}. For the two former ions, a decrease of the intensity is observed at lower pH (< 3 for copper ions), indicating the formation of complexes where the deactivation of the anthracene excited state occurs via an energy transfer process from the chromophore to metal centred states. In the case of Zn^{2+}, complex formation is revealed by an almost complete recovering of the chromophore luminescence, which, after an initial quenching on increasing pH, gradually raises in the interval between 3.7 and 6. In the first part of the intensity vs pH curve, the Zn^{2+} ion does not interact with the ligand, and the progressive deprotonation of the ammonium ions allows PET to occur. At pH above 3.7, complexation with Zn^{2+} ions is achieved and, since this process increases the oxidation potentials of the amines, the PET is not anymore thermodynamically allowed. It is worth to note that in this system, as for every ligand which contains basic atoms, complexation competes with protonation. Hence metal binding induces a shift of the apparent pK_a of the receptor which is as higher as stronger is the interaction with the metal ion. Consequently, the observed behaviour is consistent with a larger stability of the Cu^{2+} and Ni^{2+} complexes with respect to the Zn^{2+} one.

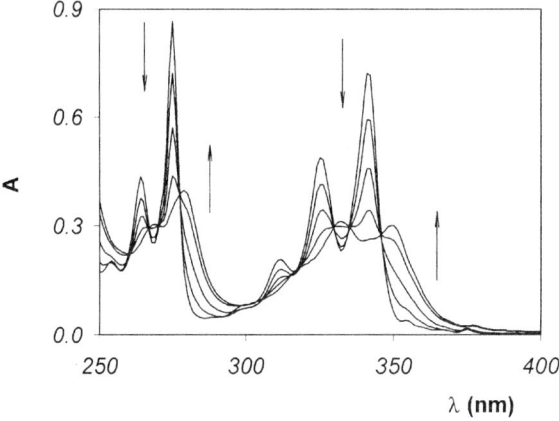

Figure 1.3. Absorption spectra of a 1.4 x 10^{-5} M acetonitrile solution of **4** and upon addition of 0.12, 0.25, 0.38, and 0.51 molar equivalents of $Ca(ClO_4)_2$.

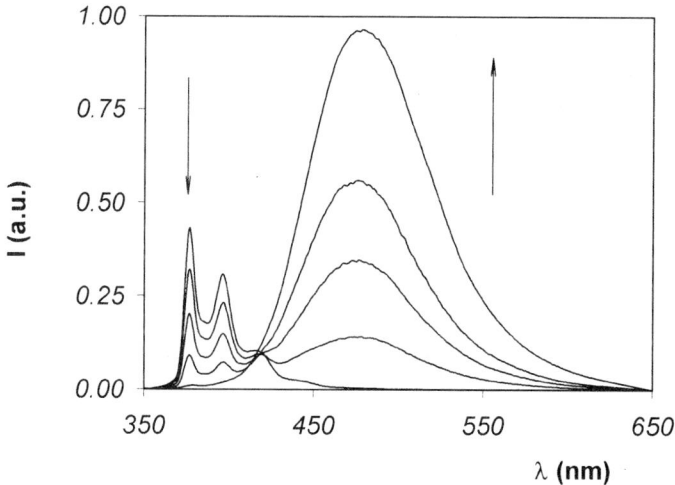

Figure 1.4. Fluorescence spectra (λ_{exc} = 330 nm) of a 1.4 x 10^{-5} M acetonitrile solution of 4 and upon addition of 0.12, 0.25, 0.38, and 0.51 molar equivalents of Ca(ClO$_4$)$_2$.

6 R = H
7 R = CH$_2$An

8

A similar molecule, **6**, synthesized by Sclafani et al.[42], shows with Zn^{2+} in pure water a comparable behaviour, although in this solvent the chelation enhanced fluorescence is lower (about 6-fold). From than onward, many other systems possessing an anthracene chromophore and a linear polyamminic chain were synthesized and studied in detail.[43-45]

In all the examples reported up to this point complexation always resulted in a modulation of the luminescence intensity. This parameter unfortunately may be influenced by factors independent from the target species concentration such as environmental or instrumental effects. As a consequence, the signal measured may be not correctly correlated only to the metal complex concentration causing an error in the analyte determination. This possibility of misreading may be minimised when a different fluorescence signal coming from the uncomplexed ligand can be detected and used as reference. This situation has been achieved in the case of **7**,[42] where two anthracene fragments are present in the same ligand unit. In the fluorescence spectrum of the free molecule there is evidence of excimer formation from a tail in the 450-600 nm region; addition of Zn^{2+} ions increases the fluorescence intensity in this part of the spectrum, indicating that ion complexation promotes the intramolecular excimer formation. The different response to Zn^{2+} at 414 nm (monomer emission) and at 495 nm (excimer emission) allows the fluorescence ratio at these two wavelengths to be used for a direct concentration determination. A similar effect was also observed[46] with the molecule **8** consisting of two anthryl groups connected by a –SCH2CH2CH2S– spacer, a species that was designed pursuing the aim of the formation of thia-anthracene receptors. The absorption spectrum of **8** (in dichloromethane:methanol, 8:2, v/v) presents the typical pattern of the anthracene moieties, indicating that almost no interaction between the two units takes place in the ground state. On the contrary, the fluorescence spectrum differs greatly from that of simple anthracene (see Figure 1.5), revealing the presence of a

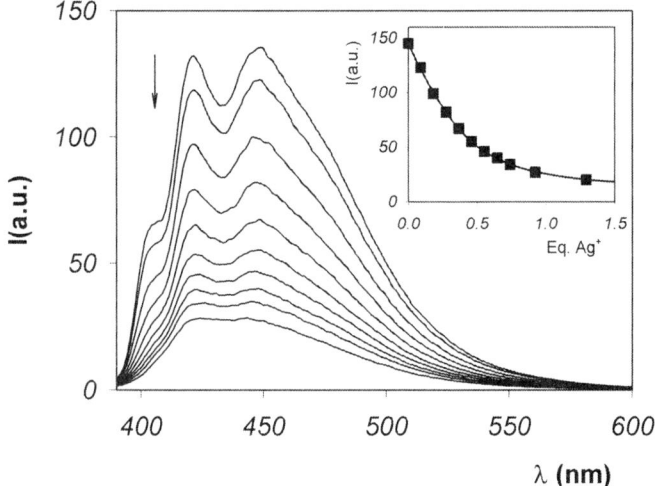

broader band superimposed on the typical structured emission of the isolated fluorophore.

Figure 1.5. Changes in the fluorescence spectra of a solution of 8 in dichloromethane:methanol, (8:2, v/v) upon addition of silver cations.

In addition, a double exponential model was needed to interpolate correctly the experimental data, giving two distinct lifetime values. It is worth to note that the above-described photophysical properties are not altered by changing the concentration in the range between 1×10^{-6} and 1×10^{-4} M. This observation clearly indicates that an intramolecular process is responsible for the double luminescence. In addition, the shape of the luminescence band clearly depends on the polarity of the solvent; in particular, the contribution of the unstructured component in MeOH/CH$_2$Cl$_2$ mixtures increases for higher methanol fractions, i.e., increasing the polarity of the mixture itself. Finally, no differences were observed between excitation spectra recorded at different emission wavelengths, all being superimposable to the absorption spectrum. The appearance of the broader, red shifted fluorescence band can thus clearly be attributed to the formation of an intramolecular excimer between the two anthracene moieties of **8**.[47] In the presence of silver ions, the above-described fluorescence spectrum of **8** undergoes remarkable changes. Titration experiments showed that a gradual increase in the concentration of silver ions causes a progressive decrease in the structured luminescence band, while the other, broader band ends by prevailing (Figure 1.5). It is worth noting that a weakening of the luminescence occurs even when only the lower-energy section of the spectrum, namely, over 500 nm, is examined, indicating that the disappearance of the structured band comes together with a decrease in intensity of the other band. Also the absorption spectrum undergoes significant changes upon complexation: a slight red shift of the lower-energy bands takes place together with a broadening of the same band. The coordination of the metal ion imposes a much more rigid structure to the ligand, and in such a geometry the two anthracene moieties are much closer than in the free ligand. This makes their interaction much easier in both the ground and the excited state. The quenching of the fluorescence of the low energy component can be attributed to an electron transfer process involving the silver ions, that is thermodynamically possible. Moreover, other possible interpretations such as the occurrence of an energy transfer process from the anthracene to the metal-centered excited states must be ruled out, since the d^{10} silver ions do not have a low-energy metal-centered state.

A deeper investigation indicated that the association process takes place in two steps, and when a solution of silver ions is added to **8**, there is the formation, first, of [Ag$^+$·**8**$_2$] and, then, of [Ag$^+$·**8**]. The association constants for the two consecutive equilibria can be estimated to be $K_1 = 2 \times 10^5$ M^{-1} and $K_2 = 8 \times 10^4$ M^{-1}. A large number of metal ions usually show a good affinity toward sulfur-containing compounds. Titration experiments with Hg^{2+}, Cd^{2+}, Cu^{2+}, Ni^{2+}, Zn^{2+}, and Co^{2+} however evidenced that **8** does not complex these species, indicating its good selectivity.

9

10

Amidic groups are usually poorly effective in metal ion coordination. Their binding tendency can be anyway widely enforced when they are inserted in a receptor structure where other anchoring centres co-operatively take part to the complexation. Fabbrizzi et al. have recently shown[48-50] that a well known category of ligands, dioxo-tetraamines (and in particular dioxo-2,3,2-tet, 1,4,8,11-tetraazaundecane-5,7-dione) can be successfully employed for signalling the presence of Ni^{2+} and Cu^{2+} if connected to chromophores such as anthracene ($9^{48,49}$) or $Ru(bpy)_3^{2+}$ (10^{50}). For these chemosensors the complexation mechanism involves the deprotonation of the two amide groups. This very endoergonic process can take place only with metal ions that profit of a large ligand field stabilisation, in these cases only Ni^{2+} and Cu^{2+}. This condition makes ligands of this kind extremely selective, and the different ligand field stabilisation, higher for Cu^{2+} with respect to Ni^{2+}, allows also to distinguish between these two cations. One major drawback of **9**, however, is the need of organic/aqueous solvent mixtures as the working media, while **10** can be dissolved in water with concentrations ranging up to 10^{-2} M.

In particular, the complexation properties of **10** towards transition metal cations have been examined by comparing the changes of its fluorescence intensity (I_f) and lifetime (τ) as a function of pH, in presence or absence of transition metal cations. When no metal cations are added to solutions containing **10**, I_f remains constant in the 2 < pH < 12 range, as it has already been observed for related systems. In this pH range, in fact, the luminescence quantum yield (0.030) and lifetime (440 ns) of **10** in aerated water solutions are very similar to those observed for the parent $Ru(bpy)_3^{2+}$ chromophore in the same conditions, indicating that the dioxo-2,3,2-tet fragment does not substantially perturb the excited state properties of the Ru core. On the other hand, when Ni^{2+} or Cu^{2+} (as their perchlorate or chloride salts) are added in 1:1 molar ratio with respect to system **10**, the I_f vs pH plot shows a typical sigmoidal profile (figure 1.6), which indicates that binding of the metal ion by the dioxo-2,3,2-tet fragment takes place in the narrow pH range of the steeply descending portion of the sigmoid.

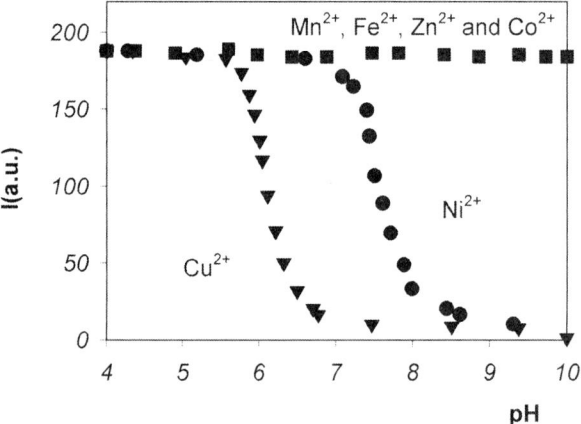

Figure 1.6. Luminescence intensity vs. pH for solutions containing ligand 10 and a metal cation in 1:1 stoichiometry. The metal species referring to each curve is indicated in plot.

Thus, from figure 1.6 it can be said that complexation by **10** (1×10^{-5} M) begins at pH 5.8 and it is complete at pH 6.8 for Cu^{2+}, while for Ni^{2+} it begins at pH 7.5, being complete at pH 8.5. In the descending I_f vs pH portion it can also be observed the appearence in the excited state decay profile of a second component with a much shorter lifetime (11 and 15 ns for Ni^{2+} and Cu^{2+}, respectively), clearly indicating an intramolecular quenching process; this component becomes the only one present at higher pH values.

From flash photolysis experiments, no evidence of the presence of Ru^I species could be detected, the only transient absorbing species being due to the excited state of the Ru^{II} chromophore. Furthermore, steady state experiments performed at 77 K showed that the quenching process was very fast also in frozen medium at low temperature. All these findings suggest that the energy transfer (ET) mechanism is the most likely explanation for the luminescence quenching of **10**.

The selectivity of this system is proven by the negligible fluorescence intensity chenges in the I_f vs pH plots in the presence of other metal centres (1:1 molar ratio), such as Mn^{2+}, Fe^{2+}, Zn^{2+} and Co^{2+}, indicating that no complexation takes place at the dioxo-2,3,2-tet fragment with these ions.

In addition, working on a solution of **10**, buffered at pH = 7.0, no variation in I_f is observed by addition of up to 2 equiv of Ni^{2+} (or other divalent first row transition metal cations), while subsequent addition of Cu^{2+} causes the expected fluorescence quenching, demonstrating that selectivity of Cu^{2+} on Ni^{2+} can be obtained by choosing the correct pH value (see figure 1.7). Finally, solutions buffered at pH 7.0 and containing system **10** at concentrations as low as 10^{-7} M revealed an easily detectable variation of I_f on addition of 1 equiv of Cu^{2+}, indicating that **10** is a suitable sensor for copper cations under analytically relevant conditions.

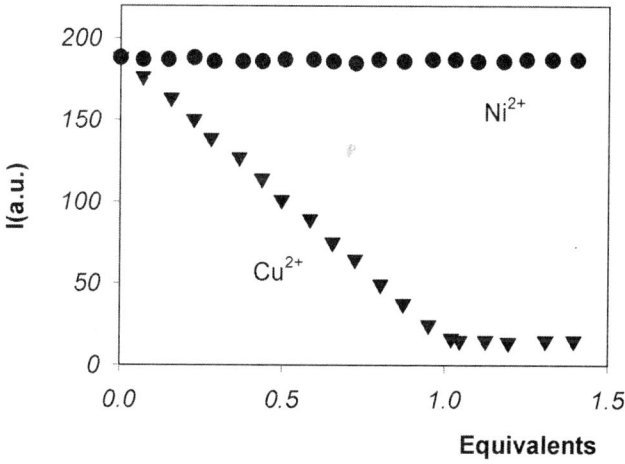

Figure 1.7. Luminescence intensity vs. equivalents of added Ni^{2+} and Cu^{2+} for ligand 10, in a solution buffered at pH 7.0.

An other family (**11-14**) of effective and selective chemosensors for transition metal ions were developed[51-51] using as receptor units the 1,2,4-thiadiazole derivative and its reduced form, the corresponding iminoyl thiourea. For **11**, a PET process from the alkylated nitrogen atom to the excited state of the chromophore is the mechanism responsible for the fluorescence quenching of the appended anthracene. Complexation with Hg^{2+} leads to a tremendous increase in the fluorescence (44-fold), while modest increases were observed upon addition of Cu^{2+} and Pb^{2+} (2-fold and 7.7-fold, respectively). The observed fluorescence increase was attributed to a generic chelation effect, rather than to a suppression of the above mentioned PET process, since addition of protons did not lead to any intensity enhancement. All of the other cations tested yielded no significant changes in emission intensity. In the case of **12**, instead, a 6-fold enhancement of the fluorescence was observed upon addition of Cd^{2+}, and, to a minor extent (3.6-fold) upon addition of Zn^{2+}. For compounds **11** and **12** the detection limit in acetonitrile, for Hg^{2+} and Cd^{2+}, respectively, is in the range of 10 μmol. **13**, closely related to **11**, displays a strong modulation of its fluorescence behaviour selectively upon addition of Cu^{2+}. In this case, the strongest fluorescence enhancement is observed for a concentration ratio 1:2 (metal:ligand). The increase of the emission intensity is time dependent, varying additionally with the concentration of the chemosensor. Immediately after addition of 0.5 eq. of Cu^{2+}, a 4-fold luminescence enhancement is obtained, and an approximately constant signal (a 46-fold intensity increase) is reached after 6 h. **14** functions instead as a chemodosimeter for Hg^{2+}, since this ion promotes the desulfurization of the iminoylthiourea leading to the corresponding urea. This process leads to a red shift and to an increase in the naphthalene emission.

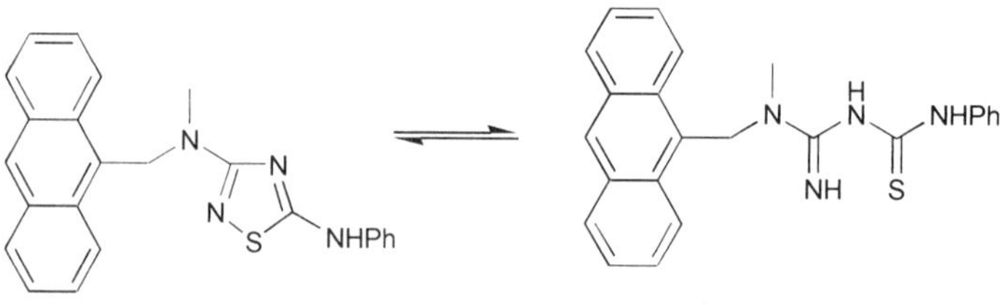

11   **12**

13   **14**

Many groups have instead choose to use derivatized polyoxyethylene chains as receptor units.[53-57] **15a-c**[55] are an example of chemosensors of this family. ^{1}H NMR, UV-visible and fluorescence studies reveal that the receptors bind to alkali and alkaline-earth metal ions to give a supramolecular complex in which the ion is nestled within the oligooxaethylene framework. The formation of the complexes induces a change in the geometry of the system toward a conformation in which the two tetrapyrrolic macrocycles tend to face together. This causes a blue shift and broadening of their Soret bands. The fluorescence quantum yield is almost completely unaffected by complexation; the lowering of the absorbance causes however a decrease also in the fluorescence intensity at low concentration. In particular, the changes observed allow the complexation process to be monitored at sub-micromolar concentrations. The association constants in chloroform/acetonitrile (1/1) lie within the range $25 - 1 \times 10^5$ M^{-1} depending on either the nature of the diporphyrinic receptors or the interacting metal ions.

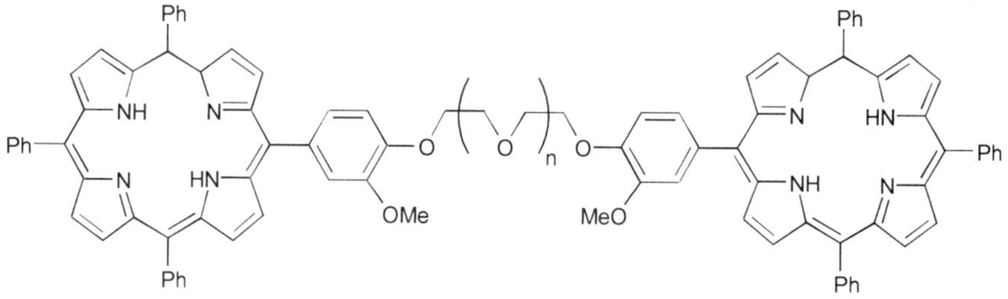

15a; n = 2
15b; n = 3
15c; n = 4

In alternative to the use of the modulation of π–π interactions as signal transduction mechanism, Valeur et al.[56] tried to change the energy-transfer efficiency between a donor and an acceptor (two different coumarins) appended at the end of an ethoxy chain (**16**), taking advantage of the changes induced in the donor-acceptor distances by metal ion complexation. Pb^{2+} is indeed complexed by **16** in acetonitrile, causing a decrease of the average distance of the two chromophores and, consequently, leading to some changes in the luminescence spectrum, due to an increase of the energy transfer rate constant among the two emitting units.

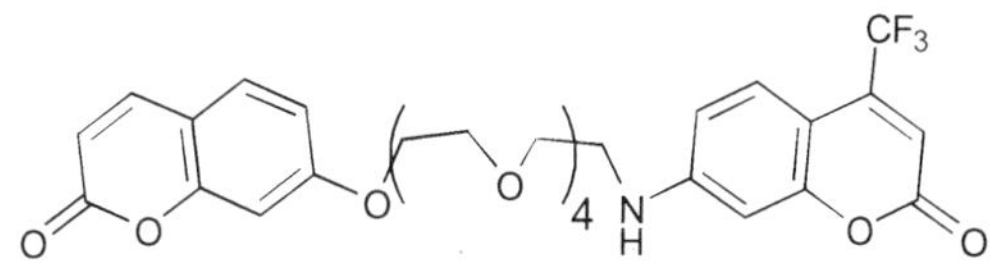

16

As a matter of fact, however, to have a complete switch off/on of an energy transfer process using flexible chain is not an easy task, as proved also by other systems.[55] A more efficient approach could be to modulate a photoinduced electron transfer reaction via conformational rearrangement induced by the binding process. This mechanism presents a much stronger dependence on the distance, if compared with Förster-type energy transfer processes, and this possibly allows to design more efficient switching systems. Following this strategy, the luminescence of the anthracene is quenched in the series **17a-f**[57] by the electron acceptor linked at the end of the ethoxy chain with an efficiency that depends on the free energy associated at the electron transfer process in agreement with the Marcus law.

17a; R = benzoyl
17b; R = 2,4-dichlorobenzoyl
17c; R = 2-Chloro-5nitrobenzoyl
17d; R = 4-Nitrobenzoyl
17e; R = 3,5-Dinitrobenzoyl
17f; R = 4-Chloro-3,5dinitrobenzoyl

18

In this context, it is worth to remind the excellent work of Roger Tsien,[58-62] a work that, started from the late 70', makes him the first pioneer in the field of chemosensors. He presented a number of very interesting examples of chemosensors; as a recent example, we will cite here two systems (**18** and **19**) based on a fluoresceine platform for the determination of intracellular concentrations of zinc ions. Both these chemosensors have excitation and emission wavelengths in the 500 nm region, association constants for $Zn^{2+} > 10^9$ M^{-1}, quantum yields around 0.9, and cell permeability, making them well suited for intracellular applications. It has been observed a 3- to 5-fold enhancement upon zinc complexation, due to the inhibition of a PET process. As stated by the authors, the primary shortcomings of these sensors are their sensitivity to protons and their relatively modest fluorescence enhancement upon binding of the zinc cations, but they represent a very interesting approach for the development of new zinc sensors suitable for the neurosciences studies, that is indubitably a very promising field.

Tripodal ligands are also often used as receptor moieties. A very interesting example is that proposed by Castagnetto and Canary,[63] who synthesised the chiroptically enhanced fluorescent sensor **20**. In this sensor, chelation with Zn^{2+} and Cd^{2+} ($K_a > 10^6$ M^{-1}) leads to an increase of the fluorescence intensity of the quinoline unit, with an observed fluorescence enhancement (378 nm, pH 7) of 30- and 6-fold, respectively. In this case, the complexation changes the nature of the lowest excited state from a $n\pi^*$ (that usually tends to give intersystem crossing processes, and thus phosphorescence) to $\pi\pi^*$ state (that usually gives a more intense fluorescence). Complexation with Fe^{2+} and Cu^{2+} did not result in a fluorescence increase. Complementary information can come looking at the exciton-coupled circular dichroism spectra (ECCD), where strong signals were observed upon complexation with trigonal bipyramidal metal ions (Zn(II) and Cu (II)), while weak signals could be detected with octahedral metal ions (Cd(II) and Fe(II)). Evaluation of both fluorescence and ECCD properties of the complexes can lead to the identification of the metal. In fact this two techniques could distinguish, for example, Zn^{2+} (strong fluorescence and ECCD response), Cu^{2+} (strong ECCD but no fluorescence), Cd^{2+} (fluorescence but not ECCD), and Fe^{2+} (neither fluorescence nor ECCD). These findings stress the principle that both isotropic and anisotropic detection may be used to maximise the information given by a single sensor molecule.

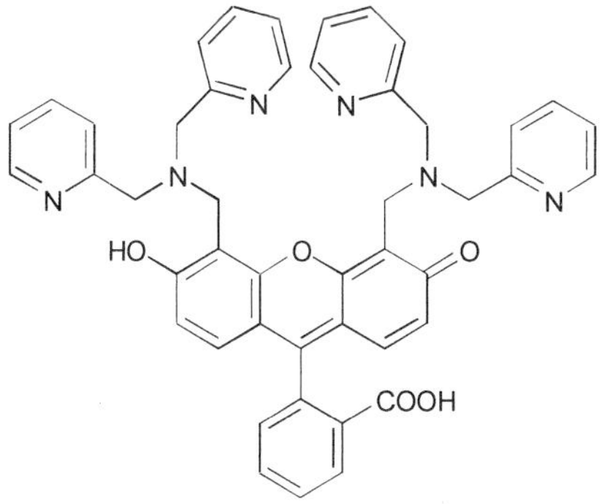

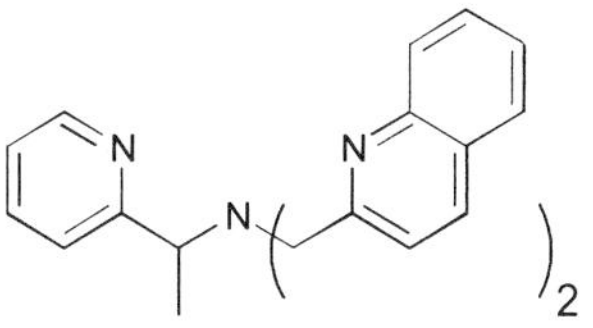

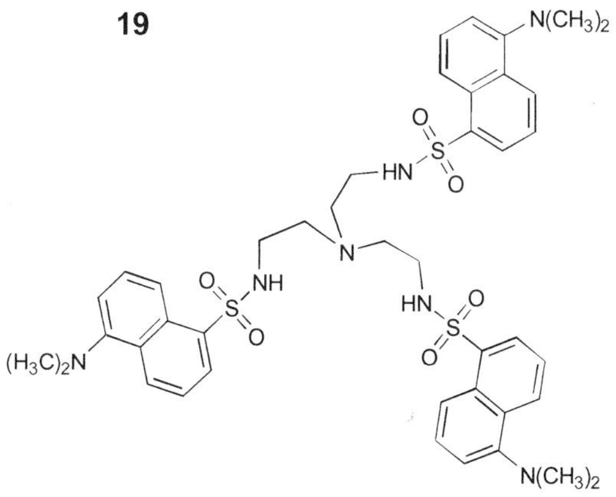

Addition to each terminal amine nitrogen atom of *tris*(2-aminoethyl)amine of a dansyl group leads to **21**.[64] The dansyl chromophore is well known to show intense and large luminescence bands in the 400-600 nm region. These bands are very sensitive to the polarity of the solvent and have considerable charge-transfer character, caused by mixing of the ^1L_a and ^1L_b states of naphthalene with a charge-transfer state arising from the promotion of a lone-pair electron on the amino group into an antibonding orbital of the naphthalene ring. The fluorescence of **21** does not depend on the pH conditions in the 3-11 pH range. However, strong changes on its absorption and luminescence properties were indeed observed upon addition of Cu^{2+}, Co^{2+}, Zn^{2+}, and Cd^{2+} at pH 9.5 (see figures

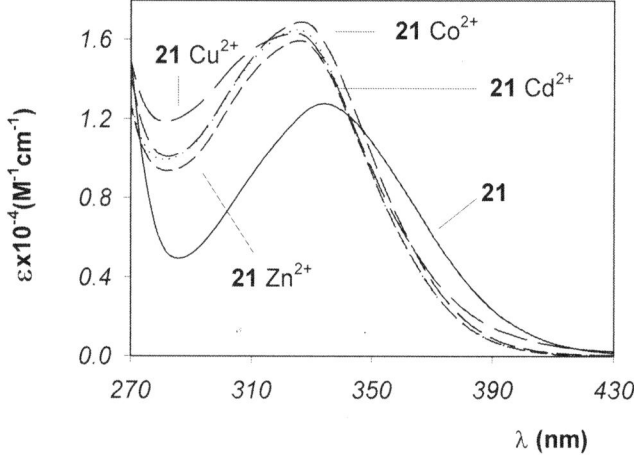

1.8 and 1.9).

Figure 1.8. Absorption spectra of 21 (9.2 x 10^{-5} M) and of its complexes with Cu^{2+}, Co^{2+}, Zn^{2+}, and Cd^{2+} in acetonitrile/water (1:1, v/v) solution at pH 9.5.

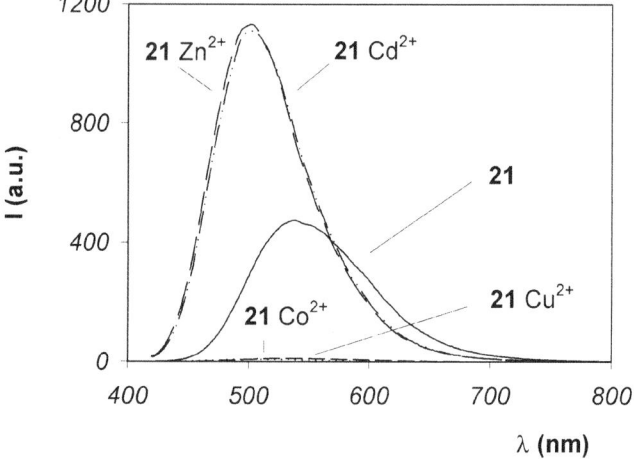

Figure 1.9. Fluorescence spectra (λ_{exc} = 338 nm) of 21 (9.2 x 10^{-5} M) and of its complexes with Cu^{2+}, Co^{2+}, Zn^{2+}, and Cd^{2+} in acetonitrile/water (1:1, v/v) solution at pH 9.5.

In particular, for the former two metal ions, a strong quenching (I_{rel} = 1%) of the fluorescence intensity was observed, while for Zn^{2+} and Cd^{2+} a blue shift accompanied by an enhancement of the fluorescence quantum yield could be detected. The complexation mechanism involves for all these metal ions the deprotonation of the sulphonamide groups. For Zn^{2+} and Cd^{2+} ions, the observed blue shift has been attributed to the increase of the electronic density on the naphthalene ring due to the deprotonation/complexation process, that moves towards higher energy the amine-to-naphthalene charge transfer state occurring in the dansyl chromophore. Cu^{2+} and Co^{2+} ions, together with this effect, introduce lower lying energy excited states centred on the metal, that deactivate the luminescent excited state centred on the dansyl chromophore. Interestingly, addition to 5x10^{-5} M solutions of **21** of up to 100 molar equivalents of Na^+, K^+, Ca^{2+}, Sr^{2+}, Ba^{2+}, Eu^{3+}, Mn^{2+}, Pb^{2+}, Fe^{2+}, Fe^{3+}, Cr^{3+}, Ni^{2+} and Hg^{2+} did not lead to any change in the absorption and emission spectra of the dansyl chromophore in the whole pH range examined (3-11).

It is noteworthy that the complexation equilibrium is pH dependent (Figure 1.10) since it involves the deprotonation of the sulphonamide groups.

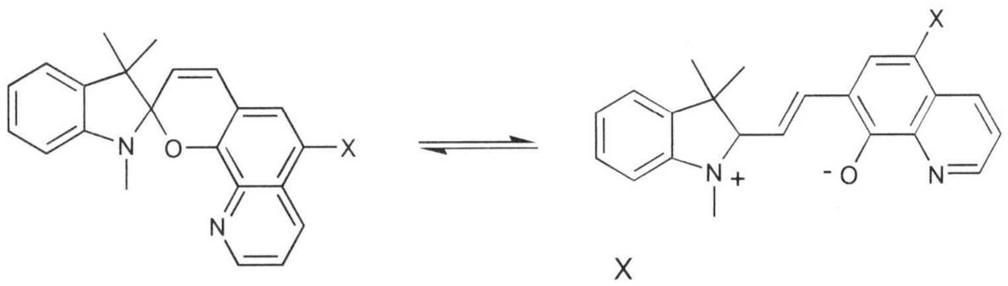

22a x = H
22b x = NO$_2$

23a x = H
23b x = NO$_2$

It was observed that, while the complexation with copper ions is complete at pH 6, the complexation with the other metal ions requires higher pH values, and this makes **21** very selective for Cu^{2+} at pH conditions close to the physiological ones. Titration experiments always indicate a 1:1 (metal:ligand) stoichiometry for all complexes. Addition of NaOH to a solution containing an equimolar amount of metal ion and ligand **21** showed that for Co^{2+}, Zn^{2+}, and Cd^{2+} the plateau of the luminescence intensity vs. added base is reached after addition of 3 equiv. of base, indicating that complexation requires the deprotonation of all the three sulphonamide groups at the same time. For Cu^{2+}, the complete quenching of the luminescence was observed after addition of only 2 equivalents of base, but further changes were also observed in the absorption spectra up to the addition of 3 equivalents, in particular for the metal centred absorption bands of the copper ion in the 450-1100 nm region. Addition of up to 2 equivalents of base to an equimolar (1x10^{-3} M) solution of **21** and Cu^{2+} leads to the appearance of a band with maximum at 620 nm, while addition of a third equivalent caused the disappearance of such a band with the concomitant formation of a band centred at 1050 nm.

A very similar behaviour as chemosensor was observed for the ligand in which only one amine of the tren moiety was reacted with a dansyl chromophore,[65] the main difference being only the formation of metal complexes at more acidic pH.

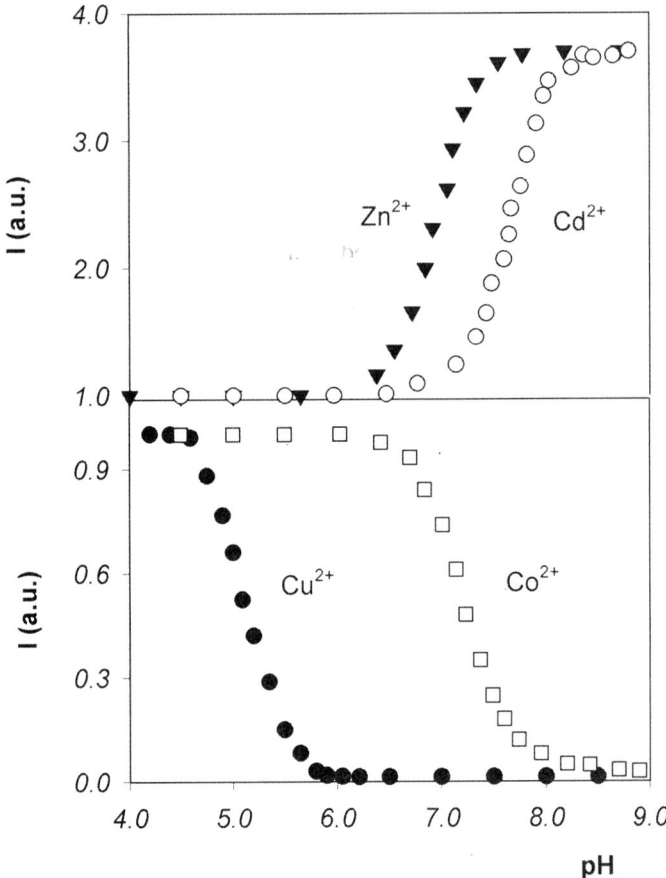

Figure 1.10. pH dependence of the luminescence intensity (λ_{exc} = 338 nm) for solutions containing equimolar amounts (9.2 x 10^{-5} M) of **21** and (i) Zn^{2+} ions (λ_{em} = 500 nm), (ii) Cd^{2+} ions (λ_{em} = 500 nm), (iii) Cu^{2+} ions (λ_{em} = 540 nm), and (iv) Co^{2+} ions (λ_{em} = 540 nm).

Systems **22a,b**[66] are very interesting since they can be interconverted into the merocyanine isomers **23** by means of light irradiation and/or metal ion complexation. In particular the form **22a** is the predominant one in absence of added metal ions or UV light, and has a very low fluorescence quantum yield. Addition of Zn^{2+} to an ethanol solution of **22a** leads to a 14-fold increase of the luminescence at 610 nm. This enhancement is due to the formation of the **23a** isomer engaged in the metal ion

PROBES AND SENSORS FOR CATIONS

complexation. Interestingly, metal free **23** are also not luminescent, since proton transfer can occur in the excited state, following non radiative paths. This process is prevented by metal ion complexation. The sensitivity of **22a** for Zn^{2+} in ethanol/water solutions was found to be around 3 ppb. Some enhancement was also found upon addition of Cd^{2+} and Mg^{2+}, while not luminescent complexes were observed with Cu^{2+}, Ni^{2+}, Co^{2+}, and Fe^{3+} monitoring the absorbance increase in the long-wavelength maximum, typical of the merocyanine form. For these complexes, irradiation of the metal chelate did not cause the release of the metal ion nor regeneration of **22a**. The insertion of the nitro group in the quinoline moiety to give **22b** induces two important changes in the system, namely (i) the shift of the equilibrium towards the open form even in absence of added metal ions and (ii) a decrease in the complexation ability of the quinoline moiety. For **22b**, a 9-fold increase of the fluorescence intensity was observed upon addition of zinc ions, only a 3-4-fold increase upon addition of magnesium or cadmium ions and, very promisingly, a 20-fold increase upon addition of one equivalent of mercury ions. For this system, expulsion of the metal ion to give the closed form was observed upon irradiation with visible light.

1.3. CHEMOSENSORS WITH CYCLIC RECEPTOR

The most frequently adopted classes of cyclic metal ion receptors are by far polyazamacrocycles and crown ethers. It is to note, however, that the first examples reported in the literature of the use of polyazamacrocycles as ligand units in fluorescent chemosensors is relatively recent. The first paper in fact appeared in 1990[67] by Czarnik and his group that described a family of fluorescent chemosensors, **24a-e**,[67-69] bearing polyazamacrocycles as receptor and an anthracene as the active luminescent unit.

24a (x = 1), **b** (x = 2), **c** (x = 3), **d** (x = 4), **e** (x = 5)   **25**

The luminescence of the appended anthracene in these systems depends on the pH. This finding is not surprising since an electron transfer process from the unprotonated amines to the anthracene group is possible, leading to a low fluorescence under basic conditions. A strong enhancement of the fluorescence is obtained when the above-mentioned quenching process is prevented by protonation of the amines. The same effect, however, can be also obtained when these groups are engaged in the complexation of metal ions. The strongest fluorescence enhancement in water was observed upon complexation by **24c** and **24e** of both Zn^{2+} and Cd^{2+} ions. When **24e** complexes Hg^{2+} ions, the fluorescence is in turn quenched, probably because of the heavy atom effect. It is very interesting to note that the complex **24d**-Cd^{2+} gives rise not only to an enhancement of the fluorescence intensity, but that its emission has different maximum and band shape respect to that one of the free ligand. This feature can be very important for a future application, since it can allow discriminating between Zn^{2+} and Cd^{2+} ions, which usually present a very similar behaviour. This difference has been attributed to the occurrence, in this system, of a ground state Cd(II)-π complexation.

A similar fluorescent chemosensor is compound **25**,[70,71] which presents a very interesting selectivity towards Cu^{2+} and Hg^{2+}. In water solution at pH 7 **25** displays chelation-enhanced quenching effects only with these two ions, with overall emission changes of 18-fold ($K_a > 10^6$ M^{-1}) and 4-fold ($K_a = 1.8 \times 10^4$ M^{-1}), for Cu^{2+} and Hg^{2+}, respectively. As for the complex **24d**-Cd^{2+} discussed above, a ground state Hg(II)-π complexation has been proposed in order to explain the high association constant and the strong quenching effect observed for **25**-Hg^{2+}.

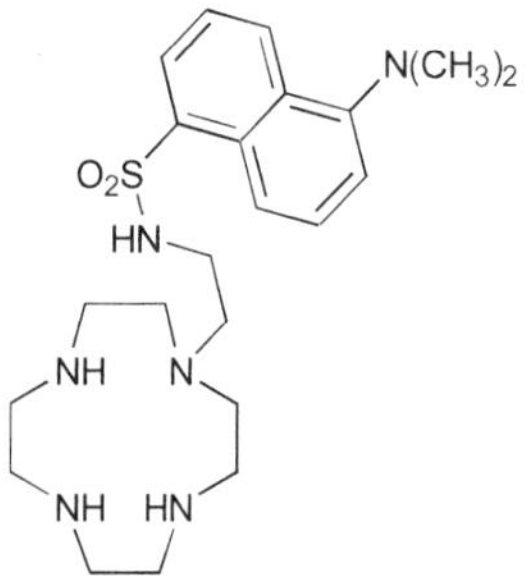

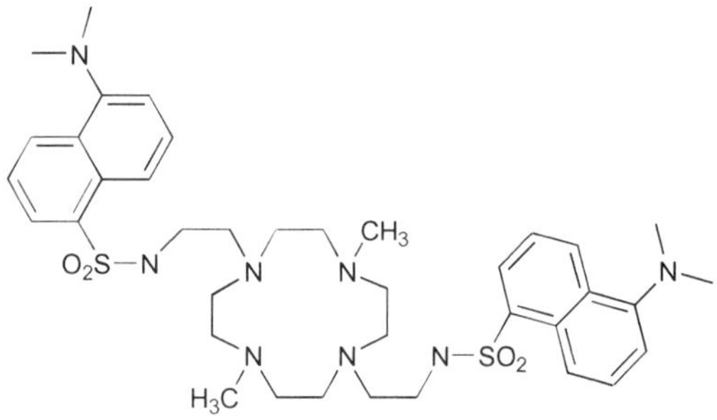

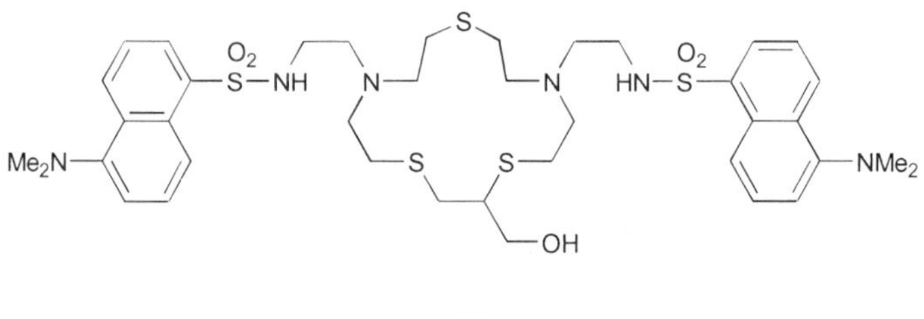

The cyclen macrocycle is the active ligand unit of the series of compounds, where it is connected with one or more chromophoric units, such as naphthalene derivatives.[72,73] A very interesting chemosensor that takes advantage of the high affinity of cyclen towards zinc ions is **26**. It was developed by Koike et al.[74] who joined the suitable properties of the cyclen macrocycle with the strong affinity towards aromatic sulphonamides of the Zn-Cyclen complex, and the good luminescence properties of the dansyl chromophore in one single species. Since the metal ion complexation requires the deprotonation of the sulphonamide group, the association constant is pH dependent, and it is very high at physiological conditions (e.g. 7 x 10^9 M^{-1} at pH 7.0 and 2 x 10^{12} M^{-1} at pH 7.8). At this pH complexation with zinc ions induces a blue shift of the peak from 582 to 540 nm, and it increases the emission intensity by 5-fold at 540 nm and by 10-fold at 490 nm. Co^{2+}, Pb^{2+}, and Hg^{2+}, on the contrary, cause some quenching of the luminescence intensity, while a complete quenching can be obtained with Cu^{2+}. This last ion tends to form very stable complexes with **26**; the use of bovine serum albumin can mask however its presence, leading to the possibility to monitor only zinc concentration.

A similar system (**27**), but with two dansyl units have been reported afterwards.[75,76] Addition of Cu^{2+}, Zn^{2+}, Cd^{2+}, and Hg^{2+} ions to a methanol/water (1/1) solution of **27** buffered at pH 9.5 leads to pronounced changes in the UV-Vis spectra of the ligand. In particular, the absorption band of the ligand at 331 nm shifts to shorter wavelengths (314, 327, 313, and 321 nm for Cu^{2+}, Zn^{2+}, Cd^{2+}, and Hg^{2+}, respectively) and increases in intensity. The intensity increase is particularly evident upon addition of Cu^{2+}, where the absorbance almost doubles. Blue shifts are also observed for the luminescence band (502, 515, 492, and 513 nm for Cu^{2+}, Zn^{2+}, Cd^{2+}, and Hg^{2+}, respectively, vs 527 nm of the free sensor). This behavior had been observed also for **21**, for which the complexation process was concomitant to the deprotonation of the sulfonamide group. As discussed above, in that case the blue shift was explained by an increase of the electronic density on the aromatic rings caused by the deprotonation process, that moves the amine-to-naphthalene charge transfer state centered on the dansyl chromophore to higher energy. In the fluorescence spectrum of **27**, an increase of the luminescence intensity was observed only upon addition of Zn^{2+} (I_{rel} = 115%, τ = 8.3 ns), while for Hg^{2+} (I_{rel} = 46%, τ = 6.9 ns) and Cd^{2+} (I_{rel} = 60%, and a double excited state lifetime, τ_1 =0.7 and τ_2 =7.0 ns), a modest decrease of the fluorescence quantum yield was monitored. Addition of Cu^{2+} led instead to an almost complete quenching of the fluorescence of the ligand (I_{rel} = 2%, τ = 0.4 ns), as expected because complexation of this ion makes energy- and electron-transfer processes accessible, providing a fast deactivation route to the ground state. The same paper[76] reported also on the photophysical properties of **28** (which still have two dansyl moieties, but with a different macrocycle). For this ligand, addition of Cd^{2+} led again to a blue shift of the luminescence bands (497 nm), but no changes in the fluorescence maximum were observed with the other metal ions. Again, a modest decrease of the fluorescence quantum yield was observed with Hg^{2+} (I_{rel} = 29%, τ = 3.8 ns) and Cd^{2+} (I_{rel} = 68%, and a double excited state lifetime, τ_1 = 0.7 and τ_2 = 6.2 ns), and an almost complete quenching with Cu^{2+} (I_{rel} = 2%, τ = 0.4 ns). In contrast to what observed with **27**, no changes were observed upon addition of up to 10 equiv. of Zn^{2+}. Furthermore, only small effects were found on the absorption and luminescence spectra upon addition of up to 10 equiv. of Zn^{2+} to a solution containing equimolar amounts of **28** and Cu^{2+}, Cd^{2+}, or Hg^{2+}. This result, obtained both by direct and by competition experiments, indicates that Zn^{2+}, in this case, does not significantly interfere with the complexation of Cu^{2+}, Cd^{2+}, and Hg^{2+}. This different complexation behavior between **27** and **28** was indeed expected, since the S atoms are known to bind Hg^{2+} preferentially over Zn^{2+}. These results are however important because they suggest that, taking advantage of their different complexation properties and their different response upon complexation, an appropriate set of ligands can be exploited as an array for the development of efficient multisensory systems. This approach, which would lead to the development of the so-called electronic tongues, could allow real time determination of the concentration of several transition metal ions at once. It is to note that it is in general still not well documented in the literature, and much less followed compared to the similar approach for the design of sensors for gaseous species that leads to the electronic noses.[77] We believe, however, that this approach is very promising for the development of efficient sensory devices.

Chemosensor **29** presents a different receptor unit, a dioxocyclam that shows a similar behaviour compared to the open-chain analogue species **9**, complexing both Cu^{2+} and Ni^{2+} ions, and exhibiting the same qualitative behaviour.[41] Sensors **29** and **9** however differ in those aspects which depend on the cyclic or non cyclic structure of the receptor that is to say: (i) the titration profiles for **29** with both cations are shifted to lower pH values, due to the thermodynamic macrocyclic effect, and (ii) the separation of Cu^{2+} and Ni^{2+} profiles is narrower for **29**, since the latter ion profits from the macrocyclic effect to a larger extent than the greater Cu^{2+} cation. This last effect causes an important disadvantage, since it reduces the pH range available for selective titration of copper in the presence of nickel ions.

Interestingly, chemosensors based on polyaminic cyclic receptors have also been very recently developed via a combinatorial approach,[78] an approach that, we believe, will be followed by many other researchers in the next future. In particular, **30,** a substituted triazacyclononane presenting a dansyl chromophore appended, has been synthesised in this way. **30** is very selective towards copper ($S_{cu/Co}$ > 500, $S_{cu/Ni}$ = 180, $S_{cu/Fe}$ = 40), revealing the presence of this ion via the quenching of the dansyl fluorescence.

Another huge class of chemosensors strictly related to the previous one has been designed and realised taking advantage from the high and easily tuneable selectivity of crown ethers towards different metal ions. One of the first and more active groups working in this direction is the one of Bouas-Laurent and Desvergne in Bordeaux, who synthesised the species **31a** and **31b**.[79-81] In low polarity solvents, their fluorescence spectra exhibit a strong emission attributed to the formation of an exciplex involving the nitrogens and the anthracene ring. In methanol these two compounds, although structurally very similar, present very different luminescence characteristics. The quantum yield of **31b** dramatically decreases due to the formation, via exciplexes intermediates, of non-fluorescent radical ions while **31a** exhibits a strong, monomer-type emission, since the nitrogen lone pairs, forming hydrogen bonds with solvent molecules, are oriented in this system outside the cavity, thus preventing exciplex formation.

Addition of alkali or alkaline earth cations to methanol solutions of both species induced an increase of the fluorescence intensity and the total disappearance of the exciplex emission. Interestingly, these two species form quite efficiently complexes also with Ag^+ ions (K_a > 10^9 M^{-1}); in this case, however, the complexation does not cause the same effect on both chemosensors. For **31a**, a strong fluorescence quenching has been observed, with the evidence of two different kinds of ground-state complexes. The major complex presents the silver ion inside the cavity, forming a non fluorescent charge-transfer complex, while the minor one has the ion residing in the external part of the ring, too far from the anthracene chromophore to influence its emission. As far as **31b** is concerned, its fluorescence upon addition of Ag^+ ions presents a different shape compared to the one observed upon addition of alkali metal ions, having a non structured component peaking at 460 nm. This contribution has been attributed to an exciplex formed by Ag^+ and the anthracene chromophore.

33 (n = 0(a), 1(b))

34 (n = 0(a), 1(b))

More recently, the asymmetric species **32** was also synthesised in the same laboratory.[82] The UV spectrum of such a ligand in acetonitrile is constituted of a structured absorption band in the 350-420 nm region, similar to the ^{1}La band typical of of anthracene derivatives, and of a non structured transition at 430-510 nm region attributed to an intramolecular charge transfer (ICT) absorption. This band disappears completely upon protonation, as expected if we assume that the nitrogen lone pair is responsible for the charge transfer transition, and, for the same reason, it is greatly reduced upon complexation of alkali metal ions, with the exception of Na$^+$. Such a particular behaviour of the sodium complex (whose association constant is 9500 M^{-1}) has been attributed to a steric interaction caused by the anthracene peri-hydrogens, which moves the ion on the side of the crown opposite to the nitrogen atom. The relative intensities of the ICT and LE bands can be used for the ratiometric titration of metal ions and, in particular, the peculiar and specific interaction of Na$^+$ with **32** allows the determination of its concentration in the presence of an excess of other alkali ions.

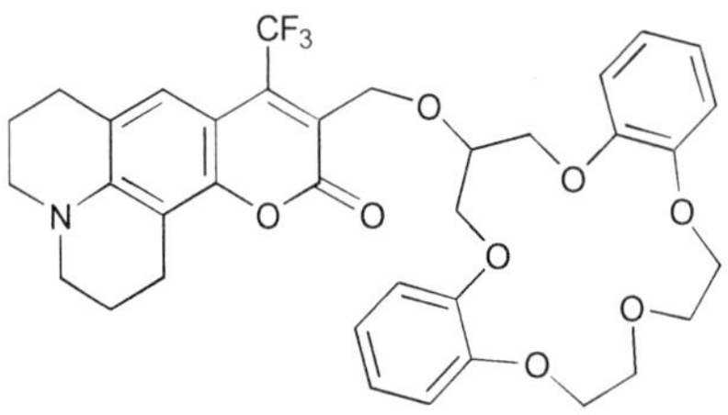

35

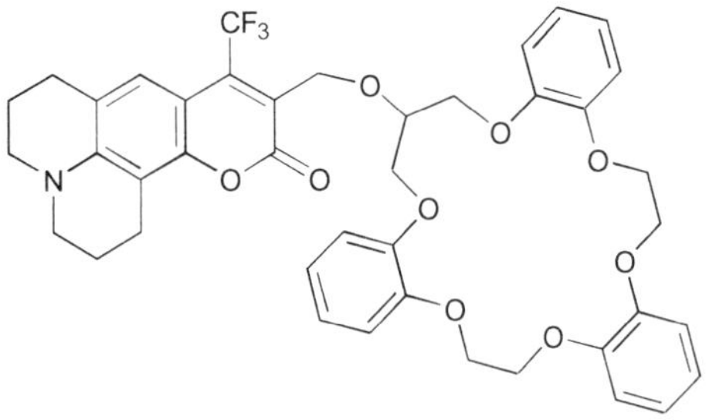

36

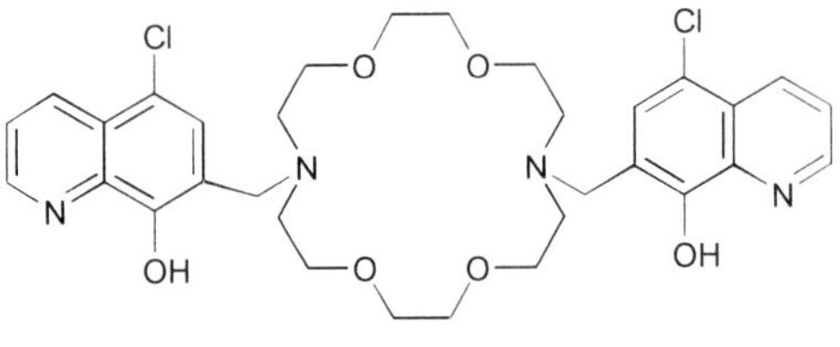

37

The species **33** and **34** studied in the Valeur's laboratory[83-85] are crown ethers derivatized with Coumarin 153, a chromophore belonging to the large family of coumarins. With these species, complexation experiments were performed with alkali, alkaline earth and transition metal ions (Ni^{2+}, Cu^{2+}, Zn^{2+}, Cd^{2+}, and Pb^{2+}). It is interesting to stress that for Coumarin 153 the dipole moment in the excited state (in which an electron is transferred from the nitrogen atom of the jololidyl ring to the carbonyl group) differs from that of the ground state as much as 8 Debyes. Since in these systems the chromophores participate to the complexation of the cations directly with the carbonyl group that acts as the electron-withdrawing unit, when a cation is co-ordinated the excited state is more stabilised than the ground state. This implies that both absorption and luminescence spectra are red shifted with respect to the free ligand. In addition to the spectral shifts, large changes in fluorescence intensities have been observed upon cation binding. As already discussed, this process reduces the efficiency of PET from the nitrogen of the macrocycle that partially quenches the coumarin chromophore in **33** and **34**. For **33**, fluorescence enhancements were observed with Ni^{2+}, Zn^{2+}, Cd^{2+}, and Pb^{2+}. The stoichiometry of these complexes is usually 1:1, with some exceptions (**33a** can form ML_2 complexes with Zn^{2+}, Cd^{2+}, and Pb^{2+}, **33b** with Pb^{2+}).

An even better selectivity towards alkali or alkaline earth metal ions was recently found for the closely related chemosensors **35** and **36**,[86] due to the higher rigidity of dibenzo- and tribenzo-crown ethers. In particular, in acetonitrile solutions both compounds showed a very high selectivity for Ca^{2+} versus Mg^{2+}, while in ethanol **35** was found to be selective for Na^+ and **36** for K^+, as expected from the relative size of the crown cavity.

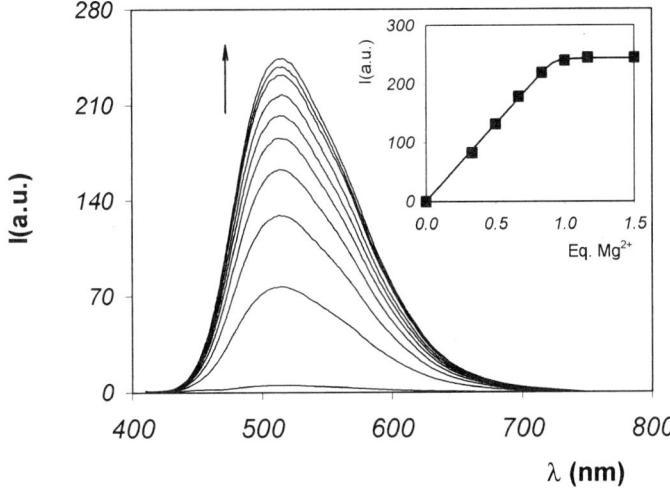

Figure 1.11. Fluorescence spectra of 37 (2.5×10^{-5} M, λ_{exc} = 360 nm) in methanol-water (1/1, v/v, pH 7.0) and upon addition of an increasing amount of Mg^{2+} ions. Inset: fluorescence intensity (λ_{exc} = 360 nm, λem = 520 nm) vs. equivalents of Mg^{2+} ions.

A large series of chemosensors combining crown ether receptors and 8-hydroxyquinoline derivatives were synthesised and studied hoping to take advantage from the interesting features of this class of chormophores. In particular, as discussed in the introduction, the idea at the base of the design of this class of sensors was to combine the capability of 8-hydroxyquinoline derivatives to modulate their fluorescence properties when complexed to suitable metal ions with the additional tool of controlling the selectivity towards the target ion given by the receptor moiety.

Compound **37**, bearing a chloride group as substituent on the 8-hydroxyquinoline showed to be a very efficient luminescent chemosensor for Mg^{2+} ions.[87] The uncomplexed species shows in MeOH-H$_2$O (1:1) a weak luminescence band centred at 540 nm. Its very low quantum yield ($\Phi < 5 \times 10^{-5}$) is consistent with the luminescence properties of the parent neutral 8-hyroxyquinoline in protic solvents, where intra- and inter-molecular excited state proton transfer processes are responsible for the high efficiency of the radiationless deactivation to the ground state. Moreover, PET processes involving the nitrogen atom of the crown offer an additional non radiative deactivation process to the ground state. Addition to a solution of this species of an increasing amount of magnesium ions leads to a large enhancement of the fluorescence band (Figure 1.11, the quantum yield of the complex is 0.042) allowing quantitative detection of this metal ion in solution with high sensitivity.

The fluorescence enhancement is associated to a process involving the deprotonation of the hydroxy groups in concomitance with the complexation of the metal ions and, for this reason, depends strongly on the pH conditions. If Mg^{2+} ions are present, the deprotonation of the hydroxy groups occurs at pH below 7, showing that **37** can indicate the presence of this ion also in physiological conditions. The other earth alkali metal ions in these conditions form less stable and not luminescent complexes, since, in their presence, deprotonation of the hydroxy groups occurs only at pH above 9. All these features, which are particular favourable, suggested us to use a very similar compound as a fluorescent probe for the intracellular detection of magnesium ions. This chemosensor could represent an improvement for this kind of analysis, also considering that commercial probes for these ions typically have stronger association constants for Ca^{2+} than for Mg^{2+}. The preliminary results obtained so far show that **37** and similar species indeed cross cellular membranes and becomes fluorescent because of the high intracellular concentration of magnesium (figure 1.12), although the quantitative relationship between fluorescent intensity and magnesium concentration is still under investigation.[88]

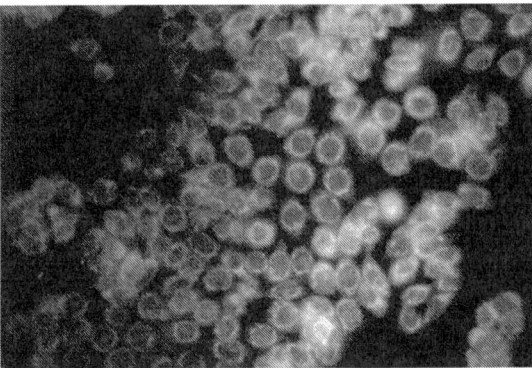

Figure 1.12. (Please see Color Inserts Section) HT29 cells (human adenocarcinoma cell line), grown on glass coverslip and stained with 37 (50 uM).

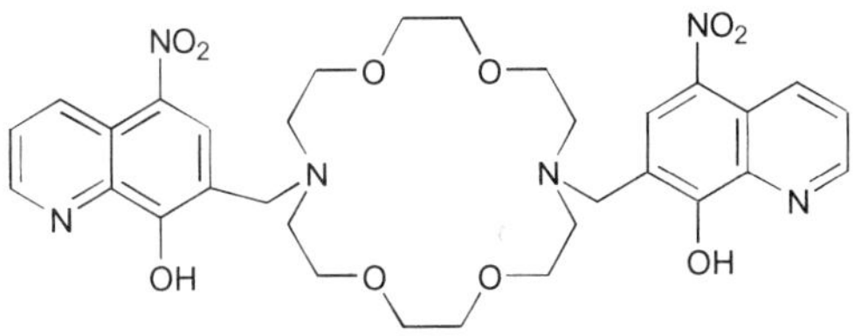

PROBES AND SENSORS FOR CATIONS

38

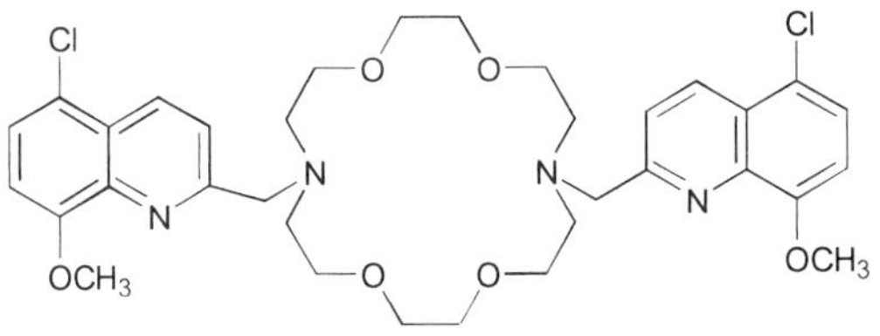

39

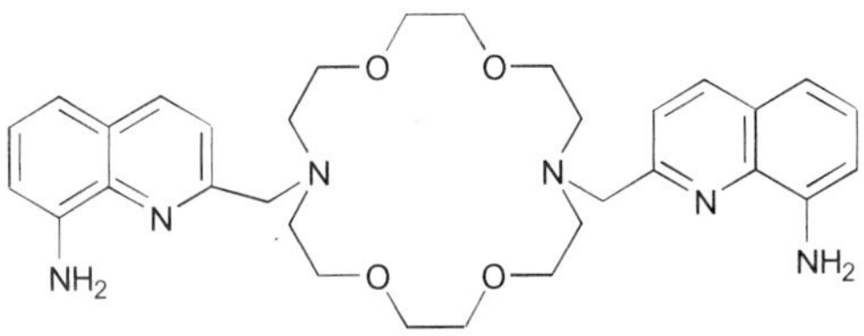

40

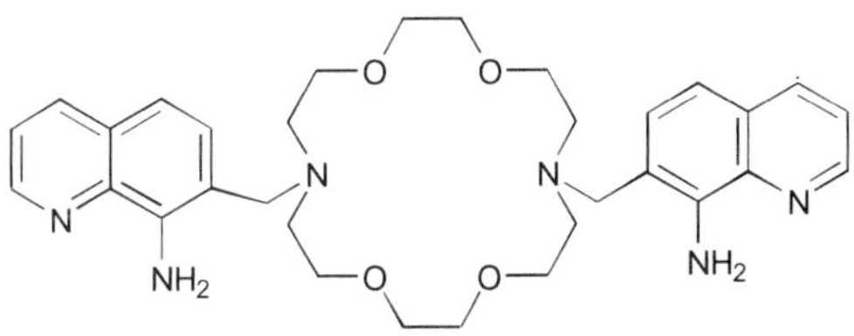

41

38, that is very similar to **37**, having in the 5 position a nitro group instead of the chlorine one, forms on the contrary stable complexes with Hg^{2+} ions.[89] The substitution with the nitro group was intended to change significantly the acid-base properties of pyridine nitrogen and of the hydroxy group, together with the complexation abilities. The absorption spectrum in methanol-water (1/1, v/v, pH 7.0), that is dominated in the visible region by a charge transfer band involving the electron withdrawing nitro groups, upon Hg^{2+} complexation shows noticeable changes (figure 1.13).

The uncomplexed macrocycle shows a luminescence band centred at ca 470 nm, and its very low quantum yield (5×10^{-5}) is consistent with the alredy discussed luminescence properties of the parent chromophore; also in this case PET processes can not be ruled out. Upon complexation with Hg^{2+} and excitation at 420 nm, the fluorescence intensity of **38** increases by a factor of 12 (figure 1.14), and the association constant was estimated to be greater than $1 \times 10^8 \text{ M}^{-1}$. The fluorescence intensity reached the maximum value when one equivalent of the metal ions was added, indicating that the stoichiometry of the complex is 1:1.

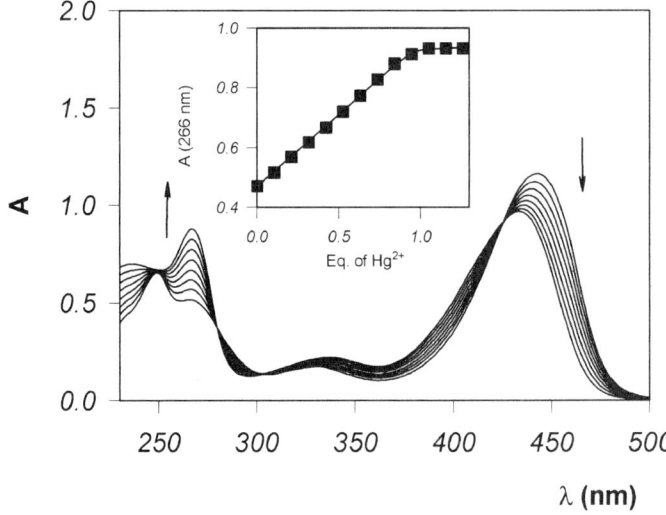

Figure 1.13. Absorption spectra of **38** (2.5×10^{-5} M,) in methanol-water (1/1, v/v, pH 7.0) and upon addition of an increasing amount of Hg^{2+} ions.

The fluorescence enhancement can again be observed only when the deprotonation of the hydroxy group occurs, in this case at pH above 3. From the point of view of possible applications, it is evident that a 12-fold intensity enhancement is not high enough to ensure the sensitivity needed for the detection of mercury in high diluted conditions. However, this is a very rare example of fluorescent enhancement upon mercury complexation since, as already discussed in previous examples, this metal ions typically causes a luminescence quenching, usually through the heavy atom effect. For this reason, we believe that this system can be a good starting point for the design of more appropriate sensors.

PROBES AND SENSORS FOR CATIONS

Another possible way to tune the sensitivity and selectivity of this class of compounds was found to be the substitution of the hydroxy group with a methoxy function as in **39**.[90] In this case, the dependence of the photophysical properties on pH is obviously less dramatic for two reasons: (i) the chromophore has lost one acid centre and (ii) intramolecular proton transfer in this system is no longer possible. On the other hand, in protic solvents intermolecular proton transfer is still active, together with PET processes involving the nitrogen of the crown moiety. **39** is selective for Cd^{2+} ions and its complexation caused the highest fluorescence enhancement among all the transition metal ions studied. The quantum yield of the complex is remarkably high (0.66 in methanol solution), and it is accompanied by high molar absorption coefficients (95000 M^{-1} cm^{-1} at 252 nm) although in the UV region. These interesting features makes **39** a promising candidate as a chemosensor for the detection of Cd^{2+}, and studies are in progress for its immobilisation on quartz or silica surfaces, a fundamental step for practical application. It is to note that also many other species were also studied in this research project; all the data obtained were of use for the design of new species that are now the object of synthetic efforts.[91]

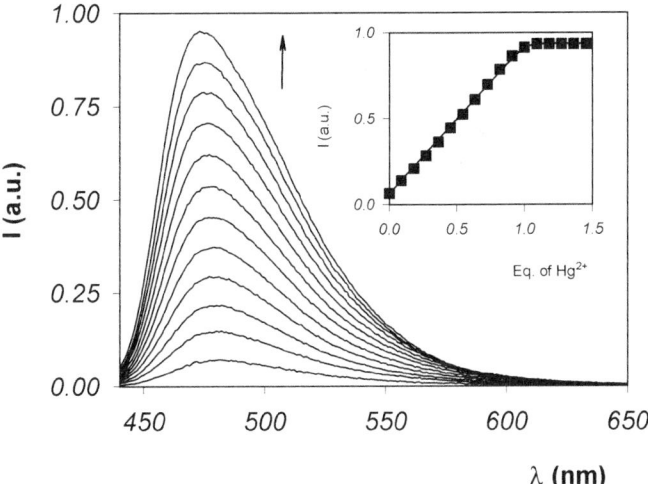

Figure 1.14. Fluorescence spectra of **38** (2.5 x 10^{-5} M, λ_{exc} = 430 nm) in methanol-water (1/1, v/v, pH 7.0) and upon addition of an increasing amount of Hg^{2+} ions. Inset: fluorescence intensity (λ_{exc} = 430 nm, λ_{em} = 480 nm) vs. equivalents of Hg^{2+} ions.

The modification of the hydroxy group in an amino one leads to the synthesis of **40** and **41**.[92] In this case, however, the addition of metal cation caused a quenching of the intense fluorescence of the ligands in acetonitrile, also when earth alkali and d^{10} metal ions were added. Interestingly, while **41** forms very stable 1:1 complexes with all the metal ions studied, for **40** this is true only for Ca^{2+}, Sr^{2+}, and Ba^{2+}, while with Mg^{2+}, Zn^{2+}, Cd^{2+}, and Hg^{2+} it forms complexes with different stoichiometries.

Appending a benzene sulphonamido group to the last two lariat ethers, Xue et al.[93] synthesised **42** and **43**, which bear the same chromophoric group of the acyclic indicator Zinquin. The responses of these two ligands to the addition of metal ions are dramatic. In particular, addition of Zn^{2+} ions to a solution of **42** or **43** leads to strong changes of the absorption spectra until one equivalent of metal ion has been added, while no changes were observed for further additions. A different pattern is in turn observed for the fluorescence spectra. The fluorescence intensity, in fact, increases in an almost linear way up to the addition of two equivalents of zinc. All these findings can be rationalized in terms of the formation in solution of two different complexes with a 1:1 and 1:2 (ligand:metal) stoichiometry, respectively, as also confirmed by X-ray data. Qualitatively, the changes observed on the absorption and luminescence spectra are very similar to those reported for TSQ and Zinquin at the same pH. For these two ligands at neutral pH, the complexation process is associated with the deprotonation of the sulfonamido group. From the changes in the absorption spectra, it is evident that the inclusion of the first metal ion leads to the deprotonation of both the chromophoric groups, while the formation of the binuclear complex causes only minor changes. The increase of the charge density close to the two chromophoric groups is however sufficient to further increase the luminescence band that, having a charge transfer character, is very sensitive to the polarity of the environment. The increase in the excited state lifetime observed on going from the mononuclear to the binuclear complex is a further support in this direction.

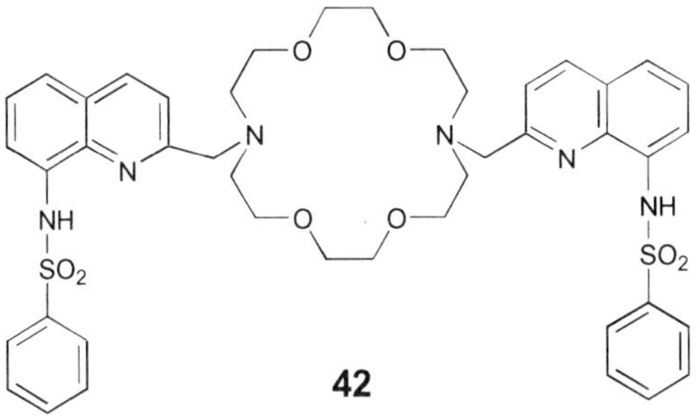

42

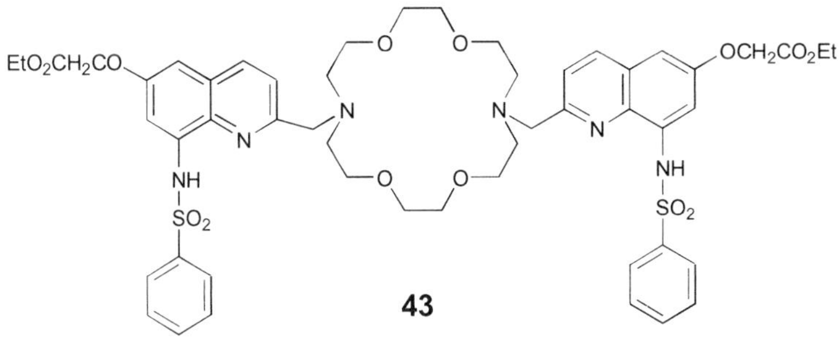

43

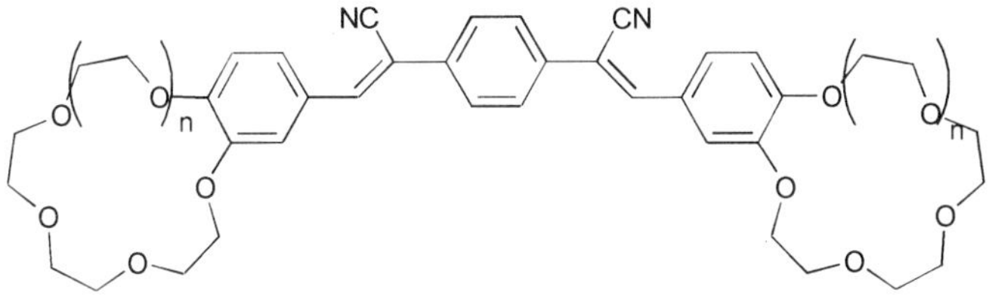

44a; n = 1
44b; n = 2

Compounds **44a,b**[94,95] are instead based on a distyryl benzene unit synthesised as Z-Z isomer. This form of the chromophore, when excited, is weakly luminescent and can give isomerisation to the E-Z form, that is not luminescent. Interestingly, when a potassium salt is added to a methanol/acetonitrile solution of **44a**,[94] the luminescence intensity increases by a factor of nearly 20. This finding is rationalised in terms of the formation of a 2:2 (metal:ligand) sandwich complex upon addition of the metal ion. In this complex, the efficiency of the deactivation route to the ground state via the isomerisation process is drastically reduced, and, as a consequence, the radiative process enhanced. A similar effect was found also for **44b** upon addition of caesium ions.[95] In this case, addition of K^+ or Na^+ has almost no influence on the fluorescence intensity.

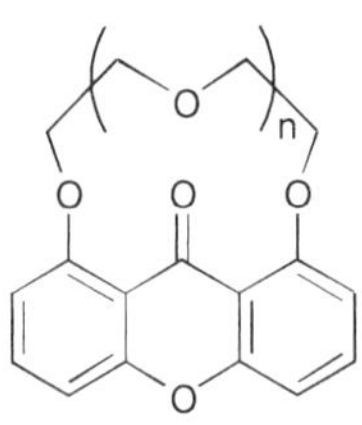

45a; n = 1
45b; n = 2
45c; n = 3

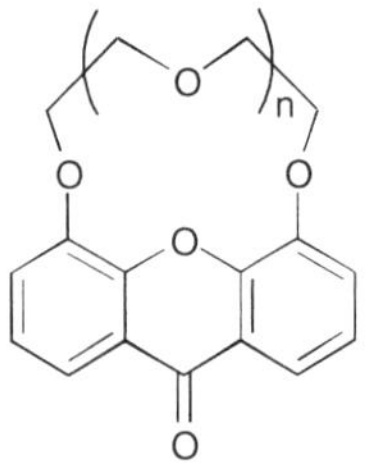

46a; n = 1
46b; n = 2
46c; n = 3

A much higher increase in the fluorescence quantum yield (more than a thousand-fold) was found for some complexes of a family of crown ethers incorporating a 1,8-dioxyxanthone residue (**45a-c**) with earth alkali metal ions.[96] The different photophysical behaviour was accounted for by the stabilisation, upon complexation, of a different ground state conformer of the chromophore (figure 1.15). This conformer presents a negative charge on the oxygen atom inside the cavity, a feature that can explain its stabilisation upon complexation with a doubly charged cation, and a planar geometry, in contrast with the belt one observed for the free ligand, whose lowest excited state is a non emissive n-π* one.

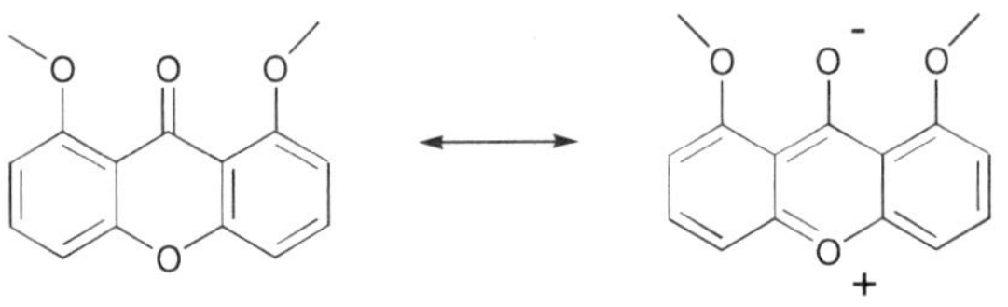

Figure 1.15. Structure and contributing canonical forms of 1,8 dioxyxanthone.

The planar geometry leads instead to a lowering in energy of the emitting π-π* level, which becomes in the complex the lowest excited state. This explanation received further ground upon the successive investigation of a series of crown ether incorporating the 4,5-dioxyxanthone residue (**46a-c**).[97] Molecular modelling has shown that the free ligand, that is highly luminescent, has in this case a planar geometry, with a positive charge pointing inside the cavity. The complexation with a positive species causes the destabilisation of this conformation, leading to the belting of the structure of the chromophore. The consequence of this phenomenon is a remarkable quenching of the fluorescence upon complexation with alkali metal ion. The earth metal ions, as demonstrated by competition and conductivity experiments, are not complexed by these sensors with sufficient affinity, most probably because of the high repulsion between the divalent cations and the positive charge inside the macrocycle. It is also remarkable that the complexes of the **45a-c** family with earth alkali metal ions show in deaerated solution at room temperature a delayed fluorescence with a lifetime in the range of 10-60 μs, a feature that could lead to higher sensitivities if time resolved spectroscopy is applied. On the contrary, for the **46a-c** family no delayed luminescence was observed, indicating that in the 4,5-dioxyxanthones the energy gap between the fluorescent lowest singlet state and the lower lying triplet state is higher, and cannot be overcome even at room temperature.

The association constants for the different metal ions depends, as expected, on the dimension of the cavity of the crown ether; in particular **45c** shows good selectivity towards Ba^{2+} ions, so that this crown ether can work as a sensor for Ba^{2+} in the presence of the alkaline and alkaline-earth metal ions Na^+, K^+, Cs^+, Mg^{2+} Sr^{2+}, and Ca^{2+}. The range of selectivities towards the alkaline and alkali earth metal ions together with the favourable photophysical properties shown by these families of crown ethers lead us to propose their use for the design of an electronic tongue which could, in principle, determine simultaneously the concentrations of Na^+, K^+, Cs^+, Ba^{2+}, Sr^{2+}, and Ca^{2+}, as a function of time. An additional relevant advantage for a possible medical or biological application in physiological conditions is that the complexation of these species does not depend on the pH of the solution at least in the range 1-12.

47a; X = O
47b; X = S

48a; X = O
48b; X = S

An other extremely sensitive fluorescence chemosensor for metal ion is **47a**.[98] This compound forms 1:1 complexes with alkali and alkaline-earth metal ions. In these complexes, coordination of the cation to the nitrogen atom of the crown inhibits a charge transfer process that is responsible for the efficient deactivation of the highly luminescent state localised on the boron-dipyrromethane (DPE) dye in the free ligand. This leads to an enhancement, in acetonitrile solutions, of its fluorescence and excited state lifetime by more than a 1000-fold. Interestingly, except for Mg^{2+} ions, two different conformers of the complexes are present in solution. The change of the donor atom in the receptor from oxygen to sulphur leads to compound **47b**,[99] that has been shown to be a very sensitive fluoroionophore for Ag^I, Cu^{II}, and Hg^{II} ions. In particular for the latter ion, that is typically known as a fluorescence quencher, a high association constant (Log K_a > 5.75) and a very large fluorescence enhancement factor (223) have been found in acetonitrile solution.

49a: X = O
49b: X = S

Rurack et al. also reported[100] on the dependence of the photophysical properties of several 1-benzothiazole-3-(4-donor)-phenyl-substituted prop-2-en-1-ones (and, among them of **48a,b**) as a function of solvent polarity, temperature, and metal ion complexation. In particular, they found that in acetonitrile **48a** shows cation induced fluorescence enhancement in the presence of alkali and earth alkali metal ions, while, as expected, its analogue **48b** binds selectively to Hg^{II}, Ag^{I}, and Cu^{II}. To be noted, also in this case, the increase of the luminescence for the latter ligand even upon coordination to Hg^{II}. A similar family of compounds, **49a,b** was studied two years later by the same group.[101] Again, a complete and elegant study of the photophysical properties was presented. The fluorescence of the free molecules is efficiently quenched via an electron transfer mechanism, which is switched off by the presence of alkali, earth-alkali, as well as transition metal cations. On the basis of these results, the same group synthetised the bi-functional fluoroionophore **50**,[102] with the aim to signal whether only one, both, or none of the binding sites are occupied by substrates. In **50** both nitrogen donor atoms are formally π-conjugated with the chromophore, The binding of Ca^{2+} by the crown ether, or of Ag^+ by the thioether counterpart, reduces the respective donor strength due to coordinative engagement of the nitrogen atom's lone electron pair, and leads to slightly blue-shifted absorption spectra. On the other hand, formation of the Ca^{2+}-**50**-Ag^+ complex largely deprives the chromophoric systems of both the nitrogen atom lone pairs, leading to a pronounced hypsochromic shift in absorption. As far as the luminescence signals are concerned, the quantum yield are 0.05, 0.02, 0.27, and 0.04 for **50**, Ca^{2+}-**50**, **50**-Ag^+, and Ca^{2+}-**50**-Ag^+, respectively. In terms of digital action, the fluorescence dependence upon the presence of the two cations represents a "INHIBIT" logic function. Rurack et al.[103] recently reported about the properties of **51**, a new chemosensor for mercury ions based on a phenoxazine scaffold, which is also at the heart of widely used polarity probes of the Nile Red family and ensures sufficient water solubility and an intense fluorescence in the visible region. In this chemosensor, that is a donor-acceptor ensemble, the carbonyl group acts as acceptor in two important processes, i.e., intramolecular charge transfer and intermolecular hydrogen bonding or, when integrated in a metal recognition unit, coordinative bond formation. In the presence of mercury ions, in acetonitrile the strong fluorescence ($\Phi = 0.65$) is almost completely quenched while in water, where a lower quantum yield of the free ligand is observed (0.08), the fluorescence maximum is blue-shifted (from 634 to 615) and the intensity is reduced to 50%. In water, a high association constant is also observed ($K = 1.20 \times 10^6$ M^{-1}).

51

52

Four heteroditopic cryptands (**52-55**) for transition metal ions have been proposed by Ghosh et al.[104-106]. These systems show the typical structured fluorescence band of the anthracene monomer along with a red-shifted broad structureless band centred at 550 nm. The authors observed a dependence of the ratio of the two intensities on ligand concentration; nevertheless they attributed the red shifted component to an intramolecular exciplex, as found for **31**.[79-81] The overall fluorescence quantum yield is much lower for **52-54** ($\Phi_{FT} \leq 0.002$) than that usually observed for anthracene derivatives, since in these systems a very efficient PET process from the aminic nitrogen to the anthracene moieties also occurs. On the contrary, the fluorescence of **55** is very similar to that of the free anthracene chromophore. The authors attributed this behaviour to the increased rigidity introduced by the carbonyl spacers, that are supposed to drastically reduce the efficiency of the PET process, since in this molecule the donor and acceptor are not be properly mutually oriented. It is to note, however, that in this case the donors are not all amine groups, as in the previous cases, but three of them are replaced by amide groups, that are known to be much less efficient in giving PET processes, because of a much lower electron donor ability. Addition of many transition metal ions (Mn(II), Co(II), Ni(II), Cu(II), Zn(II), Pb(II), and Fe(III),) to THF solutions of **52-54** leads to the formation of 1:1 adducts, in which the monomer fluorescence is increased by more than a hundred-fold for all the metal ions, and no exciplex emission could be observed. Among the different receptors, the more rigid **54** shows the lower enhancement, while the more flexible **53** gives the highest one. Small enhancements are instead observed in **55**, because of the low PET efficiency observed for the free ligand. Speaking about the different metal ions, as expected Zn(II) induces the highest fluorescence since the complexation with this species makes the PET process thermodynamically forbidden, without introducing other quenching mechanisms. The noticeably high enhancement observed with the other metal ions was explained in terms of a communication gap between the metal ion and the anthracene fragments, which would avoid efficient energy- or electron-transfer processes. Differently to the other metal ions, silver and mercury ones did not induce a fluorescence enhancement, probably because they do not fit into the cavity. The selectivity of these hosts for the different metal ions was not discussed in detail.

A very interesting and promising section of the research in the field of chemosensors is devoted to the synthesis of chemical species based of the excellent properties of calixarene derivatives as receptors for metal ions. In theory, the skeleton of many calixarenes, made by benzene derivatives, is itself luminescent, and this luminescence can be modulated by their complexation with metal ions. This is the case, for example, of 1,3-alternate calix[4]arene-crowns,[107] whose fluorescence was modulated by metal-π

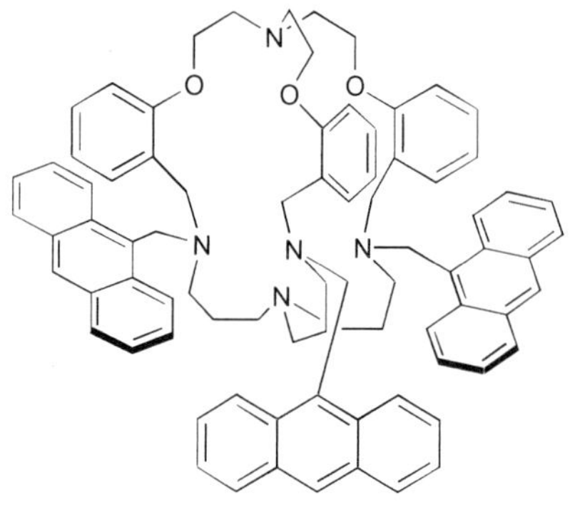

53 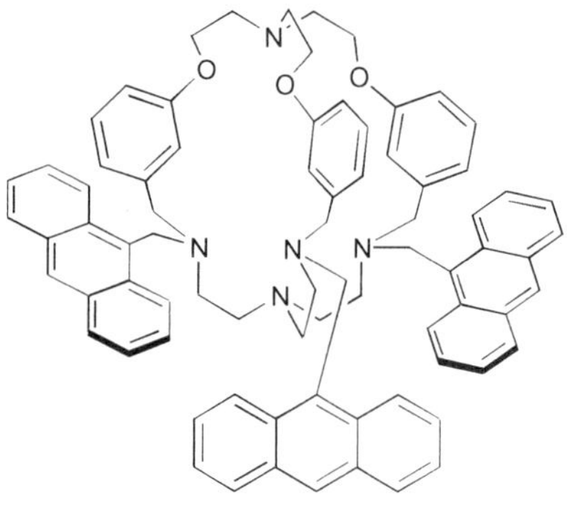   **54**

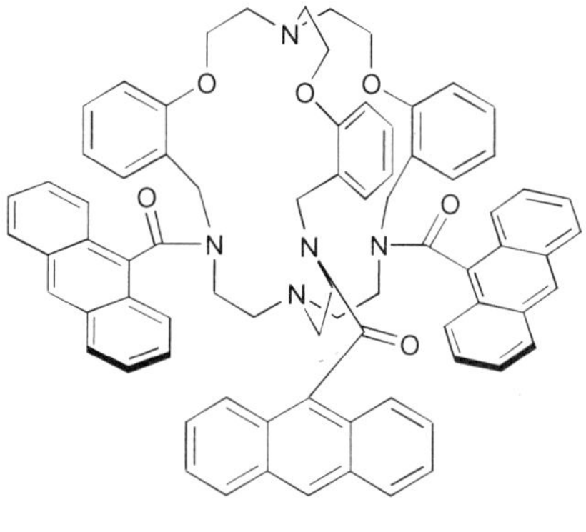

55

interactions. However, the fluorescence of calixarenes falls typically in the 300-350 nm region, and their absorption coefficients are also relatively low, so that usually one or more chromophores are inserted in the structure in order to have a more efficient signalling process. For example, in **56** and **57** a naphthalene and a binaphthyl chromophore were inserted in the polyethoxyethylene part, respectively.[108] In both these cases, an intensity decrease in the absorption and fluorescence spectra of the chromophores was observed upon metal ion complexation. This effect can be attributed to the interaction of the metal cation with the oxygen atoms directly linked to the chromophores that reduce in this way their conjugative effect to the π-systems of the aromatic rings. In addition, in **57** the complexation induces also a distortion of the angle between the two naphthalene rings, further reducing the conjugation of the system, so that the strongest changes could be observed for this compound. Among all the metal ions investigated, the highest affinity and sensitivity of both these calixcrowns were found for Cs^+ ions. It is to note that **56** shows the highest efficiency for caesium ion binding among the calix-crown-6 known so far.

56 **57**

More recently we reported[109] the synthesis and characterisation of **58** and **59**, that combine the fluorogenic properties of hydroxyquinoline derivatives with the metal ion binding ability of the tetramide of *p-tert*-butylcalix[4]-arene, that is probably one of the most studied calixarene ligands, since it binds alkali and alkaline earth metal ions with association constants comparable to or higher than those of cryptands. Contrary to what was observed with quinoline-functionalised crown ethers, no changes in the absorption or emission spectra were observed for either ligand **58** or **59** with Hg^{2+}, Mg^{2+}, and Zn^{2+} metal ions. In addition, it is to note that the fluorescence of these two ligands is relatively high compared to that one of **37-39**. In these systems, in fact, two of the main mechanism for the non radiative deactivation of the excited state present in the azacrown derivatized systems, namely intramolecular proton transfer and PET, are not present. Alkali and alkaline earth metal ions complexes of **58** usually show a bathochromic shift and a quenching of the fluorescence band with respect to the free ligand; on the contrary, the corresponding complexes of **59** present hypsochromic shifts and an increase of fluorescence intensities for alkaline earth cation complexes. NMR spectroscopic data and semiempirical calculations support the hypothesis that this difference in the photophysical properties is due to a different mode of coordination of the quinoline part to the metal ion. In the complexes of **58** both the quinoline N and O atoms are coordinated while in complexes of **59** the N atom is too far from the metal ion. Although the changes in fluorescent properties were lower than expected on the basis of the studied performed on the crown ethers **37-39**, the results were very interesting for understanding the importance of metal ligand interactions with hydroxyquinoline ligands.

58

59

60 is a calixarene bearing in the lower rim three ester groups and a 6-acyl-2-methoxynaphthalene (AMN) chromophore.[110,111] AMN is known to undergo intramolecular charge transfer upon excitation and, for this reason, to be environmentally sensitive. The fluorophore is linked to the calix[4]arene so that the carbonyl group participate in the complexation process. Consequently, cation-binding results in the enhancement of photoinduced charge transfer with concomitant marked changes of photophysical characteristics; in particular, a red shift and a large increase of the fluorescence can be observed. In water-ethanol mixture, a good selectivity was found for sodium ions, and the fluorescence quantum yield is 30 times larger than that of the free ligand. Moreover, in acetonitrile solution the calcium complex has a fluorescence quantum yield as high as 0.68. This exceptional behaviour can be interpreted in terms of the relative locations of the $n\pi^*$ and $\pi\pi^*$ levels, which depend on the charge density of the bound cation.

A sensitive and selective fluorescent sensor for Pb^{2+} was similarly obtained by the same group inserting dansyl chromophores at the lower rim of a *p-tert*-butylcalix[4]-arene to give **61**.[112] Interestingly, this sensor exhibits very efficient binding in acetonitrile-water for lead ions; the changes in the luminescence properties (a 52 nm blue shift and a 1.7-fold increase of the quantum yield were observed at pH 5.2) allow a detection limit of 2×10^{-8} M.

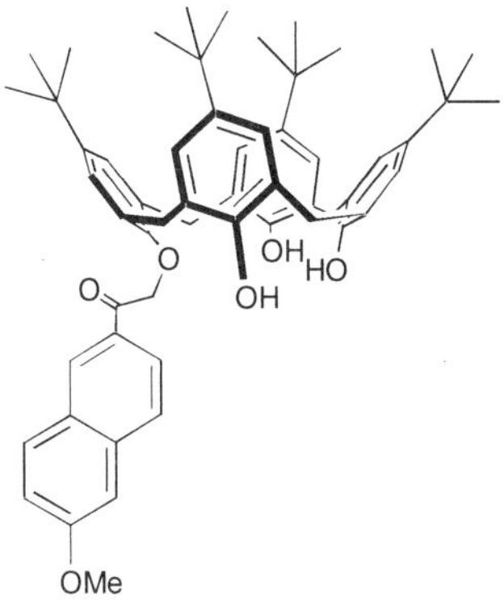

60

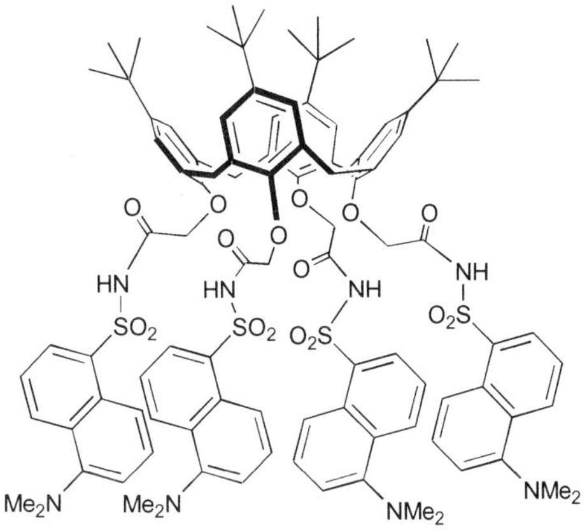

61

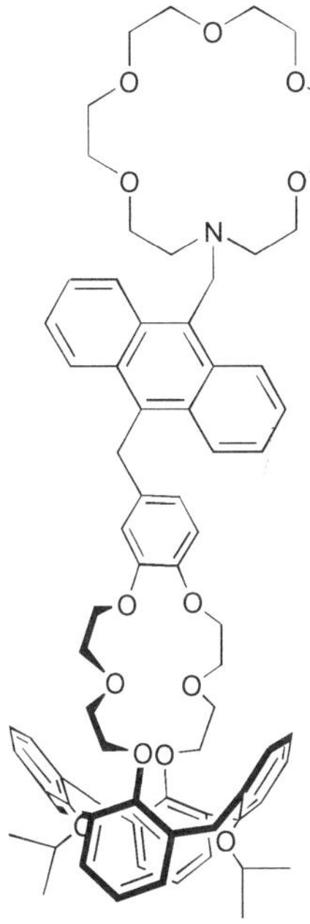

62

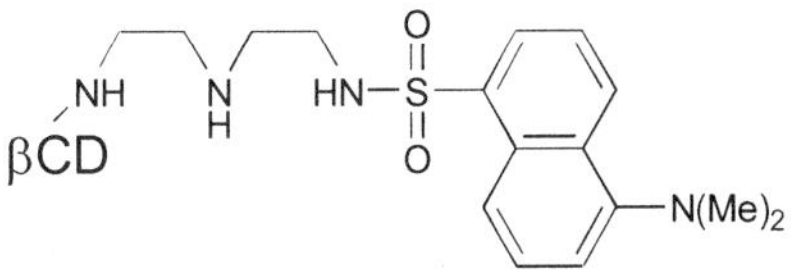

63

A more complex system, **62**, has been proposed by Ji et al.[113,114] In this molecule, the fluorescence of the anthracene is quenched via a PET mechanism involving the electron donor dimethoxybenzene moiety inserted in the crown ring. The two cavities present in the host are able to bind selectively Cs^+ ions, as already found for simple 1,3-alternate calix[4]arene-crown-6 ligand. Metal ion complexation rises the oxidation potential of the dimethoxybenzene unit, so that electron transfer can no longer occur; as a consequence a more than a ten-fold increase of the anthracene fluorescence can be observed in CH_2Cl_2-CH_3OH solution when both sites are engaged with a cation.

The dansyl moiety was also used in compound **63**,[115] in which this chromophore was appended to a β-cyclodextrin via a diethylenetriamino spacer. **63** is highly luminescent also in water, where usually the dansyl chromophore is poorly emissive, since its inclusion in the cyclodextrine cavity provides it a sufficient apolar environment. When copper ions are added to a solution of **63**, efficient quenching of the luminescence can be observed. This effect is promoted by the complexation of the ion by a donor group adjacent to the dansyl moiety, with subsequent abstraction of the sulphonamide hydrogen, leading to the observed quenching. Copper complexes of similar systems containg a β-cyclodextrin and a spacer containing a phenylalanine connected with a dansyl group were than used as probles for the chiral recognition of aminocids, with very interesting results in the determination of the optical purity of proline.[116,117]

1.4. SENSORS WITH DENDRIMER-, PEPTIDE- AND PROTEIN- BASED RECEPTOR

Very recently, the use of dendrimers has been extended to the field of chemosensors, in particular by the groups of Balzani and Vögtle.[118-120] They have synthesised and characterised the dependence of the fluorescence intensity of two different families of dendrimers containing dansyl units upon addition of different metal ions. They have demonstrated that the fluorescence of **64** is completely quenched when even only one Co^{2+} ion is complexed by the system, and that, for the largest species, up to 32 dansyl units are quenched simultaneously. This is a very nice example of signal amplification, a mechanism that results in an increase of the sensitivity of the whole system. This feature makes dendrimers very interesting systems for the design of sensory devices. In **65**, addition of Co^{2+} and Ni^{2+} causes a quenching of fluorescence; at low metal ion concentration, each ion quenches about 9 dansyl units; also in this case the fluorescence quenching takes place by a static mechanism involving coordination of metal ion in the interior of the dendrimer. Zn^{2+} ions cause very small changes on the fluorescence of **65**, but they are able to displace the former two ions when present, with a concomitant revival of the dansyl fluorescence.

Proteins are particularly selective and avid receptor units for divalent metal cations and many researchers are investigating this very attracting field. In fact, the analysis of transition metal ions in trace amounts is extremely important for environmental and biomedical applications. Because of the very low concentrations, it is however almost impossible their quantification without the introduction of very specific receptor systems such as indeed proteins.

B. Imperiali et al. have for example investigated the possibility of using polypeptides architectures for metal ions recognition and in particular to design fluorescent chemosensors sensitive to nanomolar concentrations of Zn^{2+} [121-124] or Cu^{2+} [125] in water at physiological pH.

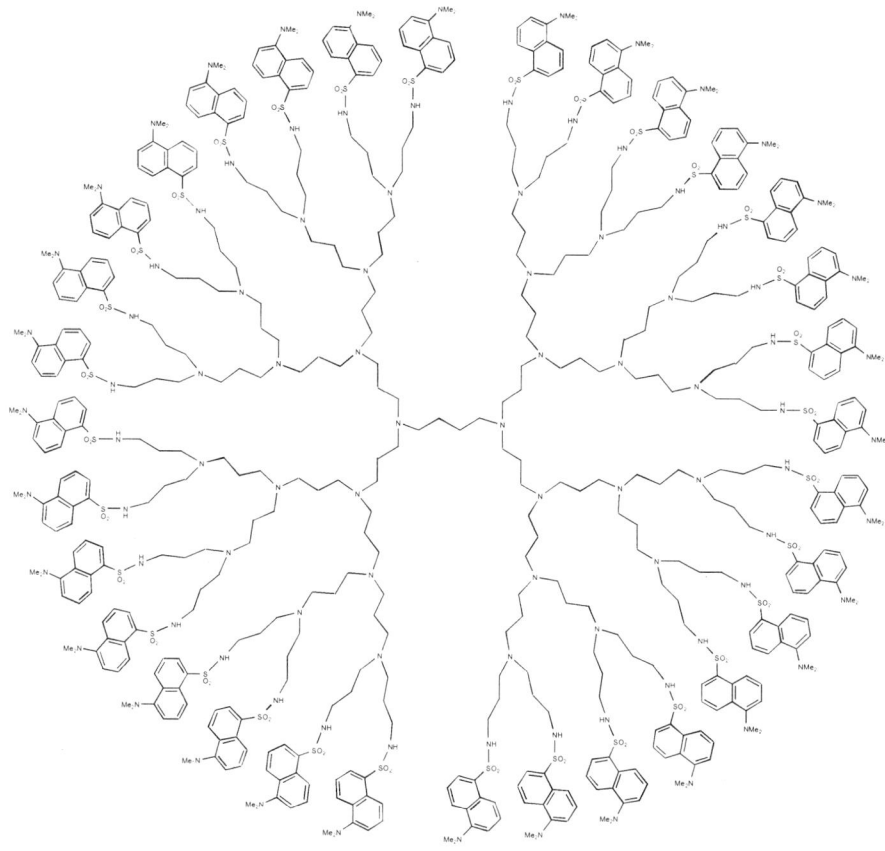

64

They reported[121] the preparation and characterisation of derivatized polypeptides acting as luminescent sensors for metal ions, undergoing a metal ion-induced conformational change that produces noticeable differences in the fluorescence properties of the peptide itself. An assembly of amino acid building blocks derivatized with a chromophore (cyanoanthracene), a metal co-ordinating site (base on bipyridine derivatives) and a donor unit (L-3,4-dimethoxy-DOPA) able to quench the fluorophore luminescence, was described. A metal ion-dependent PET luminescence quenching was observed upon titration with Zn^{2+} and/or Co^{2+} in micromolar concentrations in MeOH or in aqueous solutions and the system exhibited total reversibility upon addition of 1,10-phenanthroline.

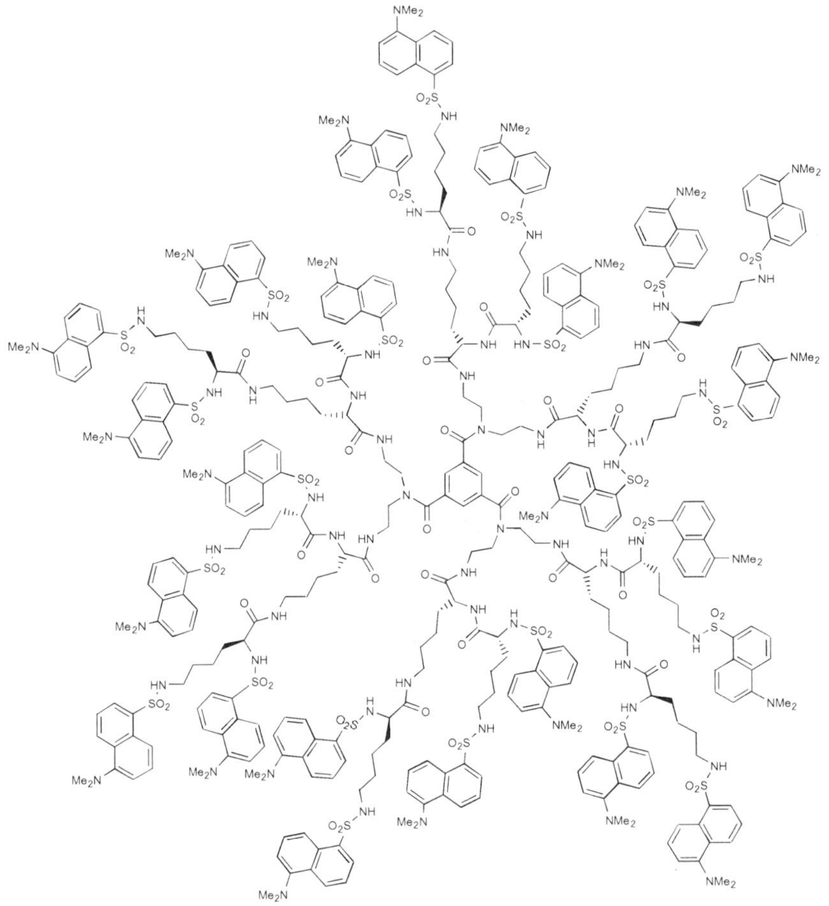

105

Following the same idea B. Imperiali and her co-workers used also zinc finger domains[122,123] as selective and sensitive receptors and replaced one of the latter residues with a dansylated β–aminoaniline amino acid derivative to monitor the metal-binding event. As already mentioned, dansyl is an environment sensitive fluorophore; the metal binding causes an enhancement of the fluorescence intensity, since it induces a peptide folding that leads to a shielding of the fluorophore from the solvent molecules. The increase of fluorescence as a function of the increasing amounts of Zn^{2+} in solution is linear in the 0.1 - 1.0 µM concentration range of the metal ion. It has been shown that Na^+, Mg^{2+} and Co^{2+} do not interfere in the analysis of Zn^{2+} even if they are present in more than a thousand-fold excess. The indispensable characteristic of reversibility is also proved by fluorescence intensity measurements before and after the addition of EDTA to a solution of the complexed chemosensor.

The peptide skeleton can be easily modified and this allowed the investigation[123] of different possible structures in order to obtain a better oxidative stability and the best luminescence signal. Other different fluorophores have been covalently attached to the peptide structure but dansyl was proved to offer the highest sensitivity. The thiolate ligand, formed by cysteine residues, is the weak site towards oxidation but its substitution is not trivial, since it would change the binding affinity for Zn^{2+}. After some different attempts aspartic acid was selected as a possible substitute of Cys^6, since it was the one that was able to mimic more closely the steric and charge neutralisation properties of the original binding site. A fluorescent chemosensor with analogues performances[122] but enhanced oxidative stability was obtained in this way.

A further step towards application in real matrixes is to support chemosensors on water-solvated resins to realise regenerable devices. To reach this goal, however, the polypeptide described above are not suitable because of their very extended structures that do not allow the realisation of a well-defined sensing device. To solve this problem it has been described[124] the synthesis of short peptides (7 residues) sensitive to sub-micromolar levels of Zn^{2+} and containing 8-hydroxyquinoline. Particular peptide sequences were designed to promote folding resulting in enhanced metal-binding properties and cysteine was incorporated to favour Zn^{2+} selectivity. A detection limit less than 250 nM was found for Zn^{2+} ions; Fe^{2+} and Cd^{2+} showed minor competition (approximately 10%) while Cu^{2+} is a strong interfering agent: approximately 40% of binding at equivalent concentrations.

The study of Cu^{2+} ions is however interesting by itself for many applications and they also studied polypeptide chemosensors selective towards Cu^{2+},[125] using the metal binding properties of amino terminal motifs found in the serum albumin, which bind both Cu^{2+} and Ni^{2+} with very high affinity. The dansyl fluorophore was appropriately covalently linked and again the metal binding intramolecular quenching signals. The fluorescence quenching was linear until the chemosensor was saturated at 1 equivalent of added metal ion and further addition of divalent metal produced no further change in the luminescence intensity. One equivalent of Cu^{2+} causes a greater change in fluorescence (7% of initial intensity) than the same amount of Ni^{2+} (~15-35% of initial intensity) and the other only interfering metal ion seems to be Fe^{2+} (~10% quenching). The selectivity towards Cu^{2+} can be increased via modification of the peptide backbone. The induced quenching is completely recovered via an addition of an excess of EDTA. Analogues species immobilised on solid phase were also prepared.

Godwin and Berg followed the same kind of strategy to obtain chemosensors for Zn^{2+}.[126] They reported the study of a new ratioable fluorescent probe that exhibits as a receptor unit the zinc finger consensus peptide (CP) that, as already reported, binds zinc tightly and selectively with consequent changes in flexibility and structure.[127] They linked to the CP unit two fluorescent moieties: fluorescein as an energy donor and lissamine as the acceptor. When the peptide is unfolded, the dyes are relatively far apart and they almost do not interact; on the contrary, Zn^{2+} ions complexation induces the folding, moving the fluorophores closer to each other. As a consequence, an appreciable intramolecular energy transfer can be evidenced. The fluorescence spectrum presents the two maxima corresponding to the fluorescein and lissamine emissions and the two intensities change differently upon addition of zinc yielding a ratioable chemosensor. This is also an excellent candidate for *in vivo* studies since the emission in the visible region is not covered by the luminescence of other cellular components.

R.B. Thompson et al. are also working in the field of optical sensors for Zn^{2+} in water using the specific ion recognition ability of the enzyme human carbonic anhydrase II.[128-131] In this case the fluorophore moiety (a dansyl derivative) is not covalently attached to the skeleton of the enzyme, but it is diffusing in solution. Dansylamide is able to bind the Zn^{2+} only when it is linked to the active site, showing, in this conditions, a fluorescence that is enhanced and shifted towards higher energies, so that the concentration of Zn^{2+} ions is proportional to the ratio of fluorescence intensity at two wavelengths. Mn^{2+} and Co^{2+} give minor interferences. A major drawback of this system is connected to the off rate of Zn^{2+} ions, that is too slow even for very low data rate applications. The application of this system to a fibre optic needs some further investigation, since the first attempts were not very successful because the membrane that should permit the exchange of the Zn^{2+} ions between the sample and the fibre optic is permeable to the dansylamide molecules as well.

Changes in fluorescence are very useful in order to monitor metal ions concentrations but sometimes some optical effects and/or other sources of error unfortunately invalidate the measurements. Methods based on fluorescence lifetime[129] measurements are for sure more precise but they present some disadvantages, especially since the instrumentation is usually quite sophisticated.

Thompson and co-workers introduced fluorescence anisotropy as a new approach for determination of metal ions in solution.[130-132] The changes in fluorescence anisotropy are caused by changes in rotational motion induced by molecular size changes. In a first attempt[130] they synthesised some *para*-substituted benzenesulfonamide which are able to bind Zn^{2+} ions when they are complexed by the enzyme. The changes in mobility and molecular size of the fluorophore result in an appreciable difference in fluorescence anisotropy able to signal nanomolar concentrations of Zn^{2+} even if non-negligible enzyme-inhibitor affinity makes the data interpretation quite difficult. For overcoming this problem, they synthesised[131] a new aryl sulphonamide ligand, which shows a large difference in affinity to the apoprotein (CAII) and holoprotein (CAII- Zn^{2+}). The new fluorescent ligand presents a very low anisotropy when free in aqueous solution that becomes high when it is bound to the Zn^{2+} in the holoprotein. This system allows the determination of Zn^{2+} in solution in the picomolar range concentration. A further development of this work[132] presents some CAII variants obtained by replacing one residue close to the binding site of the enzyme with a cysteine covalently substituted with derivatives of benzoxadiazole sulphonamide. This kind of approach opens many possibilities for applications since this system can, for example, be attached on the surface of an optical fiber. Two different variants are discussed, the first one with a high affinity for Cu^{2+} (0.3 pM) and Co^{2+} (100 nM), (although with a possible interference from Zn^{2+} and Cd^{2+} ions), and the second one for Zn^{2+} (80 pM), both studied in water at pH 7. The transition metal ions concentrations are monitored both by chenges in fluorescence intensities or in fluorescence anisotropy, this last one being particularly important (30% of increase) for the enzyme binding the Zn^{2+}. The results demonstrate that free transition metal can be quantitatively determined at trace levels.

1.5. SENSORS BASED ON SILICA NANOPARTICLES

Signal amplification is a very desirable effect in sensor technology, since in theory it allows a significant lowering of the detection limit. This effect in the case of luminescent sensors means that, as already discussed for dendrimers, the coordination of a single molecule of the target analyte is able to change the photophysical properties of a large number of luminophores.

In general, the organization of photophysically active units in broad structures typically gives rise to collective effects that can be exploited for the design of new functional materials.[1] To our opinion, modification of the surface of nanoparticles is a suitable and still almost unexplored path to constrain a set of fluorescent units into an organized network. Fluorescent nanoparticles are in fact very promising for the design of labels and sensors for the relative ease of their synthesis and for their peculiar properties. Covalent grafting is in general necessary to obtain a stable arrangement and avoid structural reorganizations due to redistribution of the dye on the surface or between the surface and the solution. Silica nanoparticles can be prepared in a very straightforward way and their surface easily modified by means of alchoxysilane derivatives. This versatility makes, in our opinion, silica nanoparticles a good choice as a scaffolding structure for a network of dye moieties, especially considering that in addition these materials are transparent to the visible light and inert as far as energy and electron transfer processes are concerned. Therefore, dye coated silica nanoparticles constitute a suitable system to characterize intermolecular photophysical processes at their surface avoiding any interference from the particle nucleus.

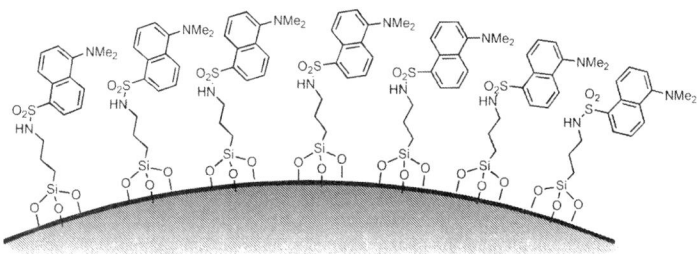

Figure 1.16. Schematic representation of dansyl covered silica nanoparticles.

In our first attempt[19] to synthesise dye coated silica nanoparticles, we choose the dansyl chromophore, because of the peculiar response of its photophysical properties to the protonation of the amino group (Figure 1. 16). As we have already discussed, the lowest excited state of dansyl has, in fact, a charge-transfer character involving the promotion of a lone-pair electron of the amino group into a π antibonding orbital of the naphthalene ring. In the protonated unit, this charge transfer state is destabilized and the π-π* transition localised on the naphthalene becomes the lowest in energy. This leads to in a strong blue shift of both absorption and luminescence bands that allow, in partially protonated polydansylated systems, to selectively excite and detect the outcoming fluorescence of the two different units. Hence, a polydansylated system represents a good example of a network composed by two different kinds of chromophores (Dns and Dns·H$^+$) whose relative composition can be controlled by changing the degree of protonation (Figure 1.17).

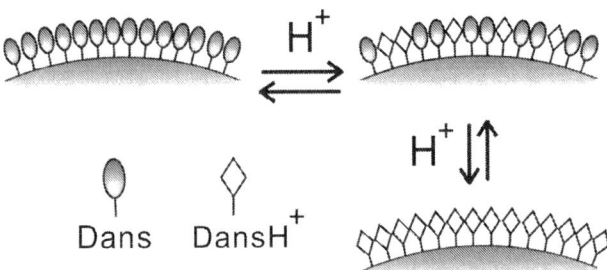

Figure 1.17. Schematic representation of the effect of protonation on dansyl covered silica nanoparticles.

We prepared the fluorescent nanoparticles by covering preformed silica colloids with silane derivatives; each nanoparticle bearing an average number of about 4000 dansyl units. The main goal of the photophysical investigation was to analyze the cooperative effects that arise from the organization of several chromophoric units into a high density pattern as at the surface of these nanoparticles.

The photophysical characterisation revealed that these nanoparticles respond to protonation in polar environment with an amplified quenching of the dansyl luminescence (up to five quenched units per added proton) due to multicomponent charge transfer processes. On the contrary, when protonated in pure chloroform they behave as an efficient antenna system, where the energy absorbed by the Dns·H$^+$ is efficiently transferred to the luminescent Dns chromophore. The different behaviour observed in the two solvents can be conveniently explained by the role of environmental polarity on the delicate balance among different photophysical processes, such as fluorescence emission and energy- and electron-transfer processes. All these results are due to the possible occurrence in an organized network, such as the one present on the surface of nanoparticles, of photophysical processes involving a large number of chromophores, leading to large signal amplification even upon small chemical inputs. The tunability of the photophysical properties of these systems makes them extremely interesting for the design of new photoactive devices especially in relation to the field of fluorescence based sensors.

Afterwards, we have also proposed[20] an alternative synthetic strategy to obtain nanoparticles in which energy transfer between the fluorophores occurs, by introducing a fluorescent alchoxysilane during the formation of the colloids. Inclusion of dyes in silica nanoparticles usually takes advantages of inverse microemulsion techniques which unfortunately do not lead to covalent grafting of the polymeric matrix with the consequent risk of dye leaching. The introduction of a triethoxysilane group directly on the dye molecule allows, instead, its covalent anchoring during the TEOS polymerization. It is interesting to note that this synthetic strategy, contrary to the one described before leaves the surface of the colloids almost unmodified and therefore available for further modification.

An amount of the fluorescein derivatized triethoxysilane (figure 1.18) corresponding to 1% (mol/mol) with respect to tetraethoxysilane (TEOS) was used in the nanoparticles preparation according to the Stroeber procedure. The formation of the colloids was confirmed by TEM images (see figure 1.19), that were graphically elaborated to evaluate the average size of the particles (23 ± 3 nm). FlFFF experiments gave definitive evidence that the fluorophores are bound to the silica matrix.

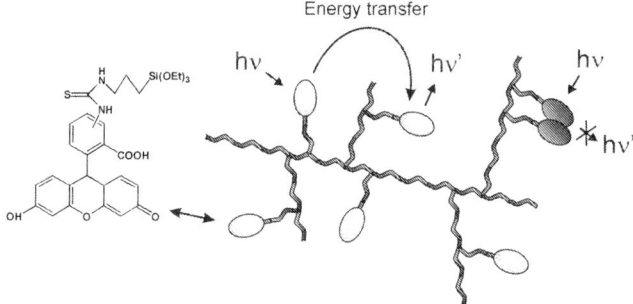

Figure 1.18. Schematic representation of fluorescein containing silica nanoparticles.

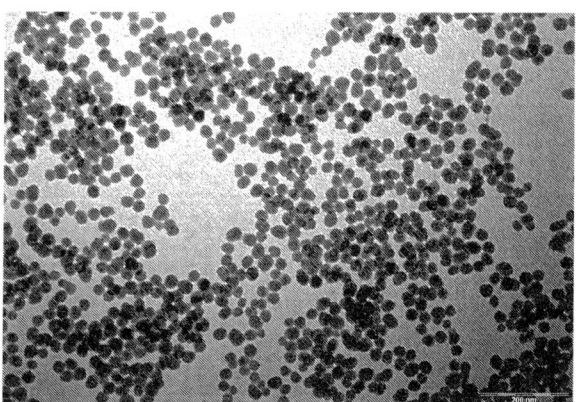

Figure 1.19. TEM image of fluorescein modified silica nanoparticles.

For all the solutions examined the fluorescence intensity at 520 nm upon excitation at 480 nm was directly proportional to the absorbance (A<0.05) at the excitation wavelength. This proportionality indicates that in the nanoparticles the average quantum yield of fluorescein is the same independently from the size of the colloids. This average quantum yield was about 50 % with respect to that of simple fluorescein. Also the excited state lifetime for the fluorescein in the nanoparticles was the same for all the nanoparticle samples. In addition, the luminescence of fluorescein in the nanoparticles is almost completely depolarized (P < 0.02). This was unexpected since the rotational mobility of the fluorophores in the colloids is strongly reduced, therefore a different mechanism for depolarization must be considered. All the photophysical results are in agreement with the occurrence of two kinds of processes which involve the fluorescein units in the nanoparticles. The first one leads to the complete quenching of some fluorophores and it is known to be due to short range interactions between fluorescein molecules. The second one leads to fluorescence depolarization and involves transfer of the excitation energy between the different dye units in the nanoparticles. The two processes are schematically depicted in figure 1.18. The possibility of having different kinds of interactions depending on intermolecular distance makes fluorescein a good probe to test the structure of the fluorophore network.

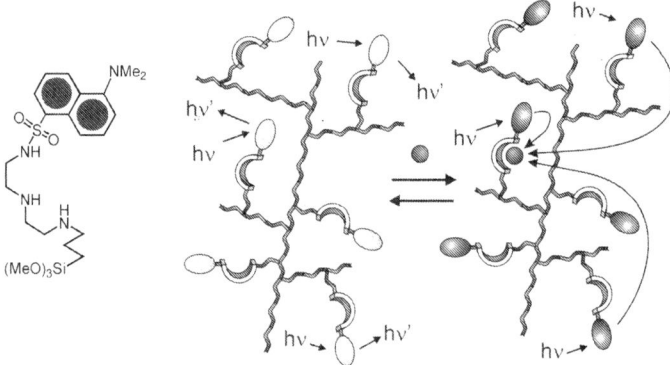

Figure 1.20. Schematic representation of the quenching processes occurring in silica nanoparticles upon copper complexation.

These results show that the local density of the dye molecules is not homogeneous in the nanoparticles. In the more dense regions the fluorescein moieties are closed enough to give self quenching causing a decrease of the overall average fluorescence quantum yield. The observation that no short components are present in the decay of the fluorescence suggests, together with the observed depolarization, that the recorded emission is due to dye molecules close enough to take part to energy transfer processes but too distant to give self-quenching. Such a discontinuous behavior implies the presence of groups of fluorophoric units completely quenched and others whose fluorescence efficiency is unmodified. This suggests some degree of pre-organization of the fluorophores during the polymerization that leads to the formation of oligomers coming from the hydrolysis and condensation of the silane derivative. The very low degree of polarization of the fluorescence rules out the presence of "isolated" fluorophore molecules too far away to transfer the excitation energy to other fluorescein moieties.

PROBES AND SENSORS FOR CATIONS 51

In our opinion, these results open up new perspectives in the design of functional nanosystems with great applicative potentialities, in particular in the fields of sensors and labels. In this nanoparticles the communication between the photoactive units, which is very efficient in the described systems, has a crucial role, and it is an essential condition for obtaining the desired signal amplification effect. We than prepared,[132] with the same synthetic strategy, nanoparticles which contain, together, a dansyl group and a linear polyamine receptor unit (figure 1.20).

Interestingly, upon addition of small amount of Cu^{2+}, Co^{2+}, and Ni^{2+} in water/ethanol solutions, the typical fluorescence of dansyl is quenched: in particular, after the addition of only 2% (compared to the number of chromophoric units) of copper ions, a more than 20% decrease of the luminescence intensity can be observed. This result is an unequivocal demonstration that the amplification process in these systems does occur. Of course, more sophisticated systems can be now appended to the nanoparticles in order to increase the selectivity but, to our opinion, these results clearly indicate that silica nanoparticles, that are typically made with cheap material and simple procedures, are very promising systems for the development of chemical sensor technology.

1.6. CONCLUSIONS

The interest devoted by the scientific community to metal ions detection can be easily understood considering that these species play an essential role in living systems and, at the same time, some of them represent an environmental concern. In particular, great attention has been focussed towards the development of sensors able to monitor the concentrations of these analytes in real time and real space.

We have reported here a large variety of fluorescence sensors for cations. One of our major purposes was to point out the principal strategies followed for metal ion recognition and the exploited mechanism for translating such a process into luminescence signal modulation. These two aspects are, in our opinion, essential to understand the effort spent up to now in the design of more and more effective systems. We believe however that for many metal ions better selectivities and sensitivities have still to be reached in order to address practical needs. A wide horizon is therefore open to further investigations and developments. In other words, as stated by Czarnik a decade ago,[135] "there is still a great demand for more and more efficient chemosensors".

1.6.1 Acknowledgments

The authors wish to thank the Ministero dell'Istruzione, dell'Università e della Ricerca, (FISR, project SAIA) and the University of Bologna (funds for Selected Topics),

1.7. REFERENCES

1. D.R. Thévenot, K. Toth, R.A. Durst, and G.S. Wilson, Electrochemical biosensors: recommended definitions and classification, *Biosens. Bioelectron.* **16**, 121-131 (2001).
2. U.E. Spichiger-Keller, *Chemical Sensors and Biosensors for Medical and Biological Applications* (Wiley-VCH, Berlin, 1997).
3. *Fluorescent Chemosensors for Ion and Molecule Recognition*, edited by A.W. Czarnik (A.C.S., Washington, 1992).
4. M. Montalti, L. Prodi, N. Zaccheroni, in: *Handbook of Photochemistry and Photobiology*, edited by M. S. A. Abdel-Mottaleb and H. S. Nalwa (American Institute of Physics, 2003), vol 3, pp. 271-317
5. J.R. Lakowicz, *Principles of Fluorescence Spectroscopy* (Kluwer Academic/Plenum Publishers, New York, 1999).
6. R. Pinalli, F.F. Nachtigall, F. Ugozzoli, E. Dalcanale, Supramolecular sensors for the detection of alcohols, *Angew. Chem. Int. Ed.* **38**, 2377-2379 (1999).
7. K.D. Schierbaum, T. Weiss, E.U. Thoden van Velzen, J.F.J. Engbersen, D.N. Reinhoudt, W. Göpel, Molecular Recognition by self-assembled monolayers of cavitands receptors, *Science* **265**, 1413-1415 (1994).
8. F. C. J. M. Van Veggel in: *Comprehensive Supramolecular Chemistry*, edited by J. L. Atwood, J. E. D. Davies, D. D. MacNicol, F. Vögtle, and D.N. Reinhoudt (Pergamon, Oxford, 1996), vol. 10, pp. 171-185.
9. L. Prodi, F. Bolletta, M. Montalti, and N. Zaccheroni, Luminescent chemosensors for transition metal ions, *Coord. Chem. Rev.* **205**, 59-83 (2000).
10. J. Janata, and M. Josowicz, Chemical sensors, *Anal. Chem.*, 70, 179R-208R (1998).
11. E. Bakker, P. Bühlmann, and E. Pretsch, Carrier-based ion-selective electrodes and bulk optodes. 1. General characteristics, *Chem. Rev.* **97**, 3083-3132 (1997).
12. P. Bühlmann, E. Pretsch, and E. Bakker, Carrier-based ion-selective electrodes and bulk optodes. 2. Ionophores for potentiometric and optical sensors, *Chem. Rev.* **98**, 1593-1687 (1998).
13. A.B. Ellis and D.R. Walt, guest editorial, *Chem. Rev.* **100**, 2477-2478 (2000).
14. R.Y. Tsien, Fluorescent probes of cell signaling, *Annu. Rev. Neurosci.* **12**, 227-253 (1989).
15. L.A.J. Chrisstoffels, A; Adronov,and J.M.J. Fréchet, Surface-confined light harvesting, energy transfer, and amplification of fluorescence emission in chromophore-labeled self-assembled monolayers, *Angew. Chem. Int. Ed.* **39**, 2163-2164 (2000).
16. N.J. van der Veen, S. Flink, M.A. Deij, R.J.M. Egberink, F.C.J.M. van Veggel, and D.N. Reinhoudt, Monolayer of a Na^+-selective fluoroionophore on glass: connecting the fields of monolayers and optical detection of metal ions, *J. Am. Chem. Soc.* **122**, 6112-6113 (2000).
17. X.X. Zhang, R.M. Izatt, K.E. Krakowiak, and J.S. Bradshaw, A thermodynamic study of complexation of alkali and alkaline-earth metal ions with low-symmetry cryptands, *Inorg. Chim. Acta* **254**, 43-47 (1997).
18. O.S. Wolfbeis, Fiber optic chemical sensors and biosensors, *Anal. Chem.* **72**, 81R-89R (2000).
19. M. Montalti, L. Prodi, N. Zaccheroni, and G. Falini, Solvent induced modulation of collective photophysical processes in dansylated silica nanoparticles, *J. Am. Chem. Soc.*, **124**, 13540-13546 (2002).
20. M. Montalti, L Prodi, N. Zaccheroni, A. Zattoni, P. Reschiglian, and G. Falini, Energy transfer in fluorescent silica nanoparticles, *Langmuir* submitted.
21. R. Krämer, Fluorescent chemosensors for Cu^{2+} ions: fast, selective, and highly sensitive, *Angew. Chem., Int. Ed. Engl.* **37**, 772-773 (1998).
22. *Molecular Luminescence Spectroscopy*, edited by S.G. Schulman (John Wiley & Sons, Inc., London, 1985).
23. R.A Bissel, A.P. de Silva, H.Q.N. Gunaratne, P.L.M. Lynch, G.E.M. Maguire, and K.R.A.S. Sandanayake, Molecular fluorescent signaling with fluor-spacer-receptor systems: Approaches to sensing and switching devices via supramolecular photophysics, *Chem. Soc. Rev.*, 187-195 (1992).
24. A.P. de Silva, H. Q. N. Gunaratne, T. Gunnlaugsson, A. J. M. Huxley, C. P. McCoy, J. T. Rademacher, and T. E. Rice, Signaling recognition events with fluorescent sensors and switches, *Chem. Rev.* **97**, 1515-1566 (1997).
25. A.P de Silva, G.D. Mc Clean, and S. Pagliari, Direct detection of ion pairs by fluorescence enhancement, *Chem. Commun.* 2010-2011 (2003).
26. *Fluorescent Chemosensors of Ion and Molecule Recognition*, NATO-ASI Series, edited by J.-P. Desvergne and A.W. Czarnik (Kluwer Academic Publishers, Dordrecht 1996).
27. O. S. Wolfbeis, in: *Biomedical Optical Instrumentation and Laser-Assisted Biotechnology*, edited by A.M. Verga Scheggi (Kluwer Acad. Publ., Dordrecht, NL 1996), p. 327.
28. L. Fabbrizzi, A. Poggi, Sensors and switches from supramolecular chemistry, *Chem. Soc. Rev.* 197-202 (1995).

29. L. Fabbrizzi, M. Licchelli, P. Pallavicini, D. Sacchi, A. Taglietti, Sensing of transition metals through fluorescence quenching or enhancement - A review, *Analyst* **121**, 1763-1768 (1996).
30. R. Bergonzi, L. Fabbrizzi, M. Licchelli, C. Mangano, Molecular switches of fluorescence operating through metal centred redox couples, *Coord. Chem. Rev.* **170**, 31-46 (1998).
31. B. Valeur and E. Bardez, Cations in control, *Chemistry in Britain* 216–220 (1995).
32. B. Valeur and I. Leray, Design principles of fluorescent molecular sensors for cation recognition, *Coord. Chem. Rev.* **205**, 3-40 (2000).
33. E. Kimura and T. Koike, Recent development of zinc-fluorophores, *Chem. Soc. Rev.* **27** 179-184 (1998).
34. A.B. Descalzo, R. Martinez-Manez, R. Radeglia, K. Rurack, and J. Soto, Coupling selectivity with sensitivity in an integrated chemosensor framework: design of a Hg^{2+}-responsive probe, operating above 500 nm, *J. Am. Chem. Soc.* **125**, 3418-3419 (2003).
35. K. Rurack and U. Resch-Genger, Rigidization, preorientation and electronic decoupling - the 'magic triangle' for the design of highly efficient fluorescent sensors and switches, *Chem. Soc. Rev.* **31**, 116-127 (2002).
36. B. Ramachandram and A. Samanta, Modulation of metal-fluorophore communication to develop structurally simple fluorescent sensors for transition metal ions, *Chem. Commun.* 1037-1038 (1997).
37. B. Ramachandram, N.B. Sankaran, R. Karmakar, S. Saha, and A. Samantha, Fluorescence signalling of transition metal ions by multi-component systems comprising 4-chloro-1,8-naphthalimide as fluorophore, *Tetrahedron* **56**, 7041-7044 (2000).
38. K.A. Mitchell, R. G. Brown, D. Yuan, S.-C. Chang, R.E. Utecht, and D.E. Lewis, A fluorescent sensor for Cu^{2+} at the sub-ppm level, *J. Photochem and Photobiol. A: Chemistry* **115**, 157-161 (1998).
39. A.P. de Silva, T. Gunnlaugsson, and T.E. Rice, Recent evolution of luminescent photoinduced electron transfer sensors - A review, *Analyst* **121** 1759-1762 (1996).
40. L. Prodi, R. Ballardini, M.T. Gandolfi, and R. Roversi, A simple fluorescent chemosensor for earth alkali metal ions, *J. Photochem. Photobiol., A: Chemistry* **136**, 49-52 (2000).
41. L. Fabbrizzi, M. Licchelli, P. Pallavicini, A. Perotti, A. Taglietti, and D. Sacchi, Fluorescent sensors for transition metals based on electron-transfer and energy-transfer mechanisms, *Chem. Eur. J.* **2**, 75-82 (1996).
42. J.A. Sclafani, M.T. Maranto, T.M. Sisk, and S.A. Van Arman, An aqueous ratiometric fluorescence probe for Zn(II), *Tetrahedron Lett.* **37**, 2193-2196 (1996).
43. M.A. Bernardo, F. Pina, B. Escuder, E. García-España, M.L Godino-Salido, J. Latorre, S.V. Luis, J.A. Ramirez, C. Soriano, Thermodynamic and fluorescence emission studies on chemosensors containing anthracene fluorophores. Crystal structure of $\{[CuL^1Cl]Cl\}_2 \cdot 2H_2O$ [L^1 = *N*-(3-aminopropyl)-*N'*-3-(anthracen-9-ylmethyl)aminopropylethane-1,2-diamine], *J. Chem. Soc., Dalton Trans.* 915-922 (1999).
44. S. Alves, F. Pina, M.T. Albelda, E. García-España, C. Soriano, and S.V. Luis, Open-chain polyamine ligands bearing an anthracene unit. Chemosensors for logic operations at the molecular level, *Eur. J. Inorg. Chem.* 405-412 (2001).
45. F. Pina, M.A. Bernardo, and E. García-España, Fluorescent chemosensors containing polyamine receptors, *Eur. J. Inorg. Chem.* 2143-2157 (2000).
46. R. Ostaszewski, L. Prodi, and M. Montalti, The synthesis and complexation studies of thia-anthracene receptors, *Tetrahedron* **55**, 11553-11562 (1999).
47. J.-P. Desvergne, F. Fages, H. Bouas-Laurent, and P. Marsau, Tunable photoresponsive supramolecular systems, *Pure Appl. Chem.* **64**, 1231-1238 (1992).
48. L. Fabbrizzi, M. Licchelli, P. Pallavicini, A. Perotti, and D. Sacchi, An anthracene-based fluorescent sensor for transition metal ions, *Angew. Chem., Int. Ed. Engl.* **33**, 1975-1977 (1994).
49. G. De Santis, M. Di Casa, L. Fabbrizzi, M. Licchelli, C. Mangano, P. Pallavicini, A. Perotti, A. Poggi, D. Sacchi, and A. Taglietti, in: *Transition Metals in Supramolecular Chemistry*, NATO-ASI Series, edited by L. Fabbrizzi and A. Poggi (Dordrecht, Kluwer Academic Publishers 1996), pp 133-152.
50. I. Costa, L. Fabbrizzi, M. Licchelli, P. Pallavicini, L. Parodi, L. Prodi, F. Bolletta, M. Montalti, and N. Zaccheroni, A $[Ru(II)(bipy)_3]$-[dioxo-2,3,2-tet] two-component system, as an efficient ON-OFF fluorescent sensor for Ni^{2+} and Cu^{2+} in water, based on an ET (Energy Transfer) mechanism), *J. Chem. Soc., Dalton Trans.* 1381-1386 (1999).
51. G. Hennrich, H. Sonnenschein, and U. Resch-Genger, Redox switchable fluorescent probe selective for either Hg(II) or Cd(II) and Zn(II), *J. Am. Chem. Soc.* **121**, 5073-5074 (1999).
52. G. Hennrich, W. Walther, U. Resch-Genger, and H. Sonnenschein, Cu(II)- and Hg(II)-induced modulation of the fluorescence behavior of a redox-active sensor molecule, *Inorg. Chem.* **40**, 641-644 (2001).

53. Y. Suzuki, T. Morozumi, H. Nakamura, M. Shimomura, Hayashita, and R.A. Bartsh, New fluorimetric alkali and alkaline earth metal cation sensors based on noncyclic crown ethers by means of intramolecular excimer formation of pyrene, *J. Phys. Chem. B*, **102**, 7910-7917 (1998).
54. J.-S. Yang, C.-S. Lin, and C.-Y. Hwang, Cu^{2+}-induced blue shift of the pyrene excimer emission: a new signal transduction mode of pyrene probes, *Org. Lett.* **3**, 889-892 (2001).
55. L. La Monica, D. Monti, G. Mancini, M. Montalti, L. Prodi, N. Zaccheroni, R. Paolesse, and G. D'Arcangelo, Synthesis, complexation properties and spectroscopic studies of the cation-induced conformational changes of some new oligooxaethylene-spacered diporphyrin arrays, *N. J. Chem.* **25**, 597-605 (2001).
56. B. Valeur, J. Pouget, J. Bourson, M. Kaschke, and N.P. Ernsting, Tuning of photoinduced energy-transfer in a bichromophoric coumarin molecule by cation binding, *J. Phys. Chem.* **96**, 6545-6549 (1992).
57. H. Jiang, H. Xu, and J. Ye, Synthesis and cation-mediated electron transfer in intramolecular fluorescence quenching of donor-acceptor podands: observation of Marcus inverted region in forward electron transfer reactions, *J. Chem. Soc., Perkin Trans. 2*, 925-930 (2000).
58. R.Y. Tsien, New calcium indicators and buffers with high selectivity against magnesium and protons: design, synthesis, and properties of prototype structures, *Biochemistry*, **19**, 2396-2404 (1980).
59. A. Minta, J.P.Y. Kao, and R.Y. Tsien, Fluorescent indicators for cytosolic calcium based on rhodamine and fluorescein chromophores, *J. Biol. Chem.* **264**, 8171-8178 (1989).
60. R.Y. Tsien, in: *Fluorescent Chemosensors for Ion and Molecule Recognition*, edited by A.W. Czarnik (A.C.S., Washington, 1992), pp. 130-146.
61. J. Zhang, R.E. Campbell, A.Y. Ting, and R.Y. Tsien, Creating new fluorescent probes for cell biology, *Nature Review, Molecular cell biology* **3**, 906-918 (2002).
62. S.C. Burdette, G.K. Walkup, B. Spingler, R.Y. Tsien, and S.J. Lippard, Fluorescent sensors for Zn2+ based on a fluorescein platform: Synthesis, properties and intracellular distribution, *J. Am. Chem. Soc.* **123** 7831-7841 (2001).
63. J.M. Castagnetto and J.W. Canary, A chiroptically enhanced fluorescent chemosensor, *Chem. Commun.* 203-204 (1998).
64. L. Prodi, F. Bolletta, M. Montalti, and N. Zaccheroni, Searching for new luminescent sensors: synthesis and photophysical properties of a tripodal ligand incorporating the dansyl chromophore and of its metal complexes, *Eur. J. Inorg. Chem.* 455-460 (1999).
65. L. Prodi, M. Montalti, N. Zaccheroni, F. Dallavalle, G. Folesani, M. Lanfranchi, R. Corradini, S. Pagliari, and R. Marchelli, Dansylated polyamines as fluorescent sensors for metal ions: photophysical properties and stability of Copper(II) complexes in solutions, *Helv. Chim. Acta* **84**, 690-706 (2001)
66. J.D. Winkler, C.M. Bowen, and V. Michelet, Photodynamic fluorescent metal ion sensors with parts per billion sensitivity, *J. Am. Chem. Soc.* **120**, 3237-3242 (1998).
67. E.U. Akkaya, M.E. Huston, and A.W. Czarnik, Chelation-enhanced fluorescence of anthrylazamacrocycle cojugate probes in aqueous-solution, *J. Am. Chem. Soc.* **112**, 3590-3593 (1990).
68. M.E. Huston, C. Engleman, and A.W. Czarnik, Chelatoselective fluorescence perturbation in anthrylazamacrocycle cojugate probes: electrophylic aromatic cadmiation, *J. Am. Chem. Soc.* **112**, 7054-7056 (1990).
69. A.W. Czarnik, Chemical communication in water using fluorescent chemosensors, *Acc. Chem. Res.* **27**, 302-308 (1994).
70. J. Yoon, N.E. Ohler, D.H. Vance, W.D. Aumiller, and A.W. Czarnik, in: *Fluorescent Chemosensors of Ion and Molecule Recognition*, NATO-ASI Series, edited by J.-P. Desvergne and A.W. Czarnik (Kluwer Academic Publishers, Dordrecht 1996).
71. J.Y. Yoon, N.E. Ohler, D.H. Vance, W.D. Aumiller, and A.W. Czarnik, A fluorescent chemosensor signalling only Hg(II) and Cu(II) in water, *Tetrahedron Lett.* **38**, 3845-3848 (1997).
72. D. Parker and J.A.G. Williams, Luminescence bahaviour od cadmium, lead, zinc, copper, nickel and lanthanide complexes of octadentate macrocyclic ligands bearing naphthyl chromophores, *J. Chem. Soc. Perkin Trans. 2* 1305-1314 (1995).
73. A. Beeby, D. Parker, J.A.G. Williams, Photochemical investigations of functionalised 1,4,7,10-tetraazacyclododecane ligands incorporating naphthyl chromophores, *J. Chem. Soc. Perkin Trans. 2* 1565-1579 (1996).
74. T. Koike, T. Watanabe, S. Aoki, E. Kimura, and M. Shiro, A novel biomimetic zinc(II)-fluorophore, dansylamidoethyl-pendant macrocyclic tetraamine 1,4,7,10-tetraazacyclododecane (cyclen), *J. Am. Chem. Soc.* **118** 12696-12703 (1996).

75. G.-P. Xue, J.S. Bradshaw, J.A. Chiara, P.B. Savage, K.E. Krakowiak, R.M. Izatt, L.Prodi, M.Montalti, and N. Zaccheroni, A Convenient Synthesis of Novel Fluorophores: Macrocyclic Polyamines Containing Two Dansylamidoethyl Side Arms, *Synlett*, 1181-1183 (2000).
76. G. Xue, J.S. Bradshaw, H. Song, R.T. Bronson, P.B. Savage, K.E. Krakowiak, R.M. Izatt, L. Prodi, M. Montalti, and N. Zaccheroni, A Convenient Synthesis and Preliminary Photophysical Study of Novel Fluoroionophores: Macrocyclic Polyamines Containing Two Dansylamidoethyl Side Arms, *Tetrahedron* **57**, 87-91 (2001).
77. A. D'Amico, C. Di Natale, and R. Paolesse, Portraits of gasses and liquids by arrays of nonspecific chemical sensors: trends and perspectives, *Sens. Actuators, B* **68** 324-330 (2000).
78. A. Singh, Q. Yao, L. Tong, W.C. Still, D. Sames, Combinatorial approach to the development of fluorescent sensors for nanomolar aqueous copper, *Tetrahedron Lett.* **41**, 9601-9605 (2000).
79. J. P. Konopelski, F. Kotzyba-Hibert, J.-M. Lehn, J.-P. Desvergne, F. Fages, A. Castellan, and H. Bouas-Laurent, Synthesis, cation binding, and photophysical properties of macrobicyclic anthracenocryptands, *J. Chem. Soc. Chem. Commun.* 433-436 (1985).
80. F. Fages, J.-P. Desvergne, H. Bouas-Laurent, P. Marsau, J.-M. Lehn, F. Kotzyba-Hibert, A.-M. Albrecht-Gary, and M. Al-Joubbeh, Anthraceno-cryptands: a new class of cation-complexing macrobicyclic fluorophores, *J. Am. Chem. Soc.* **111**, 8672-8680 (1989).
81. H. Bouas-Laurent, J.-P. Desvergne, F. Fages, and P. Marsau, Tunable fluorescence of some macrocyclic anthracenophanes, in: *Fluorescent Chemosensors for Ion and Molecule Recognition*, edited by A.W. Czarnik (A.C.S., Washington, 1992), pp. 59-73.
82. B. Witulski, M. Weber, U. Bergsträsser, J.-P. Desvergne, D.M. Bassani, and H. Bouas-Laurent, Novel alkali cation chemosensors based on N-9-anthrylaza-crown ethers, *Org. Lett.* **3**, 1467- 1470(2001).
83. J. Bourson, F. Badaoui, and B. Valeur, Coumarinic fluorescent chemosensors for the detection of transition metal ions, *J. Fluorescence*, **4**, 275-279 (1994).
84. B. Valeur, F. Badaoui, E. Bardez, J. Bourson, P. Boutin, A. Chatelain, I. Devol, B. Larrey, J.P. Lefèvre, and A. Soulet, in: *Fluorescent Chemosensors of Ion and Molecule Recognition*, NATO-ASI Series, edited by J.-P. Desvergne, A.W. Czarnik (Kluwer Academic Publishers, Dordrecht, 1996).
85. S. Fery-Forgues, M.-T. Le Bris, J.-P. Guetté, and B. Valeur, Ion-responsive fluorescent compounds. 1. Effect of cation binding on photophysical properties of benzoxazinone derivative linked to monoaza-15-crown-5, *J. Phys. Chem.*, **92**, 6233-6237 (1988).
86. I. Leray, J.-L. Habib-Jiwan, C. Branger, J.P. Soumillion, and B. Valeur, Ion-responsive fluorescent compounds VI. Coumarin 153 linked to rigid crowns for improvement of selectivity, *J. Photochem. Photobiol. A: Chemistry* **135**, 163-169 (2000).
87. L. Prodi, F. Bolletta, M. Montalti, N. Zaccheroni, J.S. Bradshaw, P.B. Savage, and R.M. Izatt, A fluorescent sensor for magnesium ions, *Tetrahedron Lett.* **39**, 5451-5454 (1998).
88. G. Farruggia, L. Masotti, unpublished results.
89. L. Prodi, C. Bargossi, M. Montalti, N. Zaccheroni, N. Su, J. S. Bradshaw, R. M. Izatt, and P. B. Savage, A fluorescent sensor for mrcury ions, *J. Am. Chem. Soc.* **122**, 6769-6770 (2000).
90. L. Prodi, M. Montalti, N. Zaccheroni, J.S. Bradshaw, R.M. Izatt, and P.B. Savage, Characterisation of 5-chloro-8-methoxyquinoline appended diaza-18-Crown-6 as a chemosensor for cadmium, *Tetrahedron Lett.* **42**, 2941-2944 (2001).
91. L. Prodi, M. Montalti, J.S. Bradshaw, R.M. Izatt, and P.B. Savage, Dependence on pH of the luminescent properties of metal ion complexes of 5-chloro-8-hydroxyquinoline appended diaza-18-Crown-6, *J. Incl. Phenom. Macr. Chem.*, **41**, 123-127 (2001).
92. G.-P. Xue, J.S. Bradshaw, N.K. Dalley, P.B. Savage, K.E. Krakowiak, R.M. Izatt, L. Prodi, M. Montalti, and N. Zaccheroni, Convenient syntheses and preliminary photophysical properties of novel 8-aminoquinoline appended diaza-18-Crown-6 ligands, *Tetrahedron* **57**, 7623-7628 (2001).
93. G. Xue, J.S. Bradshaw, N.K. Dalley, P.B. Savage, R.M. Izatt, L. Prodi, M. Montalti, and N. Zaccheroni, The synthesis of azacrown ethers with quinoline-based sidearms as potential zinc(II) fluorophores, *Tetrahedron*, **58**, 4809-4815 (2002).
94. W.-S. Xia, R.H. Schmehl, and C.-J. Li, A highly selective fluorescent chemosensor for K+ from a bis-15-crown-5 derivative, *J. Am. Chem. Soc.* **121**, 5599-5560 (1999).
95. W.-S. Xia, R.H. Schmehl, and C.-J. Li, A novel caesium selective fluorescent chemosensor, *Chem. Commun.* 695-696 (2000).
96. L. Prodi, F. Bolletta, N. Zaccheroni, C.I.F. Watt, and N.J. Mooney, A new family of luminescent sensors for alkaline earth metal ions, *Chem. Eur. J.* **4**, 1090-1094 (1998).

97. B.G. Cox, T.V. Hurwood, L. Prodi, M. Montalti, F. Bolletta, N. Zaccheroni, and C. I. F. Watt, Synthesis, Characterisation and Metal Ion Binding Properties of Crown Ethers Incorporating 4,5-Dioxyxanthones, *J. Chem. Soc., Perkin Trans. 2* 289-296 (1999).
98. M. Kollmannsberger, K. Rurack, U. Resch-Genger, and J. Daub, Ultrafast charge transfer in amino-substituted boron dipyrromethene dyes and its inhibition by cation complexation: A new design concept for highly sensitive fluorescent probes, *J. Phys. Chem. A* **102**, 10211-10220 (1998).
99. K. Rurack,, M. Kollmannsberger, U. Resch-Genger, and J. Daub, A selective and sensitive fluoroionophore for Hg(II), Ag(I), and Cu(II) with virtually decoupled fluorophore and receptor units, *J. Am. Chem. Soc.* **122**, 968-969, (2000).
100. K. Rurack, J.L. Bricks, R. Radeglia, and U. Resch-Genger, Chalcone-analogue dyes emitting in the near-infrared (NIR): Influence of donor-acceptor substitution and cation complexation on their spectroscopic properties and X-ray structure, *J. Phys. Chem. A* **104**, 3087-3109 (2000).
101. K. Rurack, J.L. Bricks, B. Shulz, M. Maus, G. Reck, and U. Resch-Genger, Substituted 1,5-diphenyl-3-benzothiazol-2-yl-Δ(2)-pyrazolines: Synthesis, X-ray structure, photophysics, and cation complexation properties, *J. Phys. Chem. A* **104**, 6171-6188 (2000).
102. K. Rurack, A. Koval'chuck, J.L. Bricks, and J.L. Slominskii, A simple bifunctional fluoroionophore signaling different metal ions either independently or cooperatively, *J. Am. Chem. Soc.* **123**, 6205-6206, (2001).
103. A.B. Descalzo, R. Martínez-Máñez, R. Radeglia, K. Rurack, and J.Soto, Coupling selectivity with sensitivity in an integrated chemosensor framework: Design of a Hg^{2+}-responsive probe, operating above 500 nm, *J. Am. Chem. Soc.* **125**, 3418-3419, (2003).
104. P. Ghosh, P.K. Bharadwaj, S. Mandal, and S. Ghosh, Ni(II), Cu(II), and Zn(II) cryptate-enhanced fluorescence of a trianthrylcryptand: A potential molecular photonic OR operator, *J. Am. Chem. Soc.* **118**, 1553-1554 (1996).
105. P. Ghosh, P.K. Bharadwaj, J. Roy, snd S. Ghosh, Transition metal (II)/(III), Eu(III), and Tb(III) ions induced molecular photonic OR gates using trianthryl cryptands of varying cavity dimension, *J. Am. Chem. Soc.* **119**, 11903-11909 (1997).
106. G. Das, P.K. Bharadwaj, M. Basu Roy, and S. Ghosh, Transition metal cryptate-enhanced fluorescence in a trianthroyl cryptand: effect of spacer on the photoinduced electron transfer process, *J. Photochem. Photobiol., A: Chemistry* **135**, 7-11 (2000).
107. L. Prodi, F Bolletta, M. Montalti, N. Zaccheroni, A. Casnati, F. Sansone, and R. Ungaro, Photophysics of 1,3-alternate calix[4]arene-crows and of their metal ion complexes: evidence of a cation-π interaction in solution, *New J. Chem.* **24**, 155-158 (2000).
108. A. Casnati, F. Giunta, F. Sansone, R. Ungaro, F. Bolletta, M. Montalti, L. Prodi, and N. Zaccheroni, Synthesis, complexation and photophysics of 1,3-alternate calix[4]arene-crowns bearing chromophoric units on the bridge, *Supramol. Chem.* **13**, 419-434 (2001).
109. A. Casnati, F. Sansone, A. Sartori, L. Prodi, M. Montalti, N. Zaccheroni, F. Ugozzoli, and R. Ungaro, Quinoline Containing Calixarene Fluoroionophores: a Combined NMR, Photophysical and Modeling Study, *Eur. J. Org. Chem.*, 1475-1485 (2003).
110. I. Leray, F. O'Reilly, J.-L. Habib Jiwan, J.-Ph. Soumillion, and B. Valeur, A new calix[4]arene-based fluorescent sensor for sodium ion, *Chem. Commun.* 795-796 (1999).
111. I. Leray, J.-P Lefevre, J.F. Delouis. J. Delair, and B. Valeur, Synthesis and photophysical and cation-binding properties of mono- and tetranaphthylcalix[4]arenes as highly sensitive and selective fluorescent sensors for sodium, *Chem. Eur. J.* **7**, 4590-4598 (2001).
112. R. Métivier, I. Leray, and B. Valeur, A highly sensitive and selective fluorescent molecular sensor for Pb(II) based on a calix[4]arene bearing four dansyl groups, *Chem. Commun.* 996-997 (2003).
113. H.-F. Ji, G.M. Brown, and R. Dabestani, Calix[4]arene-based Cs+ selective optical sensor, *Chem. Commun.* 609-610 (1999).
114. H.-F. Ji, G.M. Brown, and R. Dabestani, Synthesis and sensing behavior of cyanoanthracene modified 1,3-alternate calix[4]benzocrown-6: a new class of Cs+ selective optical sensors, *J. Chem. Soc., Perkin Trans. 2* 585-591 (2001).
115. R. Corradini, A. Dossena, G. Galaverna, R. Marchelli, A. Panagia, and G. Sartor, Fluorescent chemosensor for organic guests and copper(II) ion based on dansyldiethylenetriamine-modified beta-cyclodextrin, *J. Org. Chem.* **62**, 6283-6289 (1997).
116. S. Pagliari, R. Corradini, G. Galaverna, S. Sforza, A. Dossena, and R. Marchelli, Enantioselective sensing of amino acids by copper(II) complexes of phenylalanine-based fluorescent beta-cyclodextrins, *Tetrahedron Lett.* **41**, 3691-3695 (2000).

117. S. Pagliari, R. Corradini, G. Galaverna, S. Sforza, A. Dossena, R. Marchelli, M. Montalti, L. Prodi, and N. Zaccheroni, Enantioselective fluorescence sensing of amino acids by modified cyclodextrins: role of the cavity and sensing mechanism, *Chem. Eur. J,* submitted.
118. V. Balzani, P. Ceroni, S. Gestermann, C. Kauffmann, M. Gorka, and F. Vögtle, Dendrimers as fluorescent sensors with signal amplification, *Chem. Commun.* 853-854 (2000).
119. F. Vögtle, S. Gestermann, C. Kauffmann, P. Ceroni, V. Vicinelli, and V. Balzani, Coordination of Co^{2+} ions in the interior of poly(propylene amine) dendrimers containing fluorescent dansyl units in the periphery, *J. Am. Chem. Soc.* **122**, 10398-10404 (2000).
120. V. Balzani, P. Ceroni, S. Gestermann, M. Gorka, C. Kauffmann, and F. Vögtle, Effect of protons and metal ions on the fluorescence properties of a polylysin dendrimer containing twenty four dansyl units, *J. Chem. Soc., Dalton Trans.* 3765-3771 (2000).
121. A. Torrado and B. Imperiali, New synthetic amino acids for the design and synthesis of peptide-based metal ion sensors, *J. Org. Chem.* **61**, 8940-8948 (1996).
122. G.K. Walkup and B. Imperiali, Design and evaluation of a peptidyl fluorescent chemosensor for divalent zinc, *J. Am. Chem. Soc.* **118**, 3053-3054 (1996).
123. G.K. Walkup and B. Imperiali, Fluorescent chemosensors for divalent zinc based on zinc finger domains. Enhanced oxidative stability, metal binding affinity, and structural and functional characterization, *J. Am. Chem. Soc.* **119**, 3443-3450 (1997).
124. G.K. Walkup and B. Imperiali, Stereoselective synthesis of fluorescent alpha-amino acids containing oxine (8-hydroxyquinoline) and their peptide incorporation in chemosensors for divalent zinc, *J. Org. Chem.* **63**, 6727-6731 (1998).
125. A. Torrado, G.K. Walkup and B. Imperiali, Exploiting polypeptide motifs for the design of selective Cu(II) ion chemosensors, *J. Am. Chem. Soc.* **120**, 609-610 (1998).
126. H.A. Godwin and J.M. Berg, A fluorescent zinc probe based on metal-induced peptide folding, *J. Am. Chem. Soc.* **118**, 6514-6515 (1996).
127. P.S. Eis and J.R. Lakowicz, Time-resolved energy-transfer measurements of donor-acceptor distance distributions and intramolecular flexibility of a CCHH zinc-finger peptide, *Biochemistry*, **32**, 7981-7993 (1993).
128. R.B. Thompson and E.R. Jones, Enzyme-based fiber optic biosensor, *Anal. Chem.* **65**, 730-734 (1993).
129. R.B. Thompson and M.W. Patchan, Lifetime based fluorescence energy transfer biosensing of zin, *Anal. Biochem.* **227**, 123-128 (1995).
130. R.B. Thompson, B.P. Maliwal, and C.A. Fierke, Expanded dynamic range of free zinc ion determination by fluorescence anisotropy, *Anal. Chem.* **70**, 1749-1754 (1998).
131. R.B. Thompson, B.P. Maliwal, V.L. Feliccia, C.A. Fierke, and K. McCall, Determination of picomolar concentrations of metal lens using fluorescence anisotropy: Biosensing with a "reagentless" enzyme transducer, *Anal. Chem.,* **70**, 4717-4723 (1998).
132. M. Montalti, L Prodi, and N. Zaccheroni, unpublished results.
133. A.W. Czarnik, Desperately seeking sensors, *Chemistry & Biology* **2**, 631 (1995).

FLUORESCENT ANION COMPLEXATION AGENTS

David Curiel, Elizabeth J. Hayes and Paul D. Beer[1]

2.1. INTRODUCTION

Anions play numerous indispensable roles in biological and chemical processes, as well as contributing significantly to environmental pollution. For instance, the majority of enzyme substrates and co-factors are anionic[1], as is DNA itself. Excess amounts of anionic pollutants such as phosphate and nitrate lead to eutrophication and consequent disruption of aquatic life cycles[2]. Radioactive pertechnetate from the nuclear fuel cycle may also contribute to pollution[3]. Given their importance, there has obviously been much effort expended in the design of anion complexing reagents, particularly over the last two decades.

The main strategies to date have focused on cationic polyammonium, guanidinium, quaternary ammonium, expanded porphyrin host systems, and a variety of Lewis acidic containing receptors such as tin, boron, silicon, mercury and uranyl. Neutral organic receptors that bind anions solely via favorable hydrogen bonding interactions have also been recently exploited. Reviews on various aspects of anion coordination can be found in the references[4,5,6,7,8], and also elsewhere in this volume. A common approach to receptor design is to design molecules consisting of two sub-units, each performing a specific function. The general design principle involves a binding site coupled to a signaling subunit. The binding site offers a means of coordinating the anionic species under investigation. Coordination induces some sort of change in the signaling component which can be measured via a macroscopic physical response, such as a change in spectroscopic properties. Common choices for signaling component include redox-active moieties, and fluorogenic and chromogenic reagents.

Significant progress has been made in the synthesis of similar receptors for cationic species. However, the nature of anions makes receptor design more challenging. Anions are larger than isoelectronic cations, and therefore have a lower charge to radius ratio.

---

[1] Inorganic Chemistry Laboratory, Department of Chemistry, University of Oxford, South Parks Road, Oxford, OX1 3QR, UK. E-mail: paul.beer@chem.ox.ac.uk

This decreases the effectiveness of electrostatic binding. Anionic species display a wide range of geometries, and therefore a higher degree of receptor design may be required for host:guest complementarity. Also, anions are sensitive to pH conditions, and so receptors must function within the pH window of the target anion. Solvent effects are a crucial consideration, since anions are often highly solvated. Potential receptors are in effect competing with the solvent environment in which the anion recognition event takes place[6,9].

Fluorescence has proven to be a widely used analytical tool, with applications in many fields[7]. With respect to the use of fluorescence in anion sensor molecules, its major advantage is its high sensitivity, with the capability to measure concentration values of up to one million times smaller than other techniques such as absorbance.

There are several photochemical processes that are related to the sensing of anions, and while a full mechanistic explanation can be found elsewhere [10,11,12,13], they warrant some comment. Photo-induced electron transfer (PET) is perhaps the most extensively studied and utilized process for the sensing of both anions and cations. Typically fluorescence is observed when an excited electron, in for instance the lowest unoccupied molecular orbital (LUMO) goes to the highest occupied molecular orbital (HOMO), releasing the excess energy as light. Fluorogenic anion sensors are designed to exploit this effect by inducing the appearance or removal of energy levels (i.e. "alien" orbitals) between the HOMO and LUMO of the fluorophore, which subsequently leads to quenching or enhancement of the fluorescent emission.

Electronic energy transfer (EET) may also be responsible for the quenching of fluorescence in fluorogenic sensors. This process occurs when the substrate has empty or half-filled energy levels between the HOMO and LUMO of the fluorophore. A simultaneous exchange of two electrons can therefore occur, and this restores the fluorophore to its ground state, leading to quenching of the emission. Monomer-excimer formation is also an observed phenomenon, an excimer being defined as a complex formed by interaction of the fluorophore in the excited state with a fluorophore of the same structure in the ground state. Another common mechanism is the rigidity effect, commonly used to explain the fluorescence enhancement observed upon anion complexation. Coordination of the anion increases the overall rigidity of the complex, making non-radiative decay from the excited state less probable. The emission intensity is therefore increased.

This review article will provide an overview of receptor systems that incorporate a fluorogenic subunit, and exhibit a physical response to anion recognition, and also detail current developments in this field. The first section of the review deals with receptors that consist of an organic fluorophore such as anthracene and naphthalene, whereas the second deals with receptors containing metal-based fluorophores such as ruthenium(II) bipyridyl complexes.

2.2. ORGANIC ANION SENSORS

2.2.1. Sensors with Naphthalene as Signalling Unit

2.2.1.1. Recognition based on electrostatic interactions and/or hydrogen bonding.

One of the most popular approaches when designing neutral receptors for anion binding is to utilise hydrogen bonding. This is one of the best choices due to its directionality and the rich variety of functional groups that can establish this kind of interaction.

Naphthylthioureas can establish directional hydrogen bonds which enable them to be tested as florescent anion sensors.[14] As reported for compounds (1)-(3), the number of urea groups influences the sensitivity of the sensor. Compound (1), bearing three urea groups, showed an enhancement of its fluorescence in the presence of $H_2PO_4^-$ and a much weaker effect was caused by HSO_4^-. However no changes were detected for (2) and (3) which have two and one urea groups respectively. On the other hand, anions like bromide or iodide did not have any influence on the receptors' fluorescence spectra.

The increase in the fluorescence intensity upon complexation of dihydrogen phosphate or hydrogen sulphate is rationalized in terms of the inhibition of a PET (Photoinduced Electron Transfer) process occurring between the tertiary amine and the excited naphthalene.

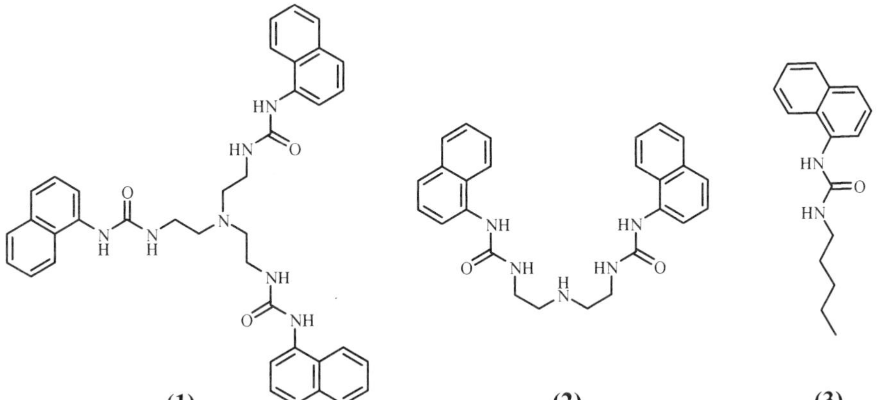

Figure 2.1. Naphthylurea receptors for the recognition of $H_2PO_4^-$.

Compounds (4)-(5) showed a very strong fluorescence enhancement in the presence of an excess (1000-fold) of some oxoanions.[15] The series of (6)-(7) follows the same trend in the anion binding selectivity in methanolic solution: $CO_3^{2-} \geq HCO_3^- >> HPO_4^- > HSO_4^-$. No changes in the photophysical properties of the receptors are detected for other anions such as chloride, bromide, iodide, perchlorate and nitrate.

The increase in the rigidity of the structure of compounds (4)-(7) upon anion binding was assigned as the cause of the fluorescence enhancement. A time dependence that affected the stability of the emission intensity was also observed and attributed to the slow formation of the anion complexes.

Regarding the structure of the receptors, it was observed that for the pairs of compounds (4) and (6), or (5) and (7), the enhancement factors were higher for (4) and (6) since they can establish a higher number of hydrogen bonds with the anion. Nevertheless, no conclusion can be drawn about the influence of the sidearms present in each compound.

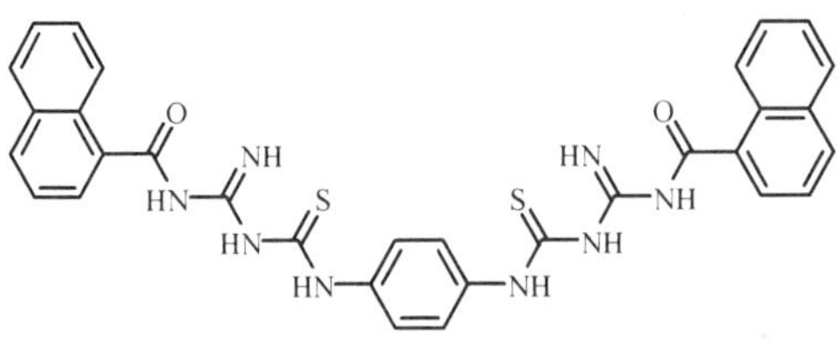

Figure 2.2. Iminoylthiourea and 1,2,4-thiadiazole-based receptors.

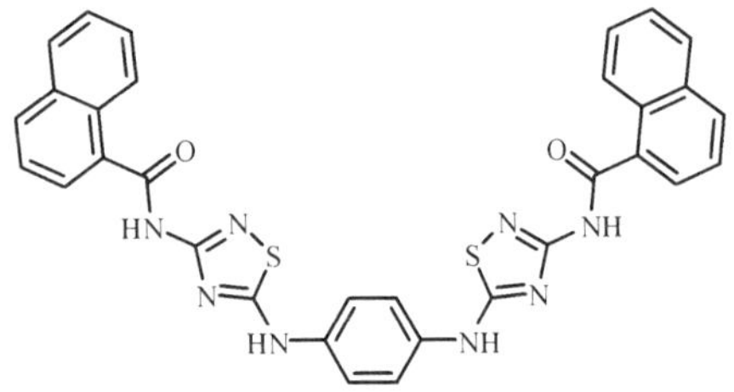

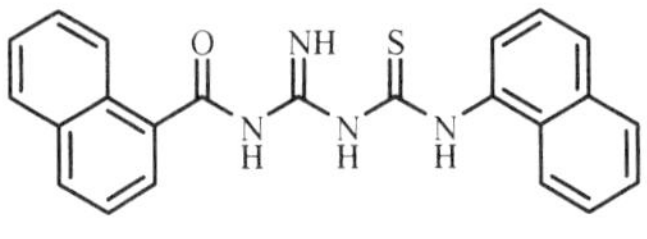

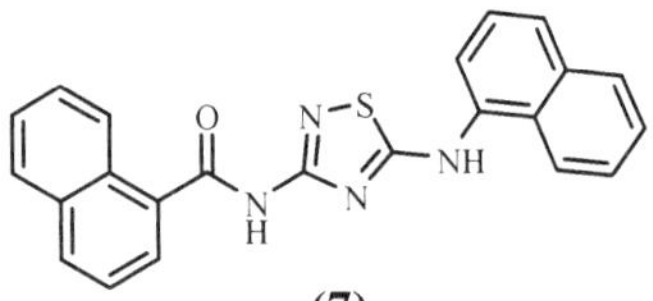

Hydroxymethyl substituted naphthylureas **(8)-(11)** were synthesised to study the influence of auxiliary groups in anion binding.[16] All the receptors showed a time dependent fluorescence enhancement when an excess of anion was present in solution. Among the different anions tested in a methanol:water (4:1, v/v) mixture, the intensity of the fluorescent response was, with minor exceptions, in accord with the basicity of the anion, showing the trend: CO_3^{2-} > HPO_4^{2-} > HCO_3^- > $CH_3CO_2^-$. The increase in emission intensity, as in similar examples, was assigned to the incremental rigidity of the naphthylureas when the anion is complexed.

Regarding the sensitivity of the fluorescent signal of compounds **(8)-(11)** towards the same anion, it seems that a direct relationship exists with the number of hydroxymethyl substituents, since the order of affinity is: **(8)** > **(9)** > **(10)** > **(11)**. Ruling out any direct interaction of the hydroxyl groups with the anion, it was proposed that these groups played a role in the preorganization of the conformation of the urea moieties. Hence, the most stable conformation for **(8)** and **(9)** was *trans-trans* whereas **(10)** and **(11)** preferred a *trans-cis* conformation. This explains how anion binding is energetically favoured in receptors XXXX and XXXX

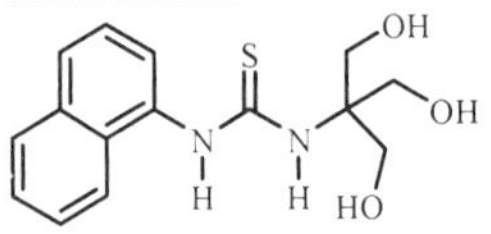

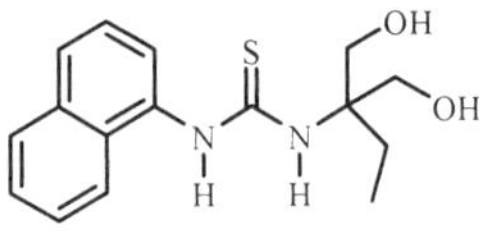

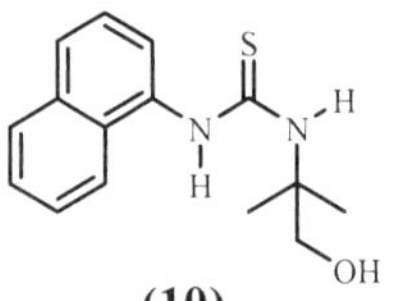

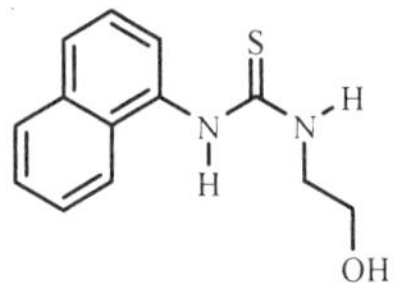

Figure 2.3 Naphthylthiourea receptors.

FLUORESCENT ANION COMPLEXATION AGENTS

The interaction of compound **(12)** with fluoride and chloride reveals a 40-fold selectivity for F⁻ over Cl⁻.[17] Upon complexation of fluoride anion a new peak appears in the emission spectrum that could provide a great advantage for detecting fluoride anions.

Based on theoretical simulations it seems that the changes in the fluorescence spectrum, observed for the ligand-fluoride interaction, are due to the large anionic character of compound XXXX when the anion is complexed.

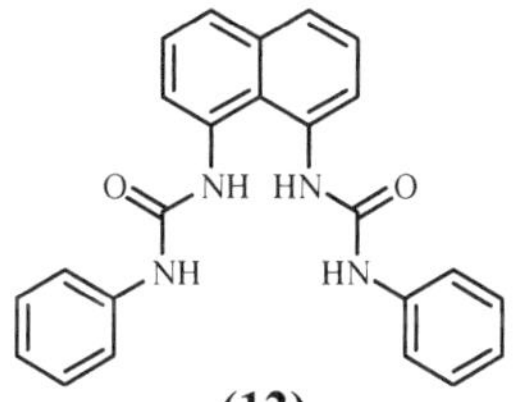

(12)

Figure 2.4. Naphthylurea selective fluoride receptor.

Inspired by the hydrogen bonding of base pairs in the nucleic acids, cytosine was linked to a naphthalene unit in order to recognize nucleotides fluorimetrically.[18] In addition to the Watson-Crick type hydrogen bonding, some of the nucleobases offer the possibility of establishing more donor-acceptor interactions through hydrogen bonds due to the functionality of their structure. This secondary hydrogen bonding can be favoured by designing a receptor with the appropriate topology. To this end the fluorophore was functionalised with a urea or thiourea group. In this way more than three hydrogen bonds can participate in the selective complexation of guanine.

The formation of a 1:1 complex in $CHCl_3$:DMSO (4:1, v/v) was confirmed both by ¹H-NMR and fluorescence titrations where a quenching of the emission intensity was detected.

When the titration experiments were carried out with anionic nucleotides, a high selectivity for 5'-GTP was again observed, as expected, because of the preferential interaction with the guanine substituted ligand.

Though both receptors **(13)** and **(14)** show the same selectivity, the one bearing thiourea has a lower binding constant presumably due to a loss of planarity caused by the steric hindrance between sulphur and naphthalene.

The use of isothiuronium groups has not been explored very much in the area of anion binding. This functional group, with an enhanced acidity of the NH protons, has turned out to be quite effective in the recognition of oxoanions over chloride anion.

Compound **(15)** showed a very weak fluorescence in acetonitrile because of a quenching process by photoinduced charge transfer. Fluorescence titrations proved that, upon addition of anions, the emission intensity remarkably increased.[19] The binding stoichiometry fitted with a 1:1 complex and was very strong for TBA (tetrabutylammonium) salts of acetate. A weaker interaction was observed with $(BuO)_2PO_2^-$ and no change was detected with Cl⁻.

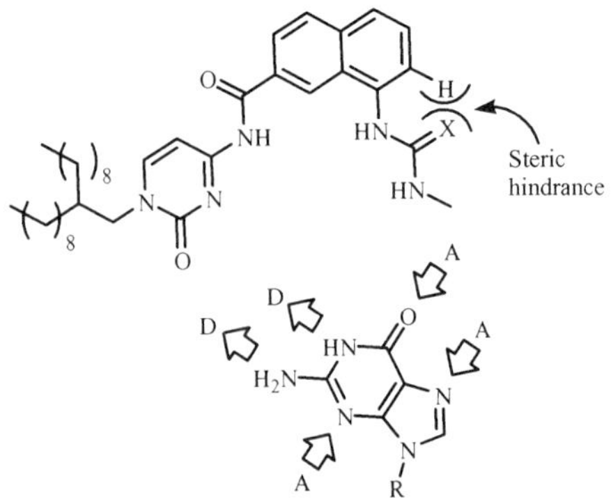

Figure 2.5. Polysubstituted naphthylurea/thiourea receptor for nucleobases.

The binaphthyl derivative (**16**)[20] was also tested as an anion sensor and the response of this compound with the same series of anions was exactly the same as the one obtained for (**15**). Nevertheless, it is worth mentioning that the emission enhancement for the binaphthyl receptor with AcO⁻ was over 4 times more intense than for (**15**).

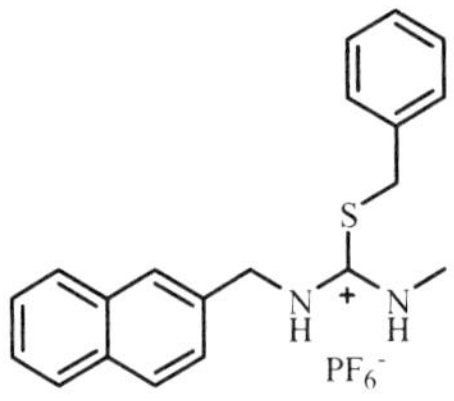

Figure 2.6. Naphthyl and binaphthylisothiouronium receptors.

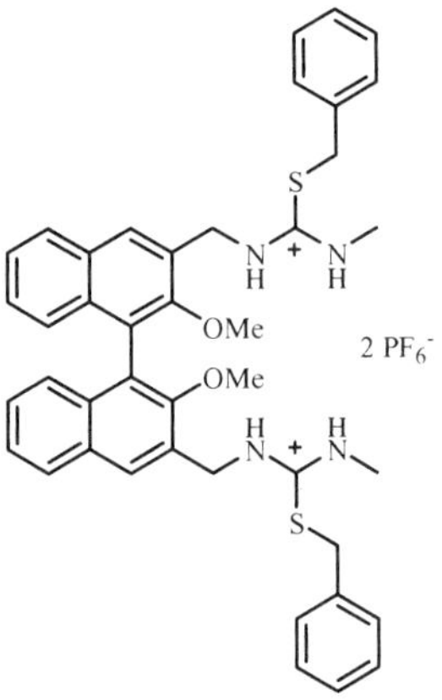

Structurally related to previous isothiouronium derivatives, compound (**17**) was synthesized with the fluorophore appended to the sulphur atom instead of being linked to the nitrogen atom.[21] Interestingly, it was observed that when the spacer between the naphthalene and the sulphur atom was a methylene group, its instability under photo-irradiation prevented its study as fluorescent chemosensor. However, good behaviour could be observed when working with an ethylene spacer.

Different anions, e.g. HPO_4^{2-}, $H_2PO_4^-$, AcO⁻ and Cl⁻, were examined in methanolic solution but only the divalent hydrogen phosphate and monovalent acetate caused changes in the emission spectra. A fluorescence enhancement in the presence of these anions was detected due to a PET inhibition, with a 2-fold increase of the emission intensity for HPO_4^{2-} compared to AcO^-.

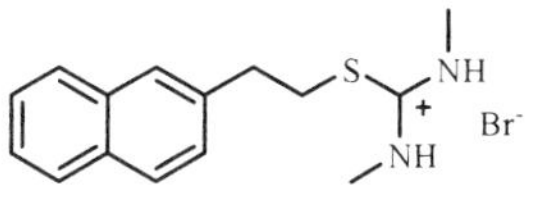

(17)

Figure 2.7. Isothiouronium receptor for HPO_4^- and AcO^-.

2.2.1.2. Recognition based on a metal coordinative interaction

Dansyl substituted β-cyclodextrins *(R)-* and *(S)-***(18)** were synthesized in order to check their enantioselectivity as fluorescent sensors for D- and L-amino acids.[22]

The formation of the copper (II) complex of both *(R)-* and *(S)-* ligands resulted in fluorescence quenching. When amino acids were added to an aqueous solution of the Cu(II) complexes, buffered to pH=7.3, an enhancement in the emission intensity was observed. The mechanism that is postulated for this photochemical change implies the formation of a ternary Cu(II)-XXXX-amino acid complex with the dansyl group being displaced from the metal by the entering anion.

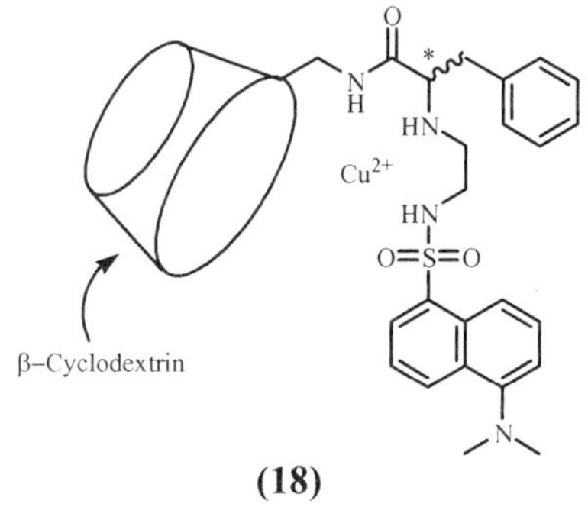

(18)

Figure 2.8. Dansyl derivative for the recognition of amino acids.

The enhancement factor depended on the identity of the amino acid and the best result corresponded to histidine for both *(R)-* and *(S)-***(18)**.

As far as the enantioselective discrimination is concerned, compound *(S)-***(18)** showed better results than *(R)-***(18)**, with Pro, Phe and Trp being the amino acids which gave the best enantiomeric differentiation with a relative fluorescent enhancement ratio ($\Delta I_D / \Delta I_L$) of 3.89, 0.33 and 0.63 respectively.

2.2.1.3. Recognition based on the formation of dative-covalent bonds

When simple phenyl and 2-naphthylboronic acids **(19)** and **(20)**[23] were titrated with KF in a mixture of $MeOH:H_2O$ (1:1, v/v) buffered to pH=5.5, both fluorophores showed a fluorescence quenching due to the hybridisation of the tetrahedral fluoride adduct.

Compound **(21)** was designed to increase the strength of fluoride binding by virtue of an additional hydrogen bonding site which is available when the amine is protonated. In this case, the titrations under identical conditions as described above also led to fluorescence quenching. It is important to mention that while changes in fluorescence could be detected with KF in the range of 5-70 mM, no change was observed with KCl and KBr until very high concentrations of these salts had been added.

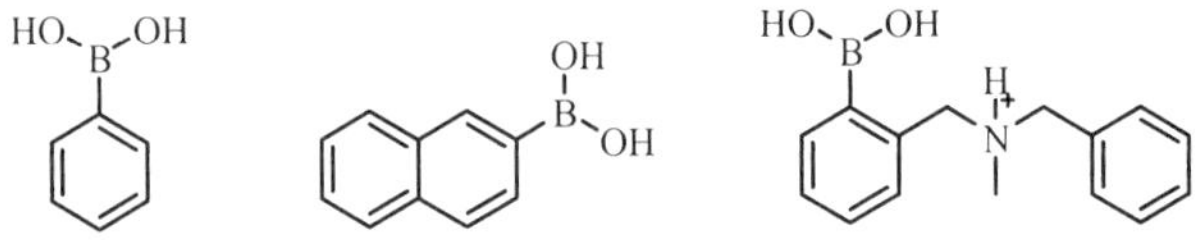

Figure 2.9. Boronic acid selective fluoride anion receptors.

2.2.1.4. Recognition based on other interactions

γ-Cyclodextrins can accommodate two molecular species in its large hydrophobic cavity. Throughout all the examples of anion chemosensors described so far, it is conceived that the anion participates as the analyte to be detected but plays a passive role at the time of interacting with the light. In the following example, the presence of γ-cyclodextrins favours the formation of an excimer involving two molecules of 2-naphthylacetate in aqueous solution buffered to pH=9.2.[24] Therefore, it is the anion itself, which acts as substrate and signalling unit showing an excimer emission band around 450 nm.

By incorporating two pyridinium substituents in the rim of the γ-cyclodextrin, compound **(22)** can establish a stronger electrostatic interaction with the carboxylate. Consequently, the exciter formation is favoured, as proven by the intensity enhancement of the peak at 450 nm. This electrostatic interaction was confirmed when upon addition of competing NaCl, a decrease in the intensity of the excimer fluorescence was observed.

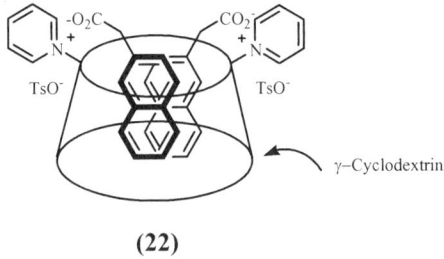

(22)

Figure 2.10. γ-Cyclodextrin-pyridinium receptor.

2.2.2. Sensors with Anthracene as Signalling Unit

2.2.2.1. Recognition based on electrostatic interactions and/or hydrogen bonding

Pioneering work in the area of fluorescent anion sensors was reported by Czarnik in 1989,[25] with compound **(23)** as the first described example of a fluorescent anion sensor.

The interaction of polyammonium salts with anionic substrates caused an enhancement in the emission intensity of the anthracene moiety. Working with aqueous solutions buffered to pH=6, the cavity defined by tris(3-aminopropyl)amine has three of the amine groups protonated, leaving the amino group by the anthracene unprotonated. This nitrogen can quench the fluorescence of anthracene by PET, but upon complexation of hydrogenphosphate, one OH group falls in the proximity of the "benzylic" nitrogen. Thus, favourable intracomplex proton transfer will then protonate that nitrogen inhibiting the intramolecular quenching and causing an increase in the fluorescence (>145%).

This Chelating Enhanced Fluorescence (CHEF) was also observed with sulphate and acetate. However, since these anions are totally dissociated at the working pH and thus cannot transfer a proton, the effect was weaker.

Bigger and more complex anions such as ATP, citrate and dimethyl phosphate were tested as well, leading to a range of different intensity increases.

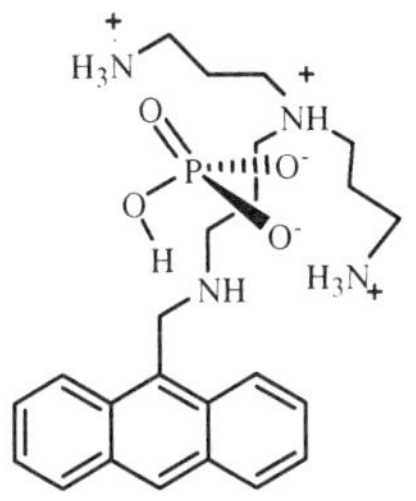

(23)

Figure 2.11. Monosubstituted anthrylpolyammonium receptor.

Mono- and disubstituted anthryl linear polyamines **(24)-(26)** were tested as fluorescent sensors for the polyanions heparin and poly-L-glutamate.[26] Titration experiments at pH=5 showed that the polyammonium probes experienced fluorescence quenching when anion was added to the aqueous solution. This result was explained as a template directed excimer formation on the surface of the anion. By increasing the amount of anion in solution, over a specific concentration, the fluorophore population can diffuse over the length of the polyanion thus avoiding excimer formation and promoting subsequent fluorescence recovery.

This behaviour was used to follow the activity of heparinase (enzyme that hydrolyzes heparin to oligosaccharide units) and pronase (enzyme responsible for the hydrolysis of polyglutamate to glutamic acid).

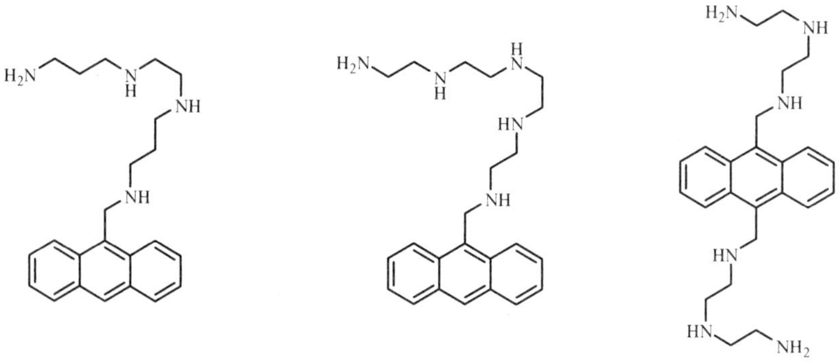

(24)   (25)   (26)

Figure 2.12. Mono and disubstituted antrylpolyamine receptors.

The structure of 1,8-disubstituted anthryl polyammonium (27),[27] allows an improvement in the range of working anion concentration and, more importantly a good selectivity is achieved based on the discrimination of the anion size. When (27) was titrated at pH=7 with pyrophosphate, a remarkable increase of the fluorescence was detected. By contrast, phosphate did not produce such an effect. Therefore a discrimination of 2200-fold could be established between these two anions. Additionally, receptor (27) was used to follow the activity of pyrophosphatase, an enzyme that hydrolyzes pyrophosphate to phosphate.

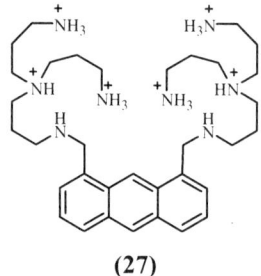

(27)

Figure 2.13. 1,8-disubstituted anthrylpolyammonium receptor for pyrophosphate.

The recognition of biologically relevant substrates is a target on which the research and design of new sensors focus its attention. Linear polyamines (24), (28)-(30) were synthesized and tested for the recognition of nucleotides by means of an electrostatic interaction between the polyammonium salt and the anionic part of the nucleotide.[28] When titration experiments in aqueous solutions buffered to pH<4 were carried out, all receptors formed complexes of 1:1 stoichiometry with ATP and showed a very similar affinities independent on the length of the chain and even the number of anthracene units.

A more detailed study of compound (24) revealed that the effective quenching of fluorescence, observed as a result of the ATP recognition, was less significant for ADP and was not detected for AMP. This proves the selectivity of this simple receptor. The assignment of the chelating enhanced quenching (CHEQ) resides presumably on an efficient π-stacking between the anthracene and the adenine protonated at position 3, since in the absence of that proton no quenching is observed. Moreover, the molecular topology of the components, with special emphasis on the number of phosphonates, plays an important role in the selectivity of the sensor.

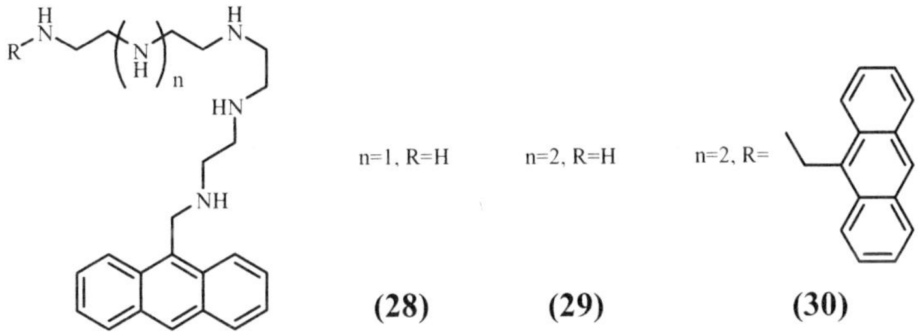

Figure 2.14. Anthracene derivative for the recognition of ATP.

As nucleotides, aminoacids are another very important group of biological substrates whose detection has attracted much attention in the area of molecular sensors. The zwitterionic nature of aminoacids, makes the approach of a ditopic receptor one of the most popular designs, allowing binding of the carboxylate anion and the ammonium cation in a cooperative way.

The combination of a triazacrown ether with two appended guanidinium groups in the presence of anthracene as signalling unit allows the formation of 1:1 complexes with γ-aminobutyric acid and lysine in a MeOH:water (2:1) solution, buffered to pH=9.5.[29] An enhancement of fluorescence by a factor of 2 is detected for both substrates. This is due to the inhibition of PET from the nitrogen atom close to the anthracene, which is participating in the complexation of the amino acid. A much weaker response is obtained for glycine, whereas no change is observed for the rest of aminnoacids tested (Ala, Phe, Val, Ser, Glu and Arg). The authors explain the selectivity of compound **(31)** according to the pKa of the substrates detected, i.e., GABA and Lys, with Lys having a higher pKa value (10.53) due to its alkyl ammonium terminal.

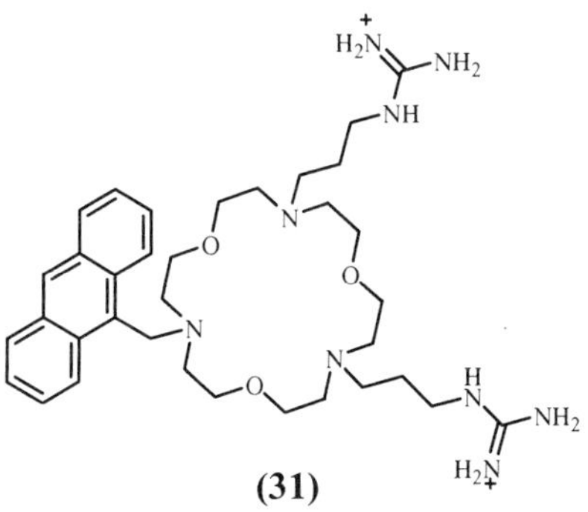

Figure 2.15. Azacrorwn ether-guanidinium receptor.

The ability of thioureas to complex anions has been thoroughly exploited in the field of molecular sensors. In the case of compound **(32)** the utilization of a tripodal structure as part of a hexasubstituted benzene leads to a more preorganized structure.[30]

The stoichiometry of the receptor:substrate complex is 1:1 according to the analysis of the titration profiles in acetonitrile. The titration experiments showed a decrease in the emission intensity that was related to a PET from the HOMO of the thiourea to the excited anthracene. Upon binding of the anion, the reduction potential of thioureas increases. This favours the electron transfer process and the subsequent fluorescence quenching.

The order of the selectivity observed for compound **(32)** was dihydrogen phosphate > acetate > chloride. Comparing this result to the one obtained for the compound **(33)** (acetate > dihydrogen phosphate > chloride), it was concluded that the tripodal nature of **(32)** increases the dihydrogen phosphate affinity over acetate.

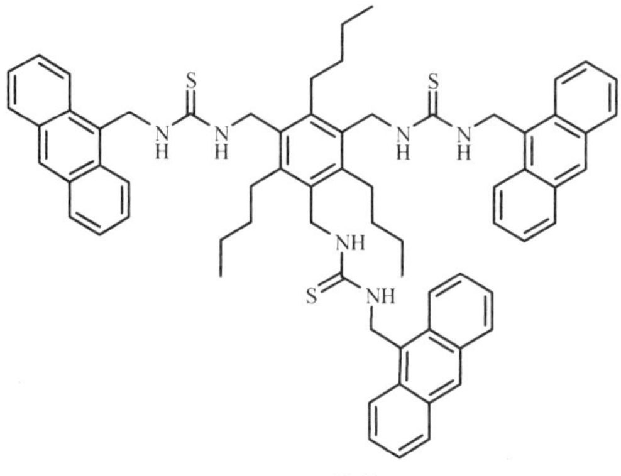

Figure 2.16. Thiourea-based receptors.

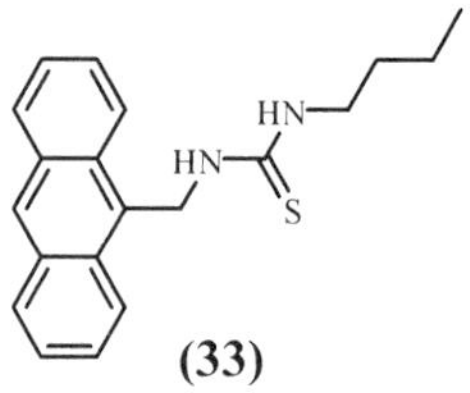

Similar anthrylthioureas **(34)**-**(36)** with different substituents on the urea function were tested with TBA salts of anions such as F^-, Cl^-, Br^-, $H_2PO_4^-$ and AcO^-, by fluorescence spectroscopy[31] in DMSO. For compound **(34)**, fluoride, acetate and dihydrogen phosphate produced 90%, 70% and 50% quenching of the fluorescence respectively. Chloride and bromide did not cause any significant change in the emission spectra. The degree of quenching matches with the selectivity of binding: F^- > AcO^- > $H_2PO_4^-$. A trend also observed for **(35)**. It is not surprising that fluoride shows the higher affinity, since its small size and high charge density enables it to form strong hydrogen bonding with the thiourea. Nevertheless, for compound **(36)** acetate and dihydrogen phosphate swap their positions and the selectivity is: F^- > $H_2PO_4^-$ > AcO^-. It is therefore worth noting how the affinity of the receptor can change with subtle changes to the structure.

Comparing these results with those described with those described previously for compound **(33)**, which differs only in the length of its aliphatic chain, it is demonstrated how the influence of the solvent could also affect the behaviour of fluorescent sensors.

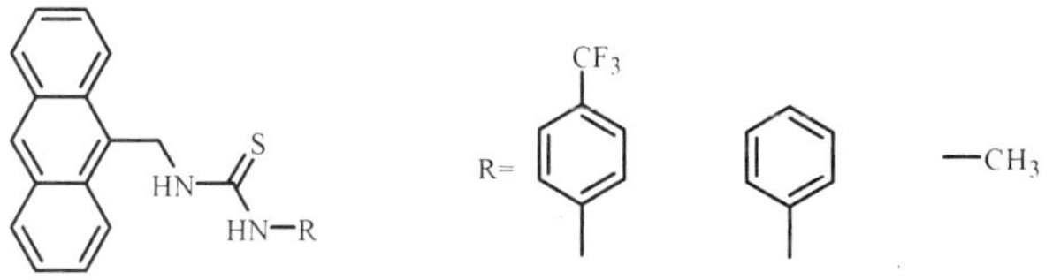

FLUORESCENT ANION COMPLEXATION AGENTS

Figure 2.17. Anthrylthiourea receptors.

The structure of anthracene offers the possibility of preparing disubstituted compounds easily. Compounds (37) and (38) bearing thiourea functions at positions 9 and 10 were titrated with TBA salts of mono- and bidentate anions in DMSO solutions.[32] AcO⁻ and $H_2PO_4^-$ caused emission quenching (70-95%) with no other changes detected in the spectra. Of the spherical anions such as F⁻, Cl⁻ and Br⁻, only fluoride caused a significant decrease (98%) in the emission intensity. This behaviour was exhibited by both (37) and (38). From the fluorescence titration profiles a 1:2 receptor:anion stoichiometry could be determined for acetate and dihydrogen phosphate, which was confirmed by ^{1}H-NMR titrations.

Regarding the bidentate anions glutarate, malonate and pyrophosphate, both receptors were sensitive to all three as could be concluded by the quenching of luminescence. Although (37) and (38) showed similar quenching the sensitivity of (37) was slightly better. The strength of the binding was of the order glutarate > pyrophosphate > malonate, following the same trend for (37) and (38). The 1:1 stoichiometry of these complexes was confirmed by ^{1}H-NMR experiments.

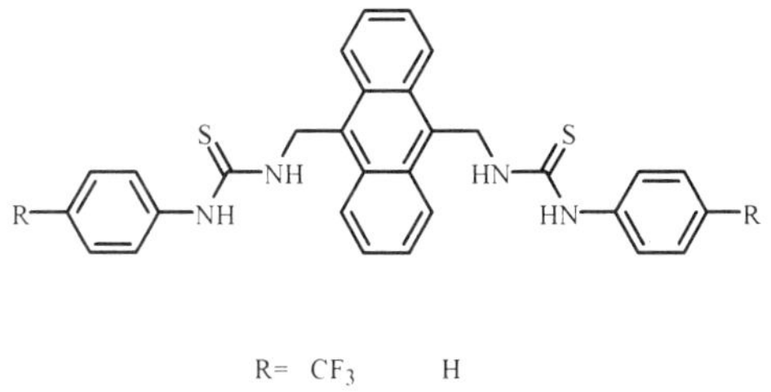

Figure 2.18. Disubstituted anthrylthiourea receptors.

Compound (39), consisting of 1,8-disubstituted anthracene,[33] with two urea groups as binding sites, was titrated with the series of halides (F⁻, Cl⁻, Br⁻ and I⁻) as tetraethylammonium salts, in a mixture of acetonitrile:DMSO (9:1, v/v). Fluoride was the only anion that caused a noteworthy decrease in the fluorescence of XXXX by PET. No significant changes were observed for Cl⁻, Br⁻ and I⁻. Thus a very good selectivity for fluoride was achieved, being 120 fold over chloride.

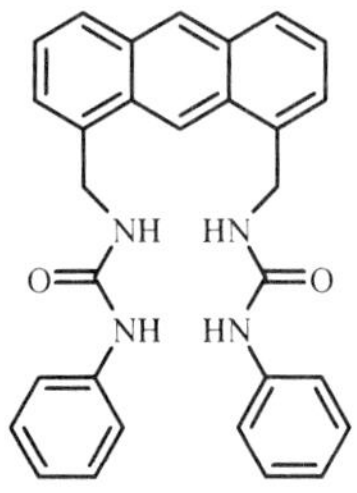

(39)

Figure 2.19. Antrhylurea receptor.

Calixpyrroles are known to be good anion binders through hydrogen bonding interaction. The attachment of this structure to an anthracene moiety allows the study of derivatives **(40)-(42)** as fluorescent anion sensors.[34] Solutions of these receptors in dichloromethane and acetonitrile were titrated with different anions. Upon addition of tetrabutylammonium salts of fluoride, chloride, dihydrogen phosphate and hydrogen sulphate, quenching of the fluorescence turned out to be quite strong for all except HSO_4^-. Good agreement was found between the efficiency of quenching and the association constants calculated either from the fluorescence data or ^{1}H-NMR.

Regarding the degree of sensitivity, it is worth mentioning that due to the different distances between the binding site and the signalling unit, compound **(40)** always behaved better as a sensor followed by **(41)** and **(42)** respectively. Besides the length of the spacer, **(40)** also has a conjugated bond pathway that could provide electronic communication between the fluorophore and the cavity where the anion is complexed.

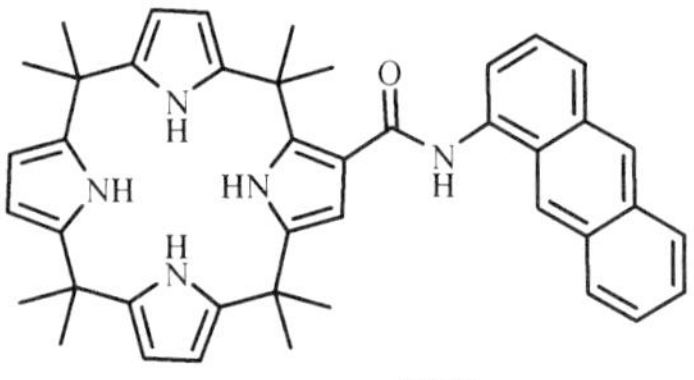

(40)

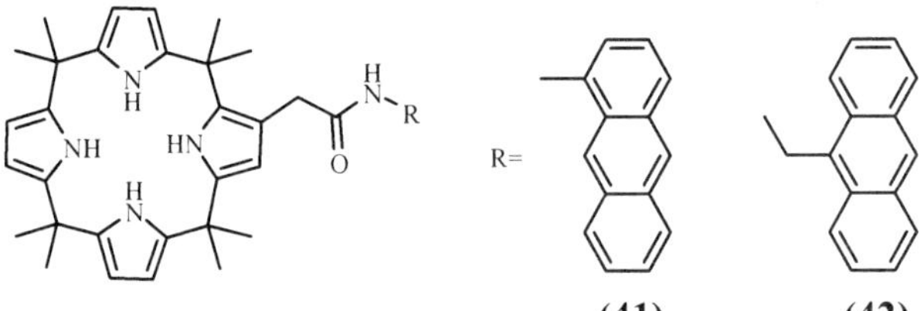

(41)   (42)

Figure 2.20. Calixpyrrole receptors.

The combination of cation and anion binding sites with a signalling unit is the most common approach to the complexation of ion-pairs. Inspired by this design, compound **(43)** was prepared to work as a logic gate[35] in which the presence of analyte represents the input condition (neither anion nor cation present; only anion present; only cation present; both anion and cation present) and the fluorescence intensity represents the output. This is only one example of how molecular recognition can evolve towards applications that go beyond the merely analytical point of view.

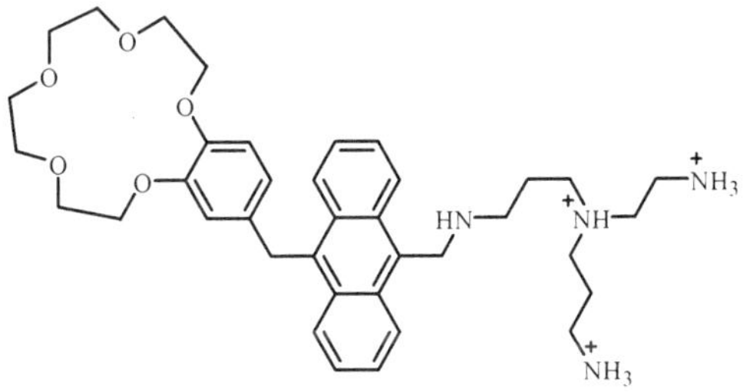

(43)

Figure 2.21. Ditopic receptor for ion-pairs.

Fluorescent anion recognition can be applied to the control of molecular motion in interlocked structures. The difference in anion binding strength can alter the self-assembling interaction between the two components of pseudorotaxane **(44)**.[36]

Taking advantage of the fluorescent properties of both macrocycle and thread in **(44)**, changes in the fluorescence spectrum can be observed depending on whether the components are assembled or apart. The fluorescence of the macrocycle enhances when components are separated and quenches when the pseudorotaxane exists. On the other hand, the fluorescence of the anthracene unit present in the thread enhances when threading happens and decreases when the system is unthreaded.

All the changes in the fluorescence are dependent upon the anion exchange in the ammonium thread. When the chloride salt is present in the ion-pair the pseudorotaxane unthreads whereas upon anion exchange to the PF_6^- salt the interlocked structure is recovered.

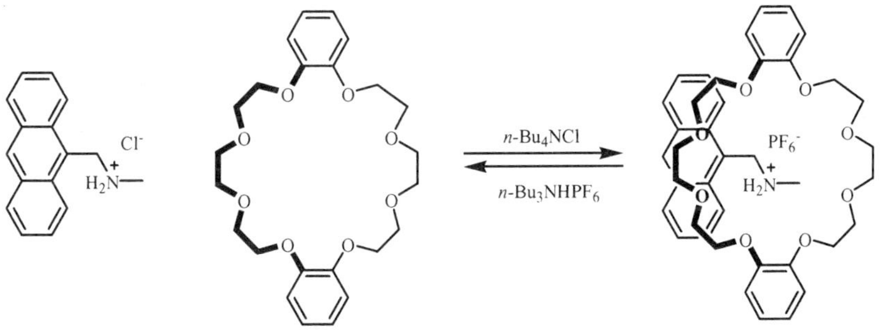

(44)

Figure 2.22. Anion-switched pseudorotaxane.

2.2.2.2. Recognition based on a metal coordinative interaction

Metal-ligand interactions can be highly energetic depending on the electrical charge of the metal, electronic configuration and the ligand field stabilisation energy that the anion coordination may induce. This higher energy in comparison with purely electrostatic interactions can be used to get better results in anion binding even in polar environments where the solvent can compete with the anion for the binding site.

Receptor design generally exploits the coordination geometry of the metal. Some of the coordinative positions can be blocked by a multidentate ligand, leaving other positions on the coordination sphere vacant and available for the incoming anion.

Using the same kind of polyamines that were described in previous sections, these ligands in their neutral state can complex metallic cations. This complexation itself is considered as fluorescent cation sensing. When the metal is complexed, the quenching of the fluorescence of anthracene by PET from the lone pair on the nitrogen atoms is no longer effective, since those electrons are taking part in the coordination of the cation. As the coordination sphere of the cation is not saturated, anionic species could occupy the available positions and the metallic complex would be working as anion sensor.

Compound **(45)**[37] favourably binds two zinc cations in the range of pH=6.5-8.5. From this pH up to pH=10, imidazole is present in its anionic imidazolate form. Working in aqueous solutions buffered to pH=9.5, this anion can bridge the two Zn(II) centres and subsequently quench the fluorescence (70%) by electron transfer from a π-orbital of the electron rich imidazolate to a π-orbital of the excited anthracene. This fluorescence attenuation was exclusive of imidazolate anion, and this fact was exploited to use **(45)** as florescent sensor for the aminoacid histidine.

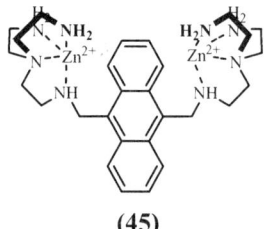

(45)

Figure 2.23. Chelated zinc receptor.

Very similar complexes can be obtained with a more preorganized cryptand structure[38]. From the distribution diagram of the ligand-metal complex it can be ascertained the optimum that pH=8.5 is for the existence of complex **(46)** in solution. Titration experiments revealed the ability of **(46)** to bind ambidentate anions in the order $NCO^- > N_3^- > NCS^-$ with a 1:1 stoichiometry. The correspondence between the length of the anion and the distance between the metallic centres dictates the selectivity of the sensor. Some other anions such as nitrate, bicarbonate, sulphate, chloride and bromide did not cause any change in the emission spectra.

In the case of N_3^- and NCS^-, their reducing tendency gives rise to an intermolecular electron transfer process which causes the quenching of fluorescence.

FLUORESCENT ANION COMPLEXATION AGENTS

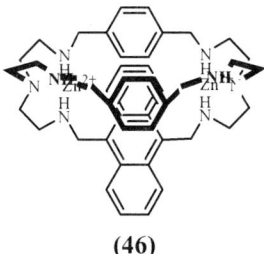

(46)

Figure 2.24. Zn(II) cryptand receptor.

The formation of Zn(II) complex (47) was favoured in dioxane:water solution (4:1, v/v) at neutral pH. This compound was initially designed to bind aminoacids through metal-anion interaction between Zn(II) and the carboxylate group of the aminoacid. The presence of three aromatic substituents in the structure of receptor (47) was intended to improve the affinity towards aminoacids bearing an aromatic substituent by means of a cooperative π-π interaction.[39]

Indeed, UV-vis titrations showed that tryptophan and phenylalanine formed quite stable 1:1 complexes. Nevertheless, when titrations were followed by fluorescence spectroscopy only Trp caused quenching of the emission. This is due to an electron transfer from the indole moiety (donor) in the tryptophan structure to the excited anthracene (acceptor). The electron transfer process was further demonstrated by reviving of fluorescence in frozen ethanolic solution. The absence of emission changes upon adding Phe was assigned to the lower electron donor tendency of its aromatic substituent.

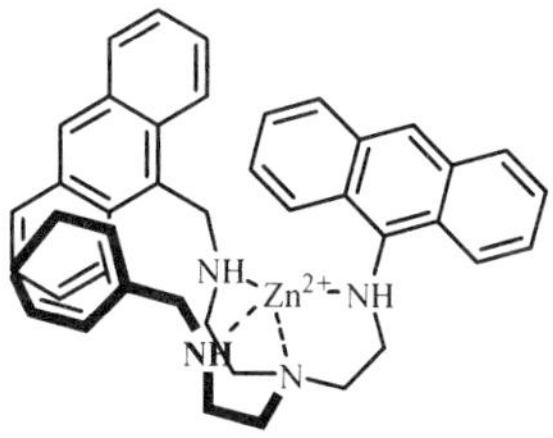

(47)

Figure 2.25. Polyanthracene receptor.

Similar modulation of the binding interaction can be achieved by changing the identity of the aromatic substituents on the tren (tris(2-aminoethyl)amine) frame. In complex (48) this is carried out by introducing two donor N,N-dimethylaniline (DMA) rings and one acceptor anthracene unit.[40] When Zn(II) is complexed in the cavity of the organic ligand, the resulting conformation permits an electron transfer through space from DMA to the excited anthracene causing quenching of the fluorescence. In the presence of sufficiently bulky carboxylates such as triphenylacetate (TPA) a significant increase of the emission intensity is observed in methanolic solution. Authors propose that upon binding of TPA to Zn(II) the aromatic substituents in the ligand are pushed apart and the electron transfer quenching is prevented. A 1:1 receptor:substrate stoichiometry was confirmed for several carboxylates (acetate, cyclohexylcarboxylate, benzoate, 1-adamantanecarboxylate and triphenylacetate). As mentioned before, only TPA produced a significant reactivation of fluorescence, which reinforces the idea of bulkiness as the main driving force in this recognition process.

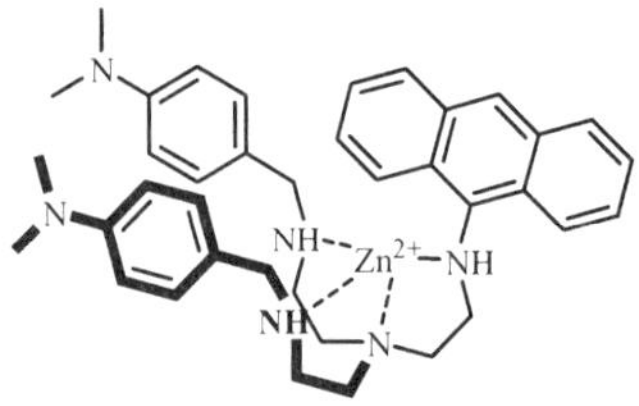

(48)

Figure 2.26. Zn(II) complex receptor for carboxylate anion.

The high affinity of TMCA (2,4,6-triamino-1,3,5-trimetoxycycohexane) for cation binding due to its preorganized structure, makes this ligand a good frame for the design of metal-anion interaction sensors.[41] Using the trimethoxy derivative and introducing two anthracene moieties, the resulting ligand selectively complexes Zn(II) over other cations at neutral pH, causing a subsequent fluorescence increase.

Titration experiments of **(49)** in aqueous solutions, buffered to pH=7.2, with analytes like nucleotides, carboxylic acids and orotic acid (vitamin B13) led to a decrease in emission intensity, which indicated a 1:1 stoichiometry.

Interestingly, when complex **(49)** was tested with a series of nucleotides, only GMP, TMP and UMP produced significant changes in the fluorescence spectra. Looking in more detail at the structure of the guests, the presence of imide or amide groups seems to be the distinguishing feature among the group of nucleobases, since a much weaker interaction between receptor and substrate is obtained for AMP and CMP. At the working pH, the coordination of the deprotonated imide or amide function to the Zn(II) brings the anthracene and the nucleobase together favouring the fluorescence quenching.

Oxalate was also selectively sensed with 40% of lessening emission. No response was obtained when titrations were effected with monocarboxylic acids or longer chain dicarboxylic acids.

Finally, orotic acid, which combines the presence of a relatively acidic amide function and a carboxylic group showed a very high affinity with **(49)**. Molecular models reveal that in the pentacoordinated complex ZnATMCA-orotate, the guest and the anthracene unit are close enough so that an aromatic interaction can be suggested as the origin of the emission decrease. The high selectivity and sensitivity towards orotate (up to 100% quenching) was proven even in the presence of competitors such as carboxylic acids, nucleotides and nucleobases.

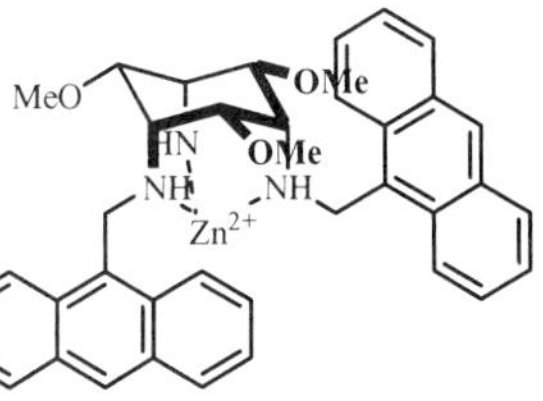

(49)

Figure 2.27. Preorganized Zn(II) complex receptor.

The discovery of fluorophore-appended dipicolylamine (Dpa) zinc(II) complex as the first artificial receptor for phosphorylated peptides, led to the design of anthracene derivative **(50)** for anion sensing.[42] Working in aqueous solution, buffered to pH=7.2, several anions (phosphate, acetate, bicarbonate, sulphate, nitrate, chloride and azide) were initially studied. Only phosphate caused a remarkable fluorescence enhancement, with bicarbonate showing a much weaker effect and azide provoking a decrease in the emission intensity. The other anions did not cause any significant change in the fluorescence spectra. Due to the selectivity of **(50)** towards phosphate, similar anionic species were fluorimetrically analysed resulting in a weak binding affinity for monomethylphosphate and no sensing of dimethylphosphate.

On the basis of these results, titration experiments were carried out with ATP in neutral aqueous solution, showing a 1:1 stoichiometry and a three-fold increase in the emission intensity. Weaker affinities were determined for ADP and AMP, which highlights the increase of the selectivity and sensitivity with the number of phosphates. By comparison with other nucleotides (GTP and CTP) a different trend in the fluorescence changes was observed which can be interpreted as a good selectivity of chemosensor **(50)** among the nucleoside triphosphates.

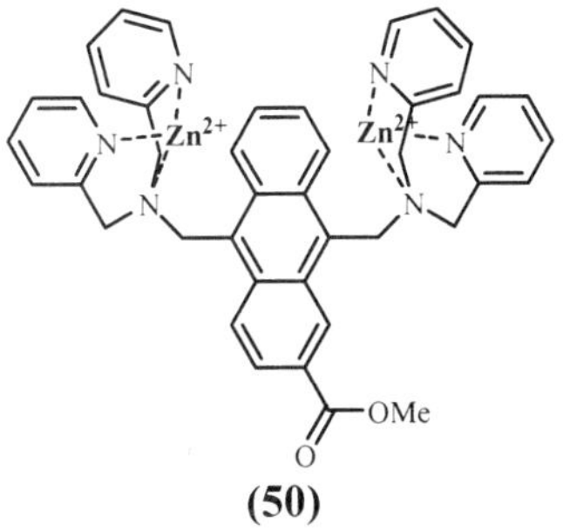

(50)

Figure 2.28. Disubstituted anthryl-Zn(II) complex receptor.

Although it is not very common to find chemosensors bearing two different signalling units, this is the case of compound **(51)**, which has ferrocene and anthracene appended to the same azamacrocycle and enables its study as an electrochemical and photochemical sensor.[43]

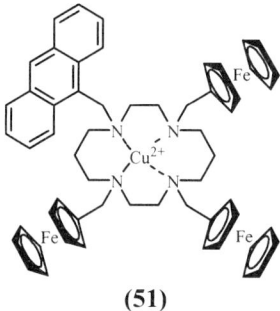

(51)

Figure 2.29. Fluorescent and electrochemical sensor.

The Cu(II) complex **(51)** was titrated with anions such as $H_2PO_4^-$, HSO_4^-, NO_3^-, F^-, Cl^-, Br^- and I^- showing a fluorescent enhancement for dihydrogen phosphate (17%), nitrate (26%) and fluoride (20%).

2.2.2.3. Recognition based on the formation of covalent bonds

Nucleotide binding studies were conducted with compound **(52)**, in neutral aqueous solution.[44] The proposed mechanism for the sensing of 5′-uridinephosphates is essentially based on the reactivity of the boronic acid with the carbohydrate integrated in the nucleotide. Additional interaction between the anionic part of the nucleotide and the polymeric ammonium chain also favours the binding process. The selectivity shown by **(50)** in aqueous solution buffered with phosphate anion was 5′-UTP^{4-} >> 5′-UDP^{3-} > 5′-UMP^{2-} and it is rationalized in terms of a competition between the analyte and phosphate buffer for the cationic polymer.

The result of the receptor-substrate interaction is an enhancement of the emission intensity. Since it is known that the nucleobase partially quenches the fluorescence of anthracene, the observed enhancement is due to a compromise between the mentioned quenching and the increase caused by the boronic acid chelation.

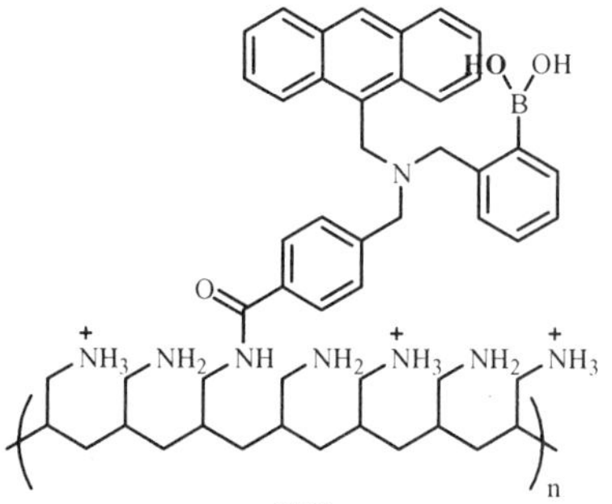

(52)

Figure 2.30. Polymeric receptor.

FLUORESCENT ANION COMPLEXATION AGENTS 79

The ditopic anthracene derivative **(53)**[45] takes advantage of the reversible formation of boronic esters and the strong interaction between guanidinium and carboxylate to work as sensor towards a series of carbohydrates in a mixture of MeOH:aqueous HEPES (1:1, v/v) (N-[2-hydroxyethyl]piperazine-N'-[2-ethanesulfonic acid]).

It is the presence of boronic acid that plays the main role in the binding of the carbohydrates, since changes in the fluorescence spectra can be observed for compound **(54)**, which does not have a guanidinium substituent. However, it is the presence of the guanidinium moiety which makes **(53)** highly selective for D-glucarate over D-gluconate, D-sorbitol, D-glucuronic acid and D-glucose. This is presumably due to an effective interaction between both of the carboxylates in the glucarate and the mentioned guanidinium.

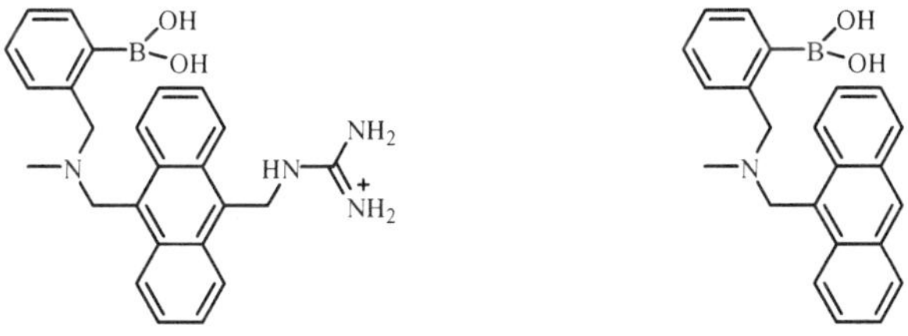

Figure 2.31. Boronic acid-guanidinium ditopic receptor.

2.2. 3. Sensors with Pyrene as Signalling Unit

2.2.3.1. Recognition based on electrostatic interactions and/or hydrogen bonding.

The very well known binding between guanidinium and phosphates was used to design compound **(55)**.[46] This simple pyrene derivative showed a high selectivity towards pyrophosphate due to an anion induced excimer formation. When compound **(55)** was titrated with pyrophosphate in methanol, the monomer structured emission band at 370-450 nm was quenched and a new band appeared at 476 nm, which was assigned to the excimer complex. The reverse was observed upon further addition of pyrophosphate. These results, which were supported by ^{1}H-NMR titration, suggested the initial formation of a 2:1 ligand:anion complex in which the anion acts as linker of two pyrenylguanidinium molecules. This complex evolved to a 1:1 stoichiometry when more than 0.5 equivalents of PPi were added.

On the other hand, a very weak excimer band was observed upon addition of HPO_4^{2-} and no change was detected in the emission spectra for other anions ($H_2PO_4^-$, AcO^-, SCN^-, Cl^-, Br^-).

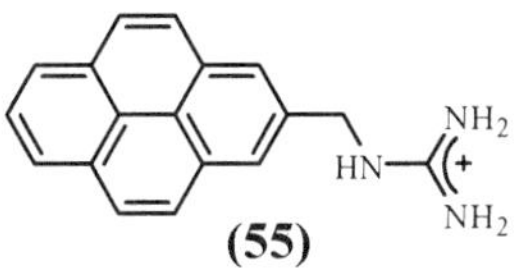

Figure 2.32. Pyrenylguanidinium receptor.

Compound **(53)** is intimately related to **(32)**. Hence, the results obtained for this pyrene derivative are basically the same as the ones described before, with a quenching of fluorescence occurring on anion binding. However, it is worth mentioning that association constants for **(53)** with dihydrogen phosphate and acetate are stronger as a consequence of the higher acidity of thioureas groups directly connected to the pyrene moieties.

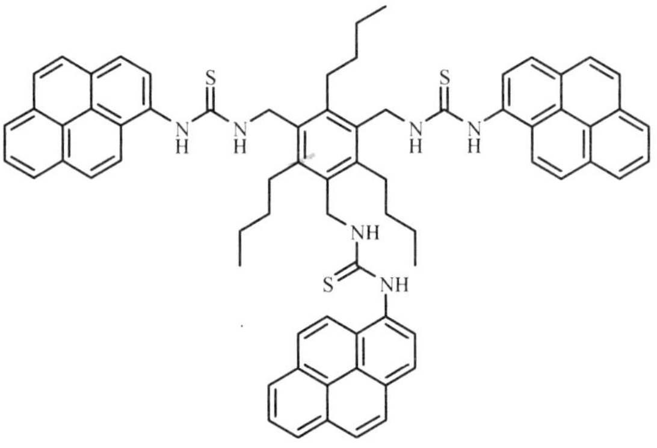

(56)

Figure 2.33. Polypyrenylthiourea receptor.

The emission bands of compound **(57)**[47] were quenched when anions such as AcO$^-$, H$_2$PO$_4^-$ and Cl$^-$ were added in acetonitrile solution. As described in previous examples, this is explained by a PET. Simultaneously, a structureless band appears at a wavelength between 480 and 500 nm. Due to the dependence of this band λ_{max} on the polarity of the solvent, this was evidence of the charge transfer nature of the emission. Therefore, this was assigned to an intramolecular exciplex formed between pyrene and thiourea.

From the analysis of the fluorescence changes a 1:1 stoichiometry could be determined, as well as the selectivity trend AcO$^-$ > H$_2$PO$_4^-$ > Cl$^-$.

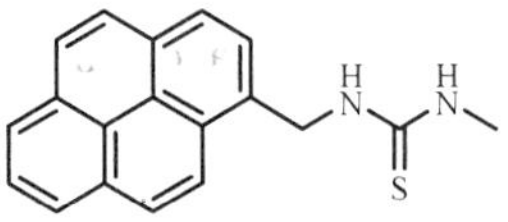

(57)

Figure 2.34. Pyrenylthiourea receptor.

With the aim of achieving a selective receptor for phosphate anion, modification of a pyridine-2,6-biscarboxamide frame by incorporating four additional amido groups was carried out in order to get a pseudotetrahedral cleft.[48] Additionally, four pyrene units were appended to act as luminescent signalling units. The fluorescence spectrum of **(58)** showed emission bands for the monomer at λ_{max}=377 and 397 nm and for the excimer at λ_{max}= 477 nm. It could be proven that this excimer corresponds to the interaction between pyrene units placed at both sides of the pyridine ring, i.e. pyrenes from the substituent at position 2 with pyrenes from the substituent at position 6.

Of all the anions titrated in THF solution (F⁻, Cl⁻, Br⁻, SCN⁻, AcO⁻, NO$_3^-$, ClO$_4^-$, HSO$_4^-$, H$_2$PO$_4^-$ and PO$_4^{3-}$) only phosphate ions caused changes in the fluorescence spectrum. The monomer bands showed an enhancement whereas excimer bands were quenched. This quenching was attributed to a separation of the pyrene units upon phosphate binding. From the analysis of titration profiles a 1:1 stoichiometry was confirmed and a better affinity of **(58)** towards PO$_4^{3-}$ than H$_2$PO$_4^-$ could be determined.

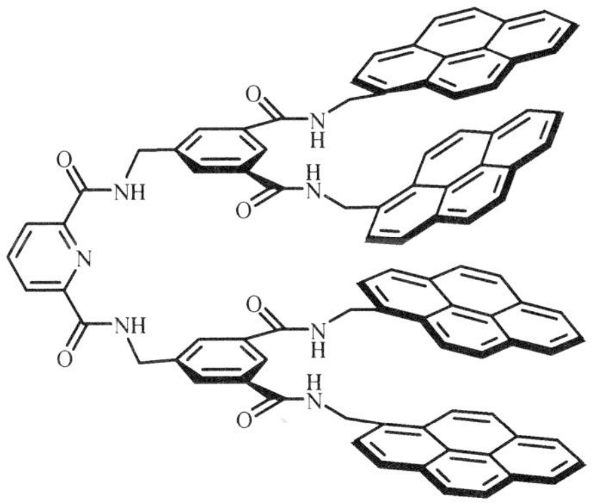

(58)

Figure 2.35. Polyamide receptor.

2.2.3.2. Recognition based on hydrophobic effect.

The pyrenophane **(59)** was synthesised to exploit the hydrophobicity of its cavity in the host-guest interaction.[49] Only one emission band, assigned to the intramolecular pyrene excimer, is observed in the fluorescence spectrum of **(59)** in water:ethyleneglycol (3:1, v/v).

When planar anions like polysubstituted naphtlene **(60)** were titrated, quenching of the excimer band was observed as a result of the π-π stacking of the analyte in between the two pyrene faces. Apart from this aromatic interaction, the hydrophobic effect of the pyrenophane cleft actively contributes as a driving force in the binding of the guest species. Similar quenching was also observed when the titrations were performed with nucleotides.

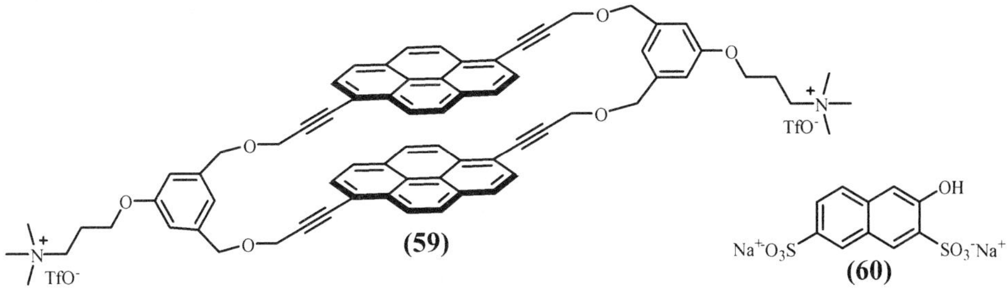

(59)   **(60)**

Figure 2.36. Pyrenophane receptor.

2.2.4. Sensors with Other Hydrocarbonated Signalling Unit

Other hydrocarbonated compounds that have been described as fluorescent anion sensors include biarylthiourea **(61)**.[50] Upon titration with F^- in $CHCl_3$, this receptor showed a fluorescence enhancement accompanied by a hypsochromic shift. However, when more than 2.5 equivalents of fluoride were added, the reverse effect was observed. This behaviour is due to the participation of two different stoichiometries. A 1:1 complex is formed in first instance, with fluorescence enhancement coming from a conformational restriction. This complex evolves to a 1:2 receptor:anion stoichiometry when excess of anion is added. Other anions such as $H_2PO_4^-$, AcO^-, HSO_4^-, Cl^- or Br^-, resulted in no changes. Thus, a good selectivity towards fluoride anion could be rationalized according to its size and its higher basicity.

A similar monosubstituted receptor, **(62)**, did not show the same evolution of the fluorescence spectrum on addition of anions, but only a broad emission increase at longer wavelengths was observed, which was assigned to the formation of an excimer.

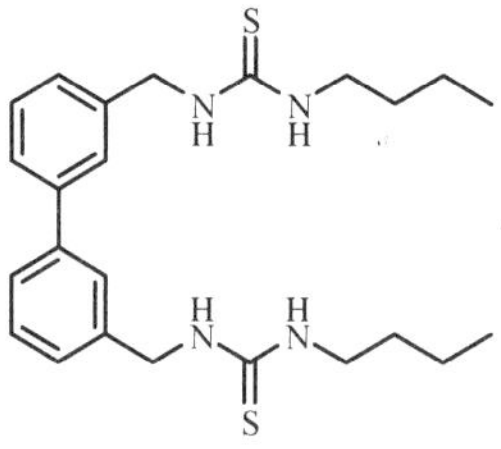

Figure 2.37. Biphenylthiourea receptors.

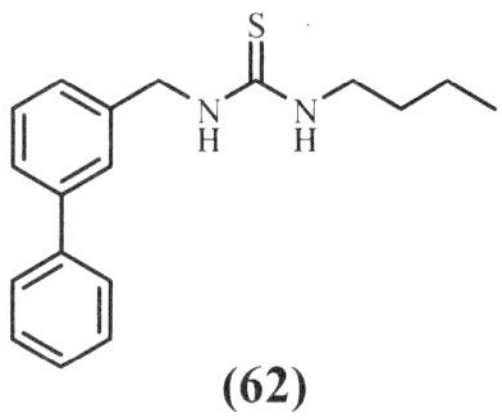

2.2.5. Sensors with Condensed or Conjugated Heteroaromatic Rings as Signalling Unit

2.2.5.1. Recognition based on electrostatic interactions and/or hydrogen bonding

As stated previously, there exists a particularly favourable electrostatic interaction between polyammonium moieties and phosphate anions. This fact has been exploited in the detection of nucleotides. In the case of **(63)** the presence of an azamacrocycle attached to an acridine unit was studied as receptor for different polyphosphate nucleotides which operates by a cooperative binding through electrostatic interaction and π-π stacking.[51]

This receptor turned out to be selective for triphosphates within the pH range 4-7.6, with only minor changes detected for ADP. This fact could be interpreted as proof of the importance of the structure and charge of the receptor in terms of binding. Interestingly, whereas ATP and CTP caused an enhancement of the fluorescence of compound **(63)**, a quenching of its fluorescence was observed in the presence of GTP. All the nucleotides tested formed a 1:1 complex, as it was also supported by NMR experiments.

Besides being studied as an anion sensor, compound XXXX was also used as a catalyst in the hydrolysis of ATP and ADP.

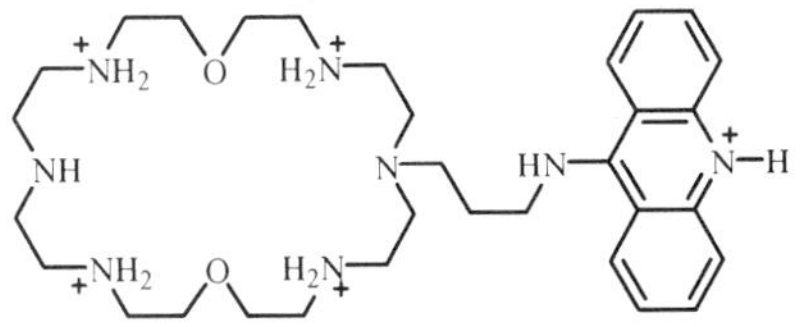

FLUORESCENT ANION COMPLEXATION AGENTS

(63)

Figure 2.38. Azacrown ether acridine receptor.

Compounds consisting of quinoxaline with appended pyridine units have proved to be useful receptors for anions.[52] Protonation of one of the pyridines enables **(64)** to participate in the binding of the anion via coulombic interactions as well as hydrogen bonding.

Despite the weak fluorescence of neutral 2,3-dipyridine-2-ylquinoxaline, when this compound is protonated an intramolecular hydrogen bond forces the structure to become flatter. This increases the conjugation of the system and consequently enhances its fluorescence intensity.

Titration experiments were carried out with anions, as their TBA salts, to detect luminescence quenching, revealing the following binding affinity: $F^- \approx H_2PO_4^- > Cl^- \gg Br^- > I^- > PF_6^-$. This change in the emission spectrum was presumably due to a loss of planarity in the molecule.

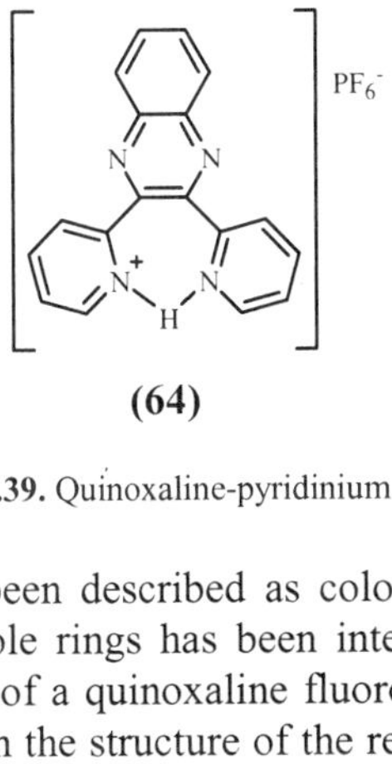

(64)

Figure 2.39. Quinoxaline-pyridinium receptor.

Dipyrrolylquinoxalines have been described as colorimetric and fluorescent anion sensors. The participation of pyrrole rings has been intensively studied in the area of anion recognition. The attachment of a quinoxaline fluorophore offers the possibility of easily modifying the substituents on the structure of the receptor, which can be exploited to tune the selectivity of the sensor.

The titration of TBA salts of different anions in dichloromethane causes a decrease in the fluorescence intensity that can be used to determine a 1:1 stoichiometry for all the complexes. The simplest example in the series, **(65)**,[53] shows a very high selectivity towards F^- over $H_2PO_4^-$ and Cl^-. The introduction of a nitro substituent increases the hydrogen bond donating character in **(66)**, which in turn increases the selectivity for F^- ($K_aF^-/K_aCl^- > 1800$, $K_aF^-/K_aH_2PO_4^- > 1400$).

In a similar way, **(67)** incorporates fluoro substituents in the structure of dipyrrolylquinoxaline.[54] As expected, an increase in the strength of anion binding is observed because of the higher acidity of the pyrrolic protons. Perhaps more important is the remarkable increase in the selectivity for $H_2PO_4^-$ over Cl^- in comparison with compounds **(65)** and **(66)**.

The expansion of the conjugated chromophore by adding aryl moieties to the quinoxaline frame, **(68)-(71)**, dramatically increases the emission intensity of the sensor.[55] All aromatic substituents showed similar values for their binding constants and as with parent compounds the strongest affinity corresponded again to fluoride. Interestingly, pyrophosphate particularly well complexed probably because of a ditopic hydrogen bonding to the two pyrrole moieties.

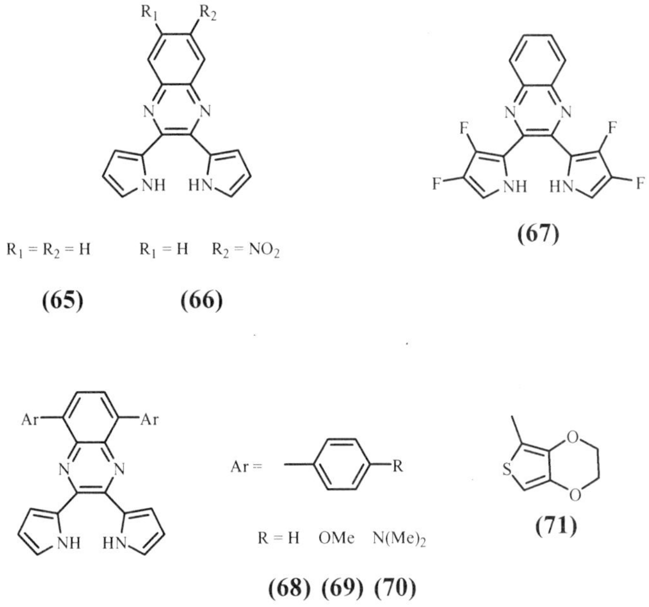

Figure 2.40. Quinoxaline-pyrrole receptors.

Biimidazole diamides **(72)-(76)**, were tested as anion sensors due to their multiple hydrogen bond options.[56] These compounds adopt a coplanar conformation, which permits the establishment of intra- and intermolecular hydrogen bonds. This last circumstance could be verified by a concentration dependence detected in the emission spectra.

Titrations with TBA $H_2PO_4^-$ and Cl^- in dichloromethane showed quenching of the fluorescence intensity in the presence of excess of anion, without very significant differences in terms of selectivity.

Structurally, the identity of the substituents seemed to better tune the affinity for Cl^- rather than dihydrogen phosphate, for which all the compounds **(72)-(76)** showed similar binding constant values.

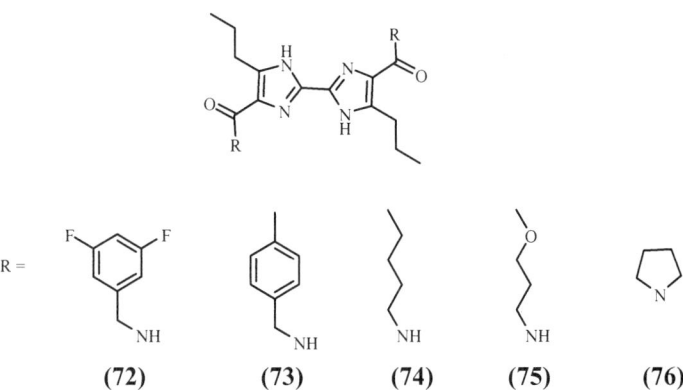

Figure 2.41. Biimidazole receptors.

2,2'-Baipyridine metallic complexes (Ru and Re) exhibit characteristic emission and have been intensely studied as fluorescent sensors (see section XXXX.XXX). On the other hand, the parent 2,2'-bipyridine is generally non-fluorescent, although it has been reported that the derivatives bearing an amino group at the 6 position exhibit fluorescence with high quantum yield. This fact allows the use of (77) and (78) as fluorophores in the study of molecular sensing.[57]

When these compounds are titrated with diphenylphosphonic acid in either cyclohexane or acetonitrile, quenching of the blue fluorescence is observed. The evolution of the spectra corresponded to that of the protonation of the bipyridine and subsequent ion-pair formation with the phosphate anion.

By analysing the relationship between the measured binding constants and the solvent polarity it was shown that for compound (77) the presence of amine protons contribute to the stabilization of the ion-pair complex in low polar solvents. However, when working in more polar solvents such as acetonitrile, the solvation of the ion-pair complex helps its stabilization and the presence of amide NH bonds is less important.

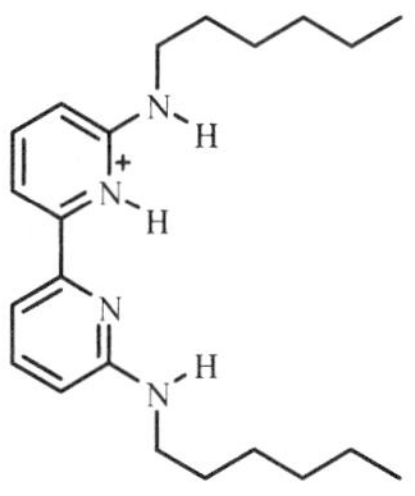

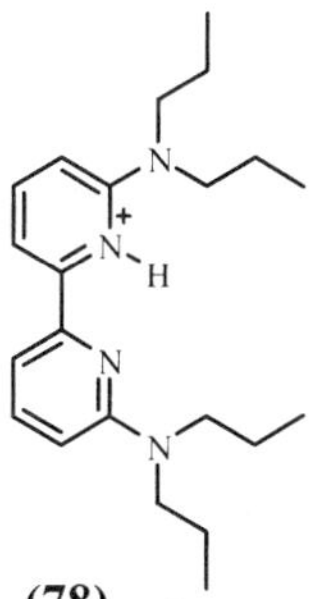

Figure 2.42. Bipyridine receptors.

2.2.5.2. Recognition based on a metal coordinative interaction

The cooperative effect of metal coordination and guanidinium electrostatic interaction was used in the design of a fluorescent sensor for citrate.[58] The fluorescence of ligand in **(79)**, with phenanthroline as a signalling unit, was quenched upon addition of $CuCl_2$ in a mixture of methanol:water (85:15, v/v) buffered to pH=7.4. When citrate was added to a solution of the Cu(II) complex **(79)**, under the same experimental conditions, an enhancement in the fluorescence was observed. This recovery of the fluorescence is due to an interaction between the metal and the anion which, in part, inhibits the quenching initially induced by the Cu(II).

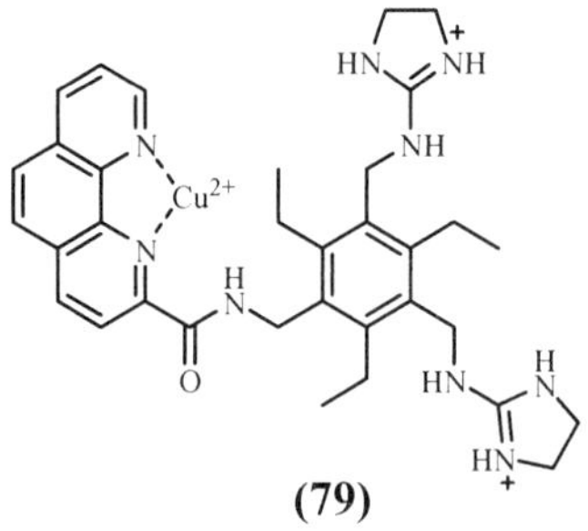

(79)

Figure 2.43. Phenanthroline receptor.

Acridine-appended cyclen **(80)** forms a stable complex with Zn(II). This system has been studied for nucleoside recognition[59] at physiological pH, proving to be selective for imide containing nucleobases. Under these experimental conditions, the imide proton is dissociated and the anionic substrate complexes the Zn(II). In addition, further stabilizing interactions have been proven such as π-π stacking between the acridine and the nucleobase and hydrogen bonding between the imide carbonyl groups and the cyclen NH groups.

Deoxythimidine presented the best binding affinity, causing quenching of the acridine fluorescence. Replacing imide containing analytes by amide functions turned out to decrease the effectiveness of the binding, as shown with deoxyguanosine. In the case of deoxyadenosine and deoxytidine changes in the fluorescence could hardly be detected in accordance with the lack of interaction with Zn(II)-**(80)**.

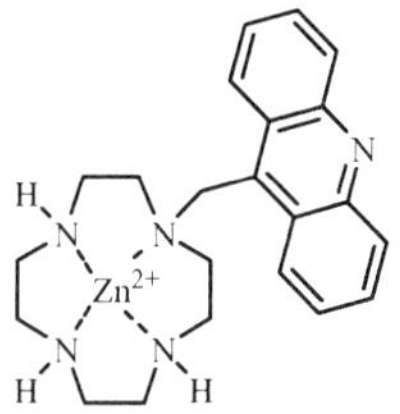

(80)

Figure 2.44. Acridine receptor.

FLUORESCENT ANION COMPLEXATION AGENTS

2.2.5.3. Recognition based on the formation of covalent bonds

Recognition of uronic acids was carried out with compound **(81)**,[60, 61] in which the boronic acid interacts with diol moieties, and the complex Zn(II)-phenanthroline is coordinated to the carboxylate group in the uronic acid. This cooperativity favours the formation of a 1:1 complex.

The recognition process can be followed by fluorescence spectroscopy and the emission intensity increased with the concentration of the tested uronic acids (D-glucuronic, D-galaturonic and sialic acid). The titration experiments were done in methanol:water (2:1, v/v) at pH=8 in order to improve the solubility and to work with the ligand in its neutral state. The emission enhancement is assigned to a more intense interaction between the tertiary amine and the boron atom upon substrate complexation, which in turn lowers the quenching ability of the mentioned nitrogen atom.

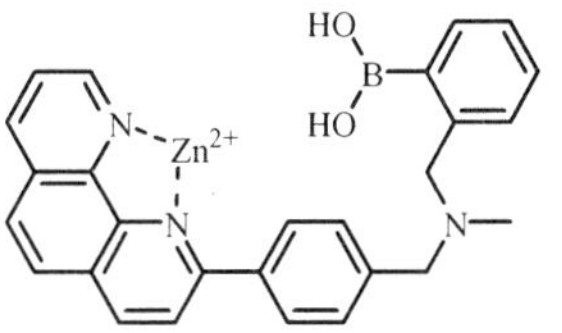

(81)

Figure 2.45. Phenanthroline ditopic receptor.

2.2.5.4. Recognition based on π-π interaction

Monocationic and dicationic 4,9-diazapyrenium derivatives **(82)-(84)** were prepared to check their response as nucleotide sensors.[62] Its large aromatic surface combined with the presence of positive charges might help to cooperatively bind nucleotides. Fluorescent titrations in aqueous solution buffered to pH=5 showed quenching of emission in the presence of nucleotides as a result of a π-π stacking interaction. It seems to be that this is the only driving force operating in the recognition process, since no distinction was made between the differently charged ATP, ADP and AMP.

All the determined binding constants are within the same range of values, with no great differences either when the receptors were mono- or dicationic or when bifunctional compounds like **(85)** and **(86)** were tested.

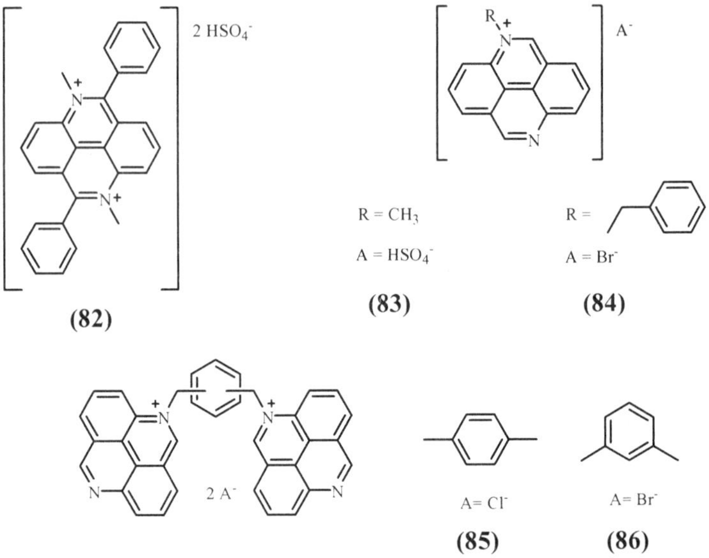

Figure 2.46. Pyrenium receptors.

Phenanthridinium is known to work as an intercalator in the helical structure of DNA and RNA. Based on previously described cyclobisintercalands receptors, phenanthridinium cyclophanes **(87)** and **(88)** were designed to bind nucleotides by means of π-π interaction.[63] In spite of the slow kinetics of the bonding process and the low luminescence of XXXX and XXXX, when aqueous solutions (pH=6.2) of these receptors with increasing concentrations of nucleotide were left to equilibrate it was possible to analyse the changes in the fluorescence spectra. It was reported that, when AMP intercalates between the positively charged phenathridium units, there may be a decrease in the acidity of receptor's amino protons and hence the quenching rate which results in emission enhancement. On the other hand, GTP and UMP did not affect the fluorescent properties of the cyclophanes.

The stoichiometry of the receptor:analyte complex varies from 1:1 to 1:2 on increasing the concentration of AMP, which might imply complexation, inside and outside the cavity of the cyclophane.

Changes within the emission spectra were also detected when the interactions with polynucleotides were tested. Single-stranded poly-A and poly-G caused fluorescence enhancement, which suggested the involvement of both phenanthridinium units in the complexation of the polynuclotide. When double-stranded polynucleotides were titrated, fluorimetric changes of **(87)** and **(88)** were found to depend on the base pair composition, with the fluorescence increase following the order: poly-A/poly-U > poly-dA/poly-dT > polyG/polyC. Moreover, a non-intercalative groove driven binding interaction is proposed for the recognition process.

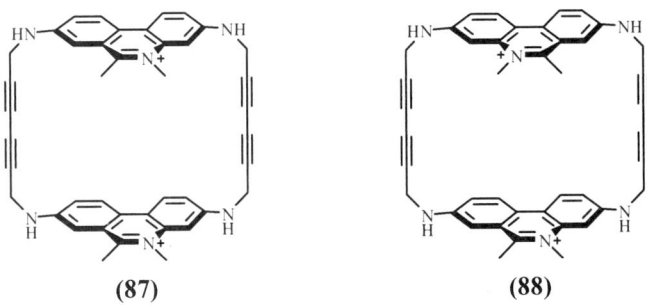

Figure 2.47. Phenanthridinium receptors.

2.2. 6. Sensors with Other Fluorescent Probes as Signalling Unit

2.2.6.1. Recognition based on electrostatic interactions and/or hydrogen bonding

Compounds **(89)-(91)** represent the second generation of calixpyrrole based fluorescent anion sensors.[64] In this instance, fluorophores that can be compatible with biological analysis conditions were chosen (dansyl, lissamine-rhodamine B and fluoresceine). A rigid spacer bearing NH groups that could assist in the complexation of the anion was incorporated into the structure of the receptor. The titration experiments were carried out in aqueous acetonitrile and, as expected, fluoride gave rise to the largest response. The order of selectivity for **(89)** and **(90)** was: $F^- \gg H_2PO_4^- > PPi \gg Cl^-$. Interestingly, compound **(91)** followed a different trend $F^- \gg PPi \gg H_2PO_4^- \gg Cl^-$. This good selectivity for pyrophosphate, over dihydrogen phosphate and chloride, was assigned to the cooperative effect of the thiourea group at the time of binding that anion.

Phenol derivatives[65] **(92)** and **(93)** show two bands in the fluorescent spectrum. For compound **(92)** it has been observed that the less energetic band changes its λ_{em} from 448 to 498 nm when the polarity of the solvent is increased, e.g. from hexane or toluene to acetonitrile or DMF. This result has been explained on the basis of the presence of several species in equilibrioum. When aliquots of F^- and $H_2PO_4^-$ were added to a solution of **(93)** in DMF, in different experiments, strong quenching was detected in the emission at 448 nm, whereas small enhancement was observed for the emission band at 354 nm. These changes are indicative of the formation of hydrogen bonds between the receptor and the analyte. Since no changes are detected during the titration with Cl^-, this represents another example of a sensor with selectivity towards fluoride and dihydrogen phosphate over chloride.

Similar results were obtained for **(93)**, but it exhibits a higher degree of selectivity for $H_2PO_4^-$, presumably due to a double hydrogen bonding interaction. ^{1}H-NMR titrations confirmed a 1:1 stoichiometry for the formed complexes.

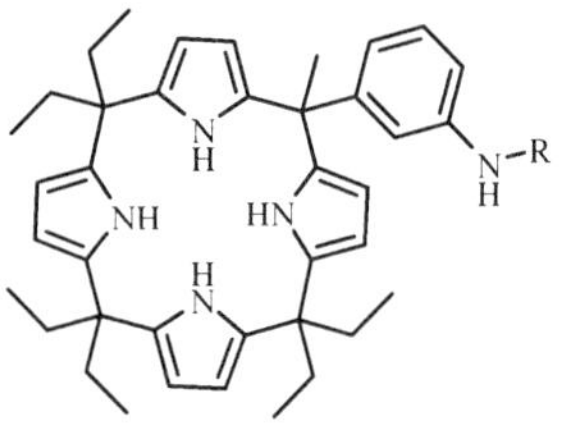

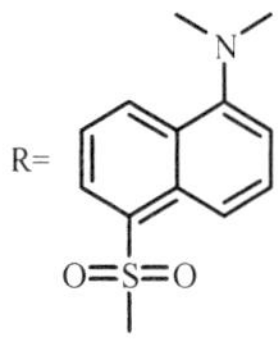

Figure 2.48. Second generation of calixpyrrol receptors.

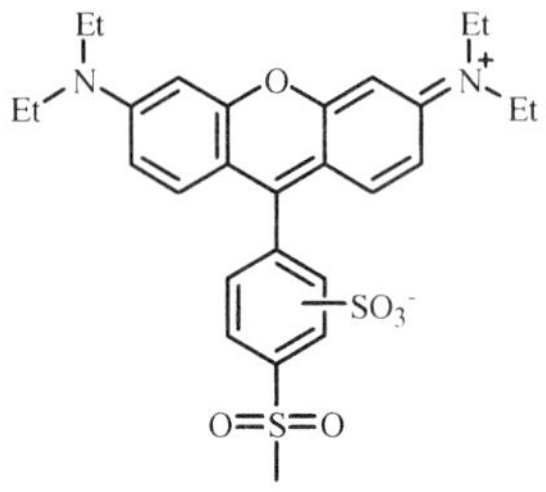

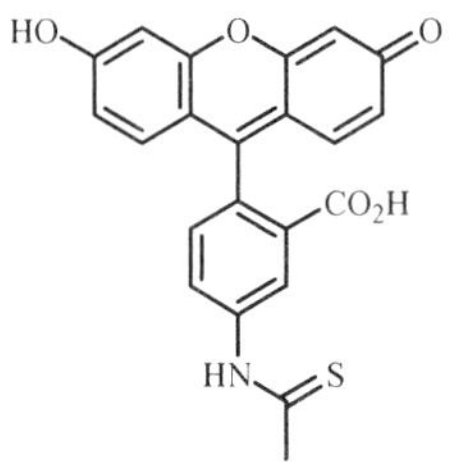

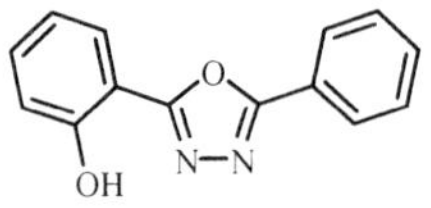

Figure 2.49. Phenol-based receptors.

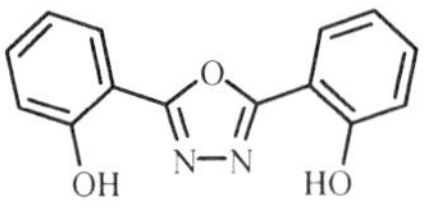

The incorporation of a fluorophore (5-aminocoumarine) into the structure of a macrcyclic receptor **(94)**,[66] in which binding of the analyte was achieved by means of hydrogen bonding, led to several changes in the fluorescence spectrum upon addition of anions. In first place, enhancement of the emission intensity was detected, presumably due to a restriction in the conformational flexibility of the receptor. Additionally, a red shift of the starting fluorescent band occurred. This was assigned to a stabilization of the fluorophore excited state. Also, the appearance of a second band in the fluorescent spectrum was observed. The origin of this new band was explained on the basis of an intermolecular excited state proton transfer (ESPT), due to its presence being related to anion basicity.

Titrations experiments were undertaken in a mixture DMSO:1,4-dioxane (1:1, v/v) and revealed a 1:1 binding mode. The following trend for the affinity between receptor and analyte could be determined: $H_2PO_4^- > PhPO_3H^- > pTsO^- > Cl^-$.

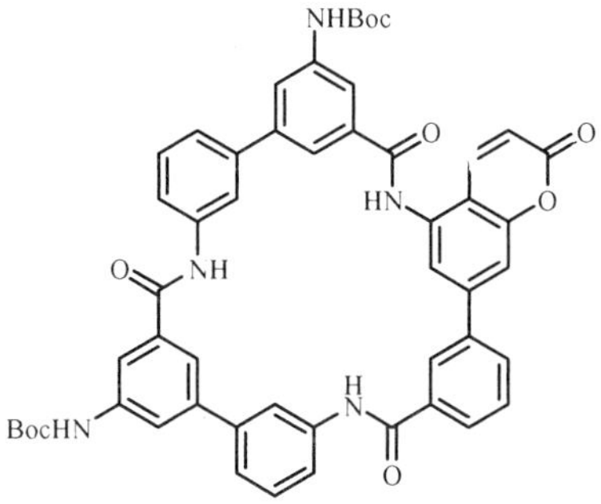

FLUORESCENT ANION COMPLEXATION AGENTS

(94)

Figure 2.50. Coumarine-based receptor.

The very well known ability of the guanidinium group to bind anions, both through hydrogen bonding and electrostatic interaction, was used in the design of compound **(95)** to sense sulphate anion.[67] The fluorophore, 4-(N,N-dimethylamino)benzoate, shows dual fluorescence originating from the simultaneous emission of the locally excited (LE) state and the twisted intramolecular charge transfer (TICT) state. Addition of TBA sulphate to a solution of **(95)** caused a response in which the LE band and the TICT band evolved in different ways depending on the number of equivalents of added analyte. In the range 0-0.5 equivalents an intramolecular hydrogen bond is broken causing a decrease of the LE emission intensity and enhancement of the TICT intensity. In the range 0.5-2 equivalents the influence of the lipophilicity of the TBA countercation causes enhancement of the LE fluorescence.

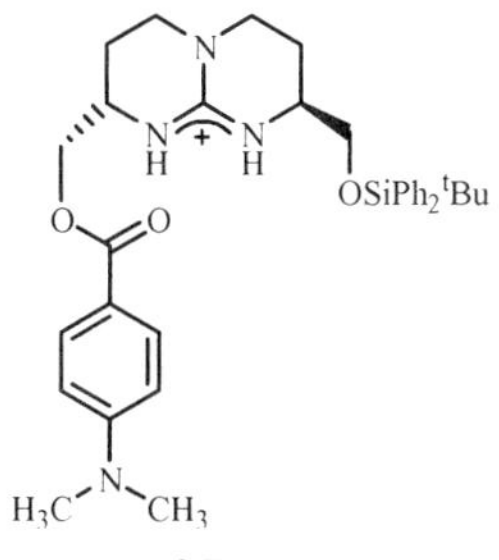

(95)

Figure 2.51. Dimethylaminobenzoate-guanidinium receptor.

All of the examples of chemosensors described in this review so far are based on the structure: binding unit-spacer-signalling unit. A different approach to the design of anion sensors takes inspiration from the antibody biosensors. In this approach, the receptor is initially complexing a fluorescent indicator, and on addition of anion, a competition for the binding cavity can be established. As long as the binding constant between receptor and anion is higher that the binding constant for receptor-indicator, the fluorophore will be displaced from the cavity of the receptor, consequently causing a change in the fluorescent properties of the solution.

Compound **(96)**[68] is known to bind citrate in water (K_a= 6.8 10^3 M^{-1}) through hydrogen bonding and charge pairing. It was also shown that, in a mixture of MeOH:H$_2$O (75:25, v/v) at pH=7.4, **(96)** could also bind the fluorescent probe 5-carboxyfluoresceine (K_a= 4.7 10^3 M^{-1}). Competition assays showed that on adding citrate aliquots to a mixture of the **(96)**-fluoresceine ensemble, under the described experimental conditions, a decrease in the fluroescence intensity of the fluorophore could be detected. This was due to its protonation after being displaced from the binding cavity.

A good selectivity towards citrate could be determined, since small changes were observed with succinate and no changes were observed with acetate or sugars. This system was therefore used for the analysis of citrate in commercial drinks, since common contaminants present in commercial beverages (malate, ascorbate, lactate, benzoate, phosphate) should not affect the competition assay.

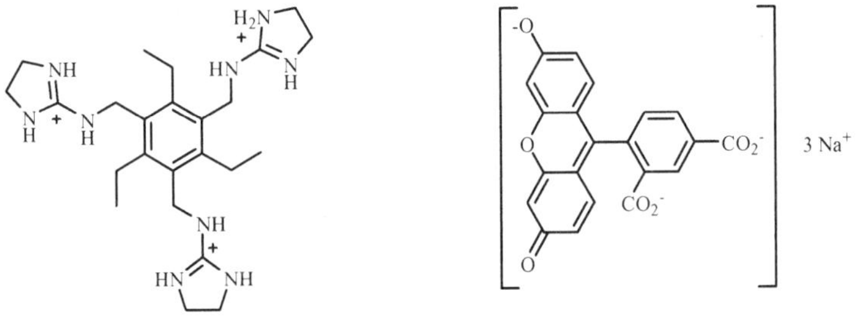

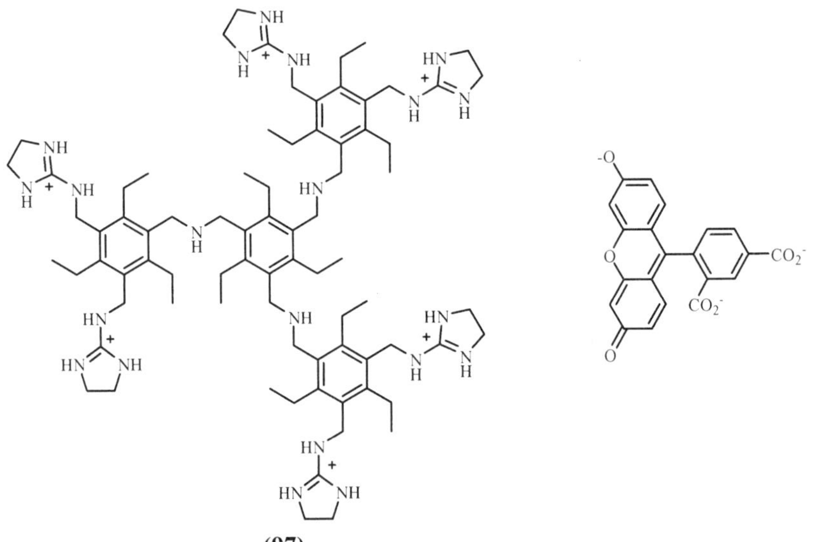

Figure 2.52. Tripodal guanidinium-carboxyfluresceine ensembles.

FLUORESCENT ANION COMPLEXATION AGENTS

Using the same methodology, compound **(97)** in combination with 5-carboxyfluoresceine showed a decrease in the emission intensity when the complexed fluorophore was replaced by inositol-1,4,5-triphosphate[69] in methanolic solution buffered to neutral pH.

Macrocyclic squaramide **(98)** presents the spatial requirements to bind tetrahedral anions[70] due to its concave cavity, in which hydrogen bonding and electrostatic interactions can be established. An ensemble with fluoresceine is formed with the fluorescence of fluoresceine being quenched presumably by a PET from the donor squaramide rings to the fluoresceine single excited state. Titration experiments of this ensemble with sulphate, in methanol:water (9:1, v/v) at pH ≥ 10, showed an increase in the intensity of fluorescence as a result of the liberation of fluoresceine. This system was applied to the real time on-line determination of sulphate in water.

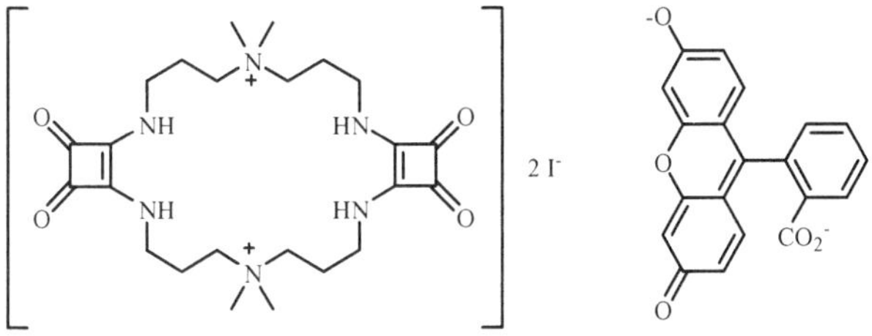

(98)

Figure 2.53. Squaramide-fluoresceine ensemble.

2.2.6.2. Recognition based on a metal coordinative interaction.

The competition approach has also been used for the recognition of anions through metal-analyte coordinative interaction. Compound **(99)**[71] binds coumarine 323 by interacting with Cu(II) atoms complexed by the cryptand and total quenching of the fluorescence is observed in aqueous solution. This quenching is assigned to an energy transfer involving the coumarine and the Cu(II) centres. When anions were added to an aqueous solution of the **(99)**-coumarine 323 ensemble (at pH=7) a total recovery of the fluorescence was detected for HCO_3^-, N_3^- and NCO^-. On the other hand, little or no changes were observed for NCS^-, NO_3^-, SO_4^-, HPO_4^{2-}, HCO_2^- and AcO^-. Due to the good behaviour of the chemosensor, this ensemble was utilized in analytical studies of commercial mineral water for the determination of bicarbonate.

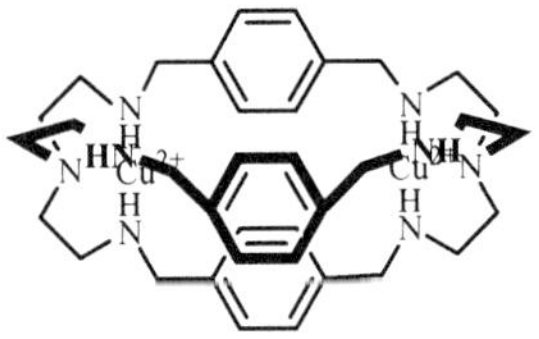

(99)

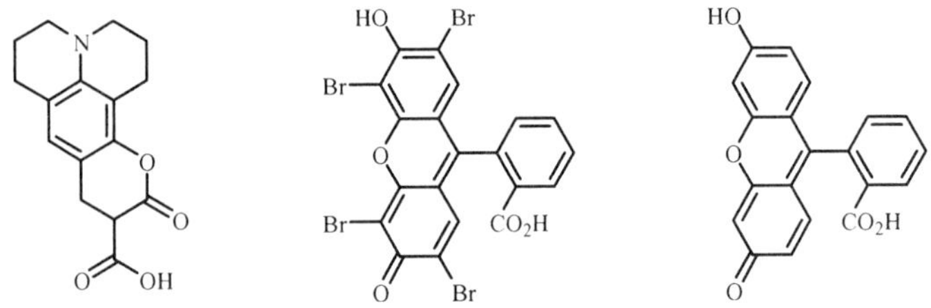

Figure 2.54. Fluorophore-Copper complex ensemble.

Similar results have been reported in the sensing of pyrophosphate with compound (**99**).[72] Three fluorophores (coumarine 343, fluoresceine and eosine Y) were complexed with total quenching of their emission. On addition of PPi to a solution of the sensing ensemble buffered to pH=7 a recovery of the fluorescence was observed for the three fluorophores. Interestingly, in the cases of fluoresceine and eosine Y, selective binding of pyrophosphate could be observed. Additionally, discrimination between PPi and other potentially interfering anions (phosphate, N_3^-, NCO^-, NO_3^-, SO_4^-, Cl^-, AcO^- and $PhCO_2^-$) was demonstrated due to the good relationship between the binding constants, following the order $K_{PPi} \gg K_{indicator} \gg K_{interferent}$.

Compound (**99**) was also used in conjunction with the same three fluorescent probes for the sensing of aminoacids.[73] A common feature of all the aminoacids is that they bear a carboxylate function that could compete for the binding cavity. Nevertheless, better selectivity can be achieved if a stronger interaction from the residue of the aminoacid can be established. This is the case with the aminoacid histidine, which is in its imidazolate form at neutral pH. This imidazolate could displace the fluorescent indicator by bridging the two metallic centres in (**99**). The highest selectivity corresponded to the (**99**)-eosine Y ensemble which could discriminate His among the rest of the aminoacids studied.

2.3. INORGANIC SENSORS FOR ANIONS

2.3.1. Ruthenium(II)- and Rhenium(I)-Bipyridyl-based Receptors

One of the most extensively investigated sensing subunits has been tris(2,2'-bipyridyl)ruthenium(II) ([Ru(bpy)$_3$]$^{2+}$), which is popular due to its chemical stability, redox properties, excited-state reactivity, and luminescent emission.[74,75] The system exhibits several absorptions, with the longest wavelength absorption being the MLCT band at ca. 456 nm, excitation of which leads to emission at ca. 670 nm. The emission in the visible region of the spectrum has obvious analytical advantages.

The bpy moiety can easily be functionalised, which leads to the possibility of a large array of substituted systems. Numerous different binding sites of varying topologies have been attached to the fluorophore. Acyclic, macrocyclic and calix[4]arene structural frameworks have been incorporated in order to produce a new class of anion receptor capable of optical and electrochemical sensing.[76, 77, 79]

Initial efforts in this area involved the attachment of simple hydrogen bonding groups such as amides to the bipyridyl moiety, thus attaching an anion binding site to a fluorophore and subscribing to the standard receptor design methodology. Similar approaches had proven successful for redox-active receptors, which consisted of amide groups attached to metallocenes such as ferrocene and cobaltocenium.[6, 78]

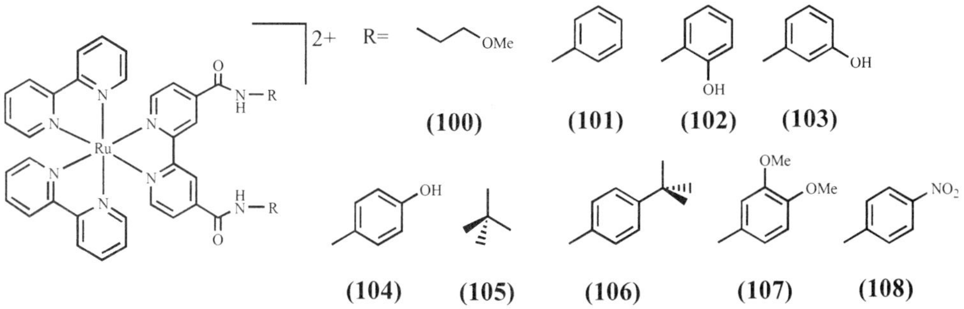

Figure 2.55. Assorted [Ru(bpy)$_3$]$^{2+}$-amide receptors

An assortment of R groups may be attached in order to tune the properties and selectivity of the receptor. For instance, molecules **(102)** - **(104)** shown in Figure 2.55 have an extra hydrogen bonding site in the form of the hydroxyl group[77]. The single crystal X-ray structure of the chloride complex of **(107)** demonstrates the importance of hydrogen bonding to the anion-binding process (Figure 2.56)[76]. Stability constants measured by NMR in DMSO-d_6 show the simple acyclic receptors **(100)** and **(107)** to form strong complexes with both Cl$^-$ and H$_2$PO$_4^-$ anions. Luminescence emission measurements in DMSO showed the receptor **(100)** to exhibit a significant blue shift in the MLCT emission band λ_{max} on addition of H$_2$PO$_4^-$. These shifts were not observed with unfunctionalised [Ru(bpy)$_3$]$^{2+}$, and addition of Cl$^-$ anion to **(100)** induced a negligible effect.

The shifts were accompanied by a large increase in emission intensity, and it was proposed that this might be due to the rigidity effect. Such a response is common with the binding of anions in [Ru(bpy)$_3$]$^{2+}$-containing receptors. Complexation of the anion leads to the adoption of a rigid conformation not observed in the free receptor. This structural rigidity decreases the probability of vibrational and rotational relaxation modes of nonradiative decay, which therefore leads to an enhancement of emission intensity.[79]

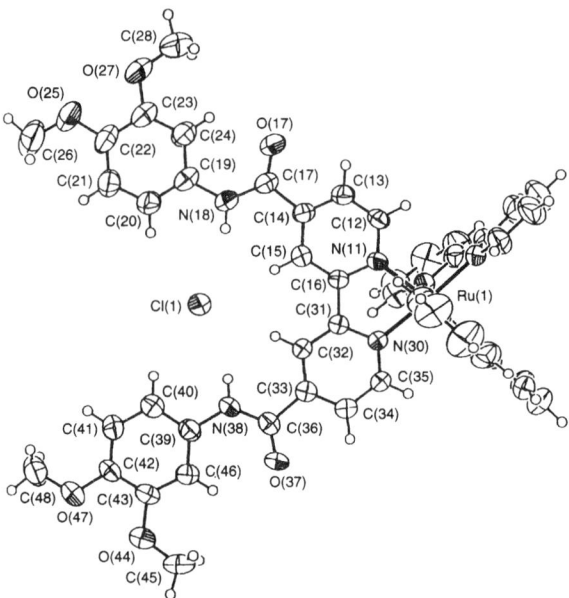

Figure 2.56. Single crystal X-ray structures of the chloride complex of **(107)**. Reproduced with permission from *Inorg. Chem.* **1996**, *35*, 5868. Copyright 1996, American Chemical Society.

Molecules **(101)** – **(106)** exhibited similar effects when studied in acetonitrile with halides. The MLCT emission band is observed in the range 630-650 nm, and shows an enhancement in all cases with chloride and bromide. The most obvious effect is seen with **(105)**, which shows a 42% increase in emission intensity of the MLCT band with chloride, as well as an 8 nm hypochromic shift. Addition of iodide leads to a decrease in emission intensity, which is attributed to the heavy atom effect[77].

The 5,5'-bipyridyl disubstituted analogues of receptors **(100)**, **(101)** and **(107)** have also been synthesised. The effects on the luminescent properties were studied with an excess of halide anions in acetonitrile. Quenching of the emission band was observed, with receptor **(107)** showing the largest effect. A quenching of 24 % and 28 % upon addition of 10 equivalents of chloride and bromide, respectively, was observed.[80]

Investigations are currently underway into the enantioselective recognition of anionic guest species using chiral receptors based on similar systems with both ruthenium(II) and rhenium(I).[81]

An acyclic [Ru(bpy)$_3$]$^{2+}$-ferrocene receptor **(109)** has also been prepared,[82] (Figure 2.57). The ferrocene units quench the emission of the ruthenium center in the free receptor, but addition of H$_2$PO$_4^-$ ions increases the emission by 20-fold in acetonitrile. This increase is not observed with Cl$^-$ or HSO$_4^-$ ions. Competition experiments in the presence of these anions gives rise to an emission increase identical to that obtained upon addition of H$_2$PO$_4^-$. It is thought that coordination of the anion may alter the energy transfer between the ruthenium and ferrocene centres, and thereby induce emission enhancement. Receptor **(109)** can therefore be considered a H$_2$PO$_4^-$-selective luminescent anion receptor.

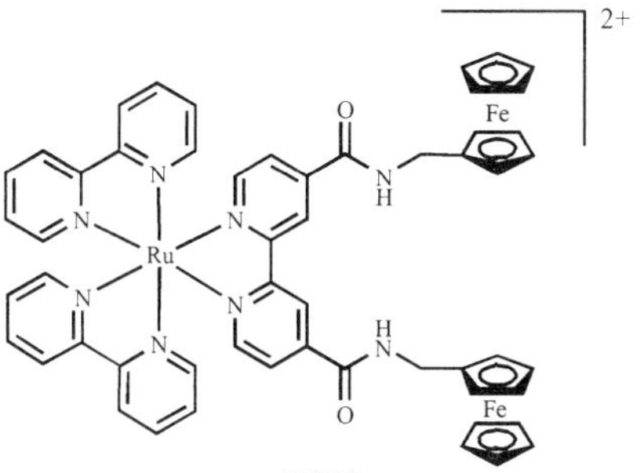

FLUORESCENT ANION COMPLEXATION AGENTS

(109)

Figure 2.57. Acyclic [Ru(bpy)$_3$]$^{2+}$-ferrocene receptor.

The acyclic water soluble ruthenium(II) bipyridyl polyaza receptors **(110) – (113)** in Figure 2.58 have been shown to bind and detect phosphate anions in aqueous media by MLCT luminescent emission quenching.[83] The protonated amines bind anions via both hydrogen bonding and electrostatic interactions. At pH 6.0, addition of both KH$_2$PO$_4$ and Na$_2$H$_2$ATP, up to 15 % quenching of emission intensity is observed.

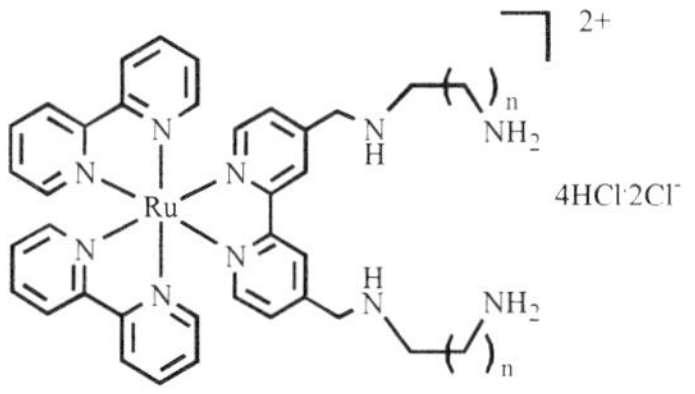

(110), n = 1
(111), n = 2
(112), n = 3

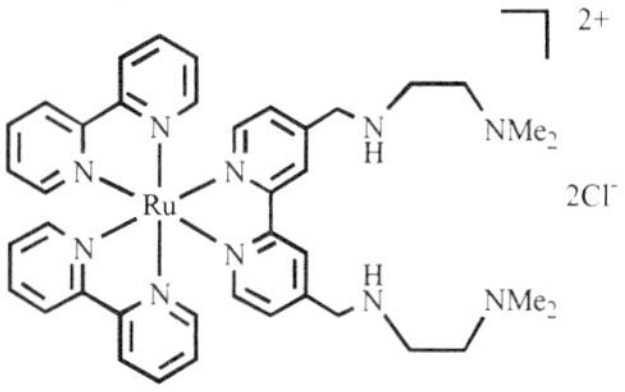

(113)

Figure 2.58. Water soluble ruthenium(II) bipyridyl polyaza receptors.

Further evidence of the tuning ability provided by assorted R groups is demonstrated by molecule **(114)**[84]. The extended aromatic amide pendant groups make the receptor selective for aromatic, linear and cyclic dicarboxylates, with binding giving rise to changes in the emission band. The most marked effect is observed with trans-1,4-cyclohexanedicarboxylate which induces an enhancement of emission intensity, ascribed to the rigidity effect. The host:guest complex is shown in Figure 2.59. The opposite effect is noticed with the cis isomer, which may be due to the opening of non-radiative decay channels via vibrational relaxation.

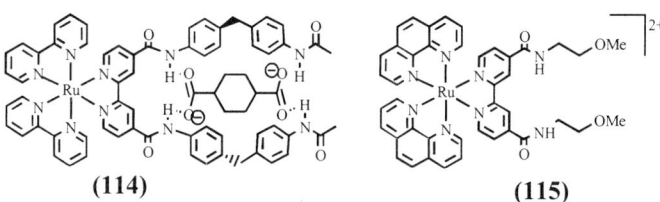

(114)   **(115)**

Figure 2.59. Stereoselective dicarboxylate receptors.

Ruthenium complexes of other pyridyl ligands can also be used as a fluorophore, as demonstrated by receptor **(115)**.[85] In this case two of the bipyridyl ligands have been replaced by phenanthroline, giving a molecule analogous to **(100)**. Addition of dihydrogen phosphate anions in acetonitrile solution resulted in a decrease in absorption maximum and an increase in emission intensity, which was due to the rigidity effect.

The $[Ru(bpy)_3]^{2+}$ moiety has also been utilized by Deetz and Smith, who have attached saccharide-binding boronic acid groups to produce molecule **(116)** (Figure 2.60(a)).[86] The phosphate group of the sugar is presumably bound to the amide protons in the cleft, while the boronic acid groups coordinate to the sugar. The presence of the phosphorylated sugars, in particular fructose-6-phosphate, in aqueous solution is detected by perturbations in the luminescent properties of the complex. Rhenium analogues of this material have previously been used to bind simple sugars.[87]

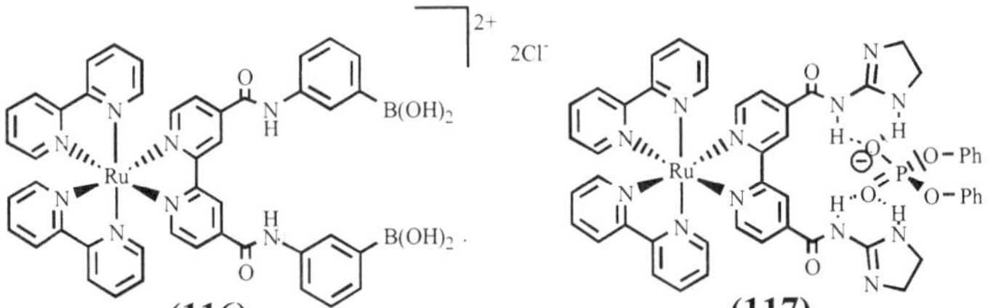

Figure 2.60. (a) Saccharide-selective $[Ru(bpy)_3]^{2+}$-boronic acid receptor (b) Imidazole-functionalised $[Ru(bpy)_3]^{2+}$ receptor + TDPP.

Watanabe and coworkers have produced imidazole-functionalised $[Ru(bpy)_3]^{2+}$ complexes such as **(117)**[88] (Figure 2.60(b)). These recognize anionic and neutral phosphodiesters with luminescent signal enhancement observed in acetone. The receptor is proposed to form 1:1 complexes with tetraethylammonium diphenyl phosphate (TDPP) or dibenzylphosphate (DBHP).

Other acyclic receptors include mixed-metal cleft-type molecules (Figure 2.61), in which d^6 metals Ru^{II}, Os^{II} and Re^{I} were used, with a variety of different bridging groups.[89] While fluorescence behaviour is unreported, NMR studies show selectivity for $H_2PO_4^-$ over chloride. The bridging unit has a considerable effect on the strength of binding. For instance, a *meta*-phenylene-bridged ligand gives a stability constant with dihydrogen phosphate in DMSO of 55 M^{-1}. The *para* analogue results in a binding constant of 4320 M^{-1}. The Os^{II} receptors exhibited larger anion stability constants, due to the efficient Lewis-acidic character of the $[Os(bpy)_3]^{2+}$ moiety.

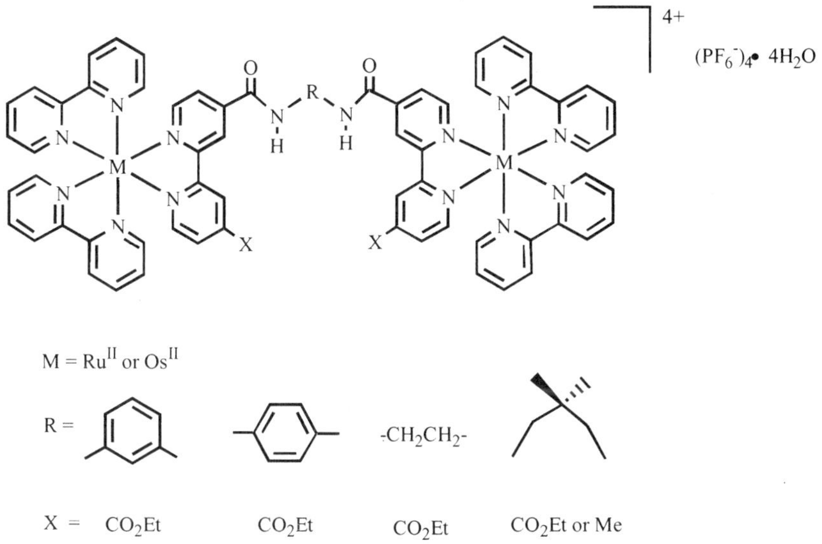

FLUORESCENT ANION COMPLEXATION AGENTS

Figure 2.61. Cleft-type mixed-metal receptors.

Macrocyclic receptors confer a greater degree of preorganisation, and therefore provide a means of achieving greater selectivity. Receptor **(118)** in Figure 2.62 provides some evidence for the explanation of the rigidity effect common to these type of receptors[76]. Upon addition of dihydrogen phosphate to this receptor, the emission intensity remains unchanged. The macrocycle provides a more rigid cavity than other acyclic receptors such as **(100)**, and therefore there is no conformational change upon coordination that leads to an increase in rigidity and subsequent change in luminescence.

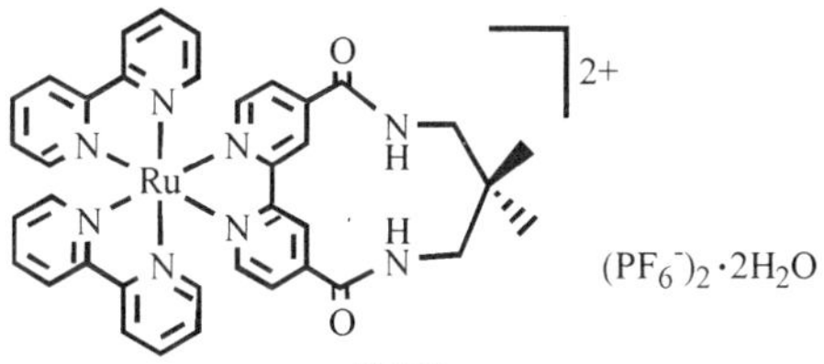

(118)

Figure 2.62. Preorganised macrocyclic receptor.

Other macrocyclic receptors include **(119) – (120)** in Figure 2.63, which incorporate, respectively, a bipyridyl and a pyridyl spacer in the macrocyclic binding site.[90] Upon addition of chloride anion to acetonitrile solutions, a blue shift of 5-10 nm is observed, along with enhancement of the emission intensity, due to rigidity effects.

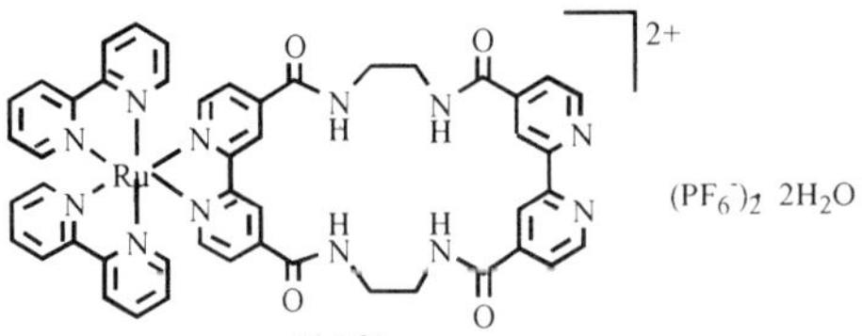

(119)

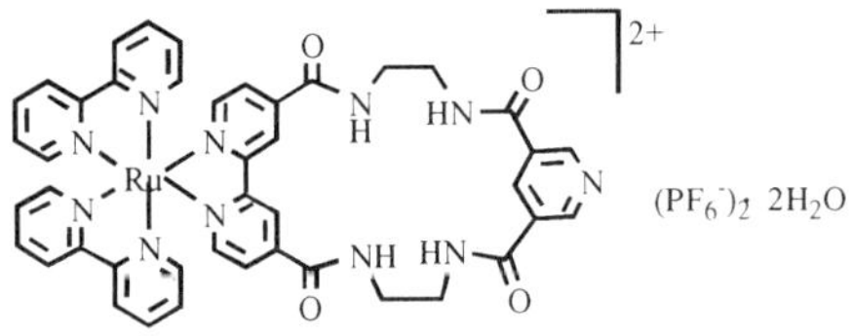

(120)

Figure 2.63. Simple macrocyclic receptors.

Particularly noteworthy are receptors **(121) – (124)** (Figure 2.64), synthesized by Beer and co-workers, which exhibit a marked selectivity for chloride ions[90]. Extremely stable 1:1 complexes are formed in DMSO, with stability constants measured by NMR of up to 4×10^4 M^{-1}. Both ^{1}H and ^{31}P NMR show little perturbation upon addition of H$_2$PO$_4^-$, and none at all in the case of **(123)**. Luminescence studies in acetonitrile show a slight blue-shift of the MLCT emission band with significant intensity enhancement in response to the addition of chloride anions. For **(121)**, **(122)** and **(124)**, chloride anions caused a decrease in the rate constant of the energy transfer process responsible for quenching of the luminescent excited state of the ruthenium bipyridyl moiety. This effect of chloride coordination indicates a decrease in the electronic interaction between the excited ruthenium subunit and the appended quencher subunit.

Increasing the size of the cavity by two or four methylene units dramatically reverses the trend in anion selectivity. The rigidity of the macrocycle is thought to contribute to this selectivity, since acyclic analogues tend to prefer H$_2$PO$_4^-$. The larger size and tetrahedral shape of this anion make it incompatible with the macrocycle's cavity.

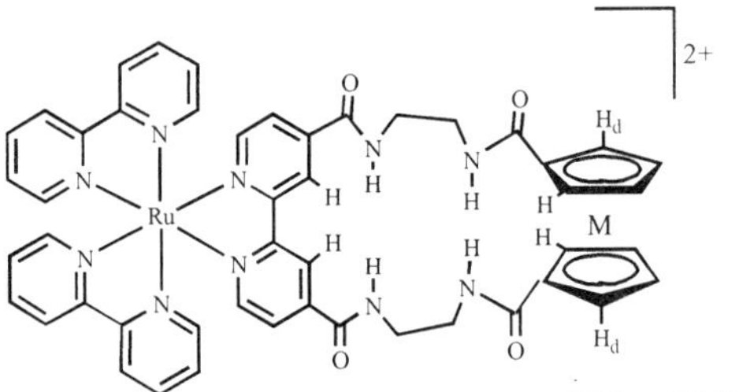

M = Fe, **(121)**
M = Co$^+$, **(122)**

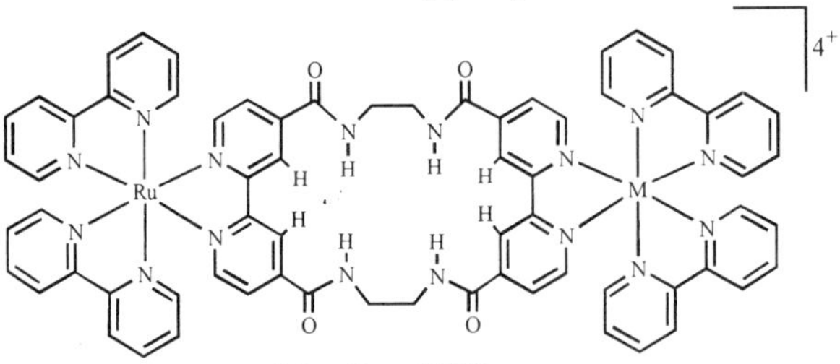

M = Ru, **(123)**
M = Os, **(124)**

Figure 2.64. Macrocyclic halide-selective receptors.

The crown ethers constitute a class of compound with a well-defined binding site, and so they have been incorporated into receptors containing the ruthenium and rhenium bipyridyl fluorophores. Compound **(125)** in Figure 2.65 consists of two fluorophores attached to an azacrown.[91] The luminescent MLCT emission band at 650 nm increases upon addition of chloride to acetonitrile solutions. The formation of a 1:2 receptor:chloride complex gives rise to increased rigidity, and hence the emission increase. Addition of potassium ions does not change the intensity enhancement, but the stability constant is enhanced.

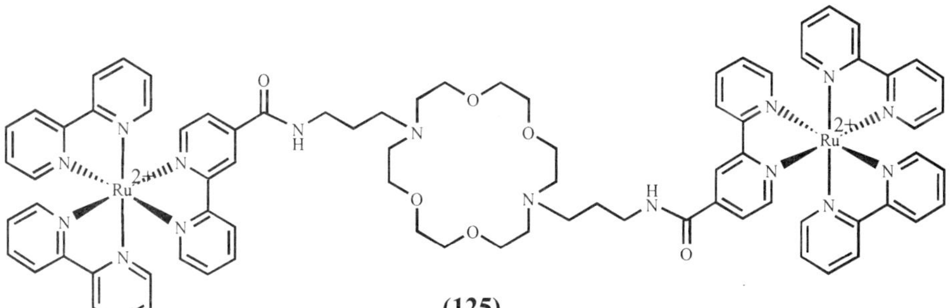

Figure 2.65. Crown ether-based receptors.

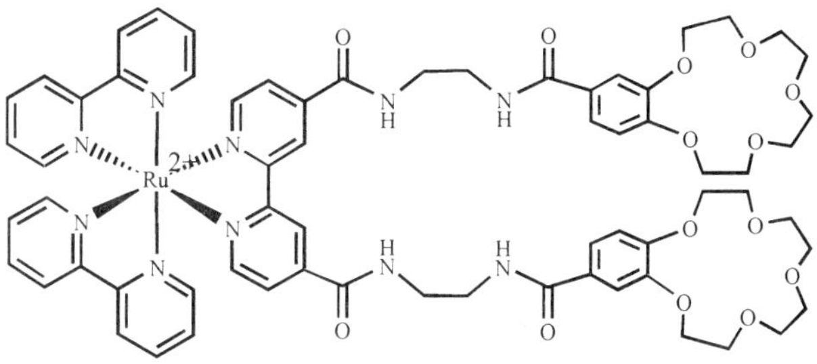

Addition of chloride and dihydrogen phosphate to receptor **(126)** in acetonitrile causes an increase in luminescence intensity of up to 140 %.[92] The MLCT emission also undergoes a hypsochromic shift of 7 nm. In the presence of potassium ions, dihydrogen phosphate has no significant effect. However, addition of chloride causes an increase in emission intensity. This is due to the formation of a potassium-crown sandwich complex, which preorganises the cavity into a psuedo-macrocyclic binding site selective for chloride over H_2PO_4.

Rhenium bipyridyl complexes have been utilized in the synthesis of ion-pair receptors. Using a similar approach to that employed for analogous cobaltocenium receptors,[93] crown ether molecules have been attached to a rhenium(I) amide bipyridine architecture (Figure 2.66).[94,95] ^{1}H NMR studies in 1:1 $CD_3CN/DMSO$-d_6 showed all receptors to be selective for acetate over chloride, with downfield shifts of the bpy protons indicative of hydrogen bonding. Titrations in the presence of potassium cations revealed a positive cooperative effect, with increases in binding constants of 80-115 %

for (127) and (129), and 40-50 % for (128) and (130). This effect may be due to the longer m-xylyl spacer in (127) and (129), which reduces the through space electrostatic interaction between cation and anion binding sites. Emission spectroscopic titrations also revealed cooperative binding effects, in particular for receptor (130) with acetate in CH$_3$CN. This suggests a preorganisation of the receptor by formation of a K$^+$ sandwich complex.

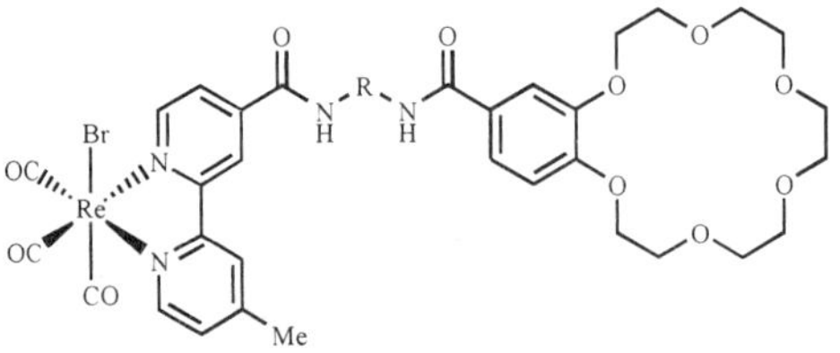

(127), R = CH$_2$CH$_2$
(128), R = m-xylyl

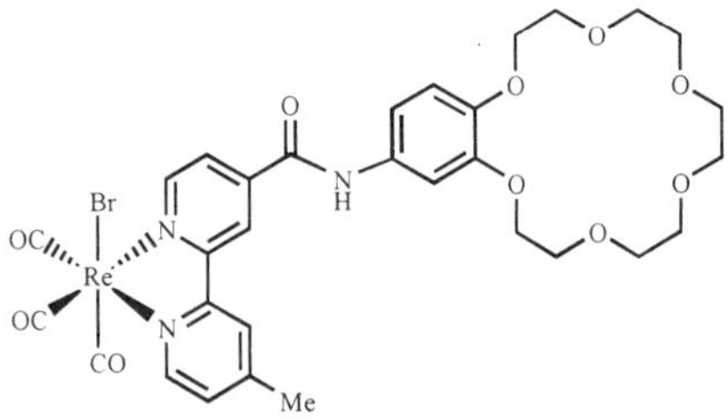

(129)

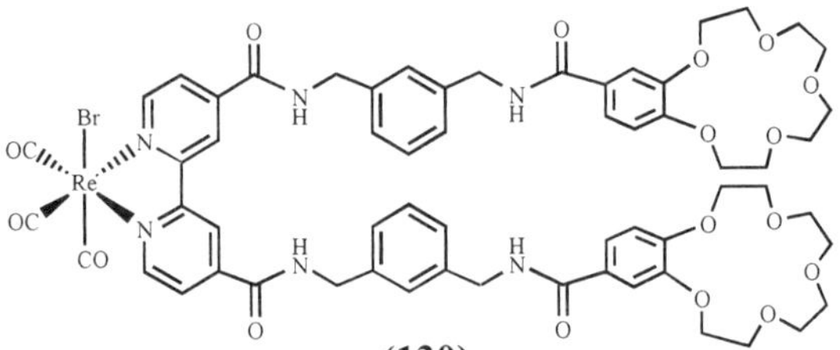

(130)

Figure 2.66. Rhenium(I) bipyridine crown ether ion-pair receptors.

Calixarenes contain a well-defined binding cavity, and as such are convenient for attachment to a fluorophore, leading to molecules such as those in Figure 2.67. Receptor (131) has been shown to form highly selective and thermodynamically stable complexes with H$_2$PO$_4^-$ ions in DMSO[76]. Electrochemical studies in acetonitrile showed substantial anion-induced cathodic perturbation of the ligand centred amide substituted 2,2'-bipyridine reduction redox couple. These shifts correlated with the stability constant values, with (131) sensing H$_2$PO$_4^-$ ions in the presence of a tenfold excess of HSO$_4^-$ and Cl$^-$ ions. Luminescence emission measurements in DMSO showed a significant blue shift (up to 16 nm) in the MLCT emission band λ_{max} on addition of H$_2$PO$_4^-$, accompanied by a large increase in emission intensity, due to the rigidity effect. A smaller shift in λ_{max} was observed with Cl$^-$, and shifts were not observed at all with unfunctionalised [Ru(bpy)$_3$]$^{2+}$.

Ruthenium(II)- and rhenium(I)-bipyridylcalix[4]diquinone receptors (132) and (133) selectively bind and sense acetate ions.[83, 96] Addition of acetate to acetonitrile solutions of (132) resulted in a 500 % increase in the emission intensity, suggesting that anion complexation inhibits the intramolecular oxidative electron transfer quenching mechanism between the ruthenium(II)-bipyridyl and calix[4]diquinone centers.

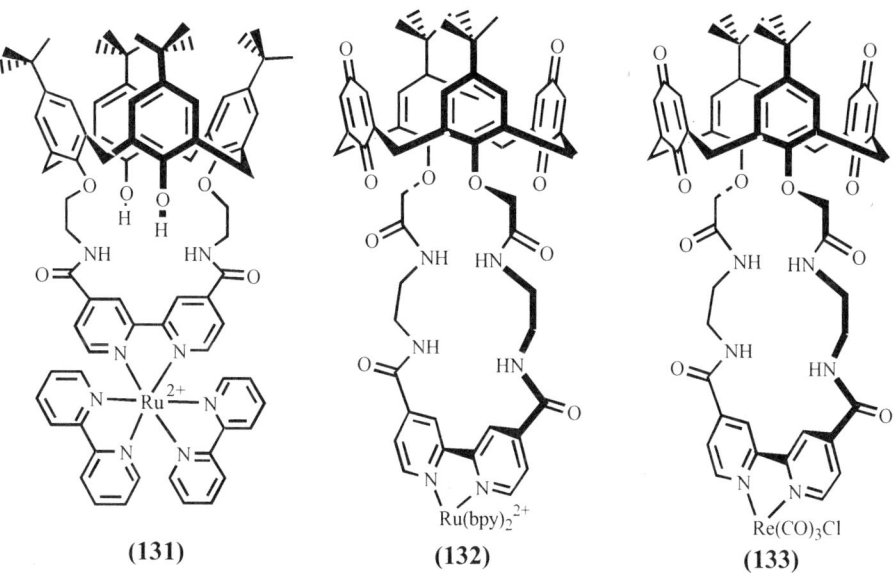

Figure 2.67. Ruthenium(II)- and rhenium(II)-bipyridylcalix[4]arene receptors.

Another class of ion-pair receptors, shown in Figure 2.68, involves attachment of a calix[4]arene tetraester cation binding site to rhenium(I) and ruthenium(II) bipyridines.[97] These molecules bind alkali metal (Li^+, Na^+)-halide (Br^-, I^-) ion pair species. Proton NMR titrations reveal that the presence of a lower-rim bound cation greatly enhances anion binding, with an increase of up to 60-fold observed with **(137)**. Halide binding enhancement is greater with the neutral rhenium(I) receptors **(134)** to **(137)** than for the charged ruthenium receptors **(138)** to **(141)**, which may be due to unfavorable electrostatic effects.

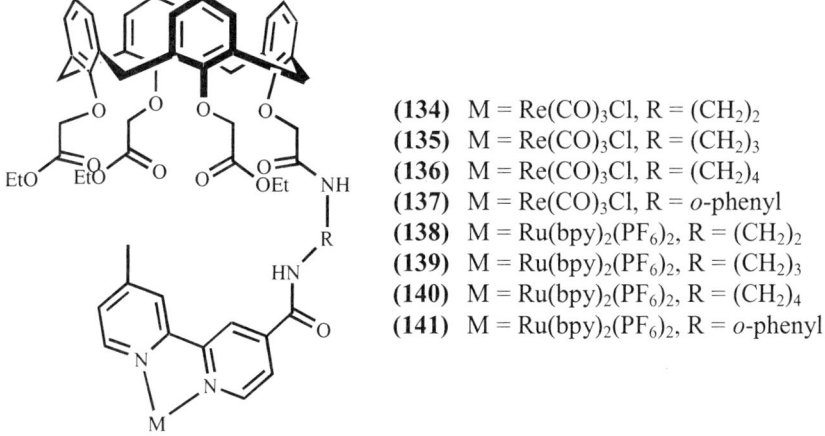

(134) M = Re(CO)$_3$Cl, R = (CH$_2$)$_2$
(135) M = Re(CO)$_3$Cl, R = (CH$_2$)$_3$
(136) M = Re(CO)$_3$Cl, R = (CH$_2$)$_4$
(137) M = Re(CO)$_3$Cl, R = o-phenyl
(138) M = Ru(bpy)$_2$(PF$_6$)$_2$, R = (CH$_2$)$_2$
(139) M = Ru(bpy)$_2$(PF$_6$)$_2$, R = (CH$_2$)$_3$
(140) M = Ru(bpy)$_2$(PF$_6$)$_2$, R = (CH$_2$)$_4$
(141) M = Ru(bpy)$_2$(PF$_6$)$_2$, R = o-phenyl

Figure 2.68. Rhenium(I) and ruthenium(II) bipyridyl calix[4]arene receptors.

The cavitand moiety has also been utilised, giving rise to receptor (142),[98] shown in Figure 2.69. Addition of acetate, chloride and benzoate to acetonitrile solutions gave rise to an increase in fluorescence emission intensity of 15, 13 and 11 % respectively.

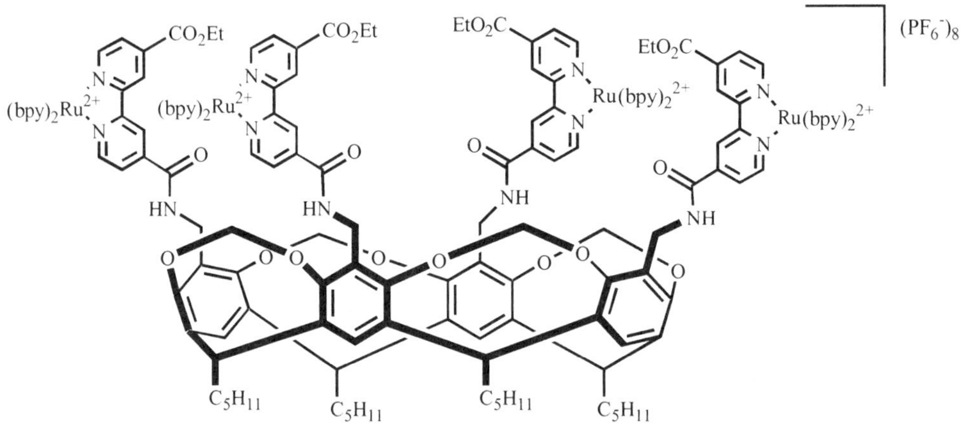

Figure 2.69. Cavitand-based anion receptor.

Receptors such as (143) make use of the hydrogen-bonding properties of the dipyrrolylquinoxaline group[99] (Figure 2.70). Addition of fluoride, cyanide and phosphate anions (98:2 dichloromethane/acetonitrile solutions) causes a decrease in the MLCT absorption band, accompanied by the appearance of a new absorption at 530 nm, which grows in intensity with anion binding. The MLCT emission is red-shifted from 594 – 610 nm, and is quenched with cyanide and fluoride, and less so with chloride and dihydrogen phosphate. The red shift is a result of the lowering of the energy of the excited state, which leads to enhancement of non-radiative decay processes. Sessler and coworkers have recently extended this methodology in the synthesis of analogous phenanthroline complexes bearing fused dipyrrolylquinoxaline anion recognition sites, such as in molecule (144).[100] These molecules possess electron-withdrawing moieties that render the pyrrole N-H protons more acidic, and thereby enhance anion-binding affinities, leading to chromogenic sensors.

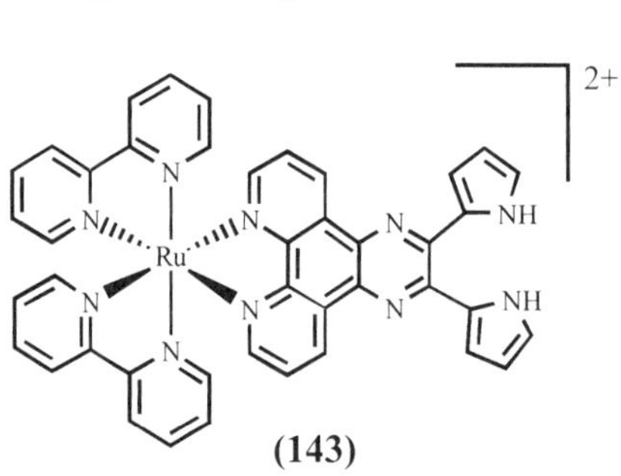

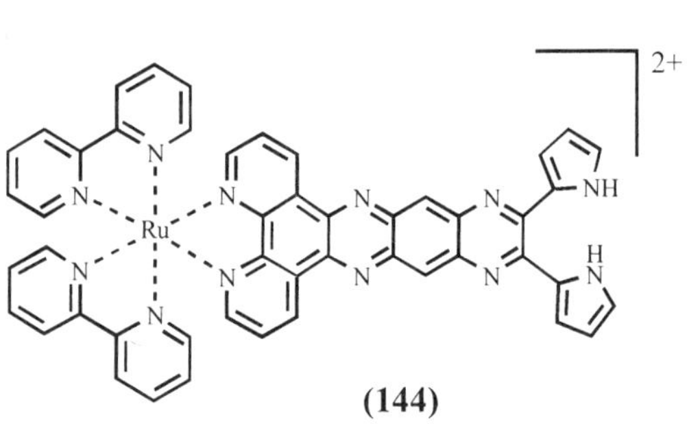

Figure 2.70. Dipyrrolylquinoxaline-based anion receptors.

2.3.2 Sensors containing other fluorophores

Self-assembled Re^I squares have also found application as anion sensors. The mixed Re^I-Pd^{II} metal macrocycle **(145)** in Figure 2.71 possesses a charged square cavity, which can bind anions by electrostatic ion-pair interactions. Excitation of the MLCT absorption bands at 320-340 nm gave luminescence emission bands at 580-630 nm. Overall quantum yields were low due to intramolecular oxidative quenching of the MLCT by palladium. The receptor binds perchlorate anion, which can be detected via MLCT enhancement in acetone.[101]

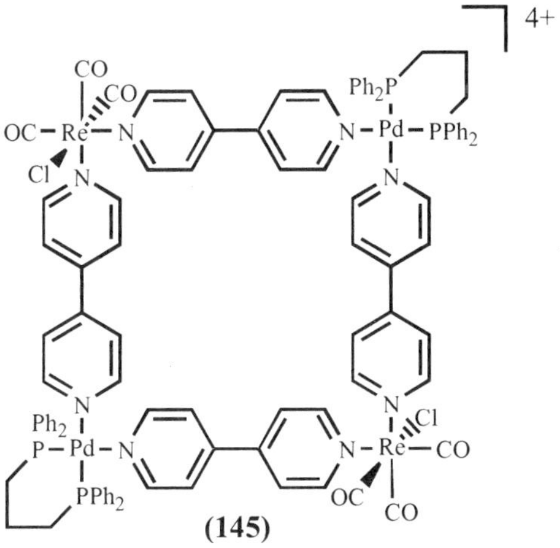

Figure 2.71. Rhenium(I)-Palladium(II) macrocycle.

Lees and coworkers have synthesised a series of heterometallic cyclophanes with both octahedral and square planar metal components (Figure 2.72).[102, 103] Molecules **(146)** to **(150)** were studied with a variety of anions, including ClO_4^-, OAc^-, OTf^-, PF_6^- and BF_4^-. Only PF_6^- and BF_4^- induced changes in the luminescence in acetone. Luminescence intensity initially decreased by a small amount, followed by an increase to a plateau. Initial anion binding lead to an increased probability of energy transfer from the MLCT to the ferrocene-based metal-centre state, and a decrease in luminescence. Addition of more anion gives rise to an increase in luminescence due to the stabilisation of the positive charge on the palladium reducing oxidative quenching.

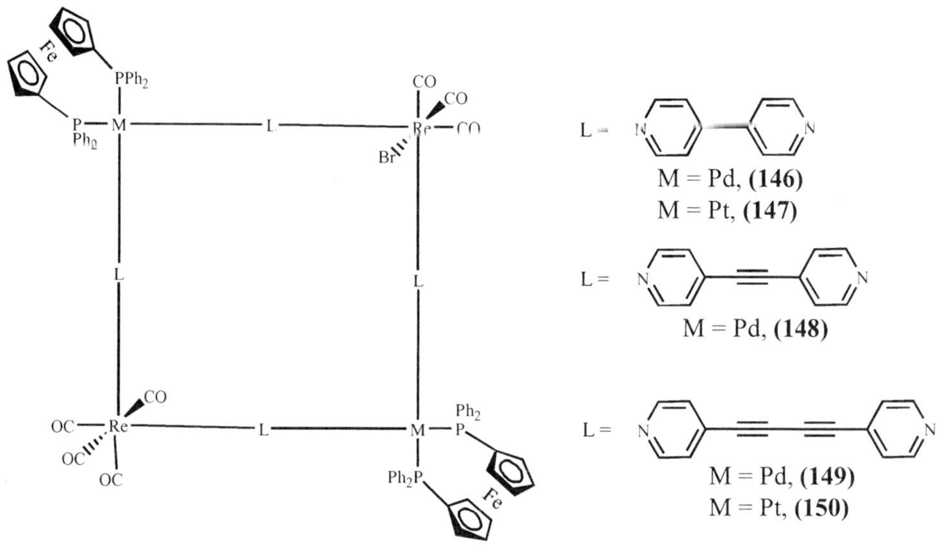

Figure 2.72. Heterometallic cyclophane anion receptors.

Receptor **(151)** in Figure 2.73 also incorporates a rhenium(I) pyridyl moiety.[104, 105] Dichloromethane solutions give intense absorption bands in the near UV, with strong luminescence at 536 nm (excitation at 360 nm). Addition of cyanide, halides, acetate, dihydrogen phosphate, nitrate and perchlorate give rise to differing degrees of quenching without any degree of selectivity. The luminescence wavelength is slightly red-shifted from 536 to 546 nm, indicating that the emission quenching is associated with the change in energy of the excited state.

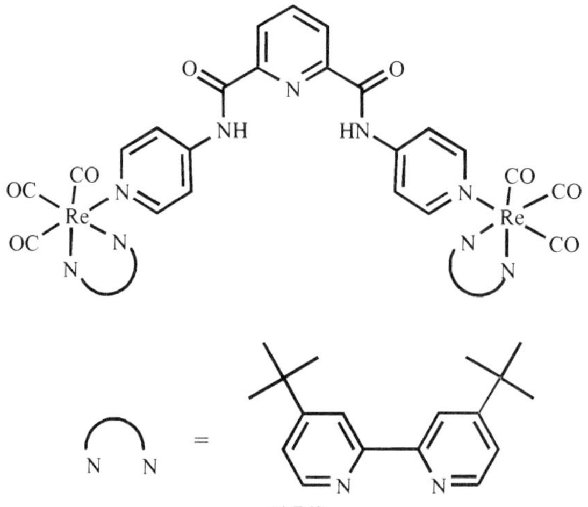

Figure 2.73. Rhenium-pyridyl cleft receptor.

FLUORESCENT ANION COMPLEXATION AGENTS

Receptor **(152)** (Figure 2.74) makes use of another bipyridyl ligand system, terpyridine. Terpyridine is less fluorescent than bipyridyl due to the low energy gap between the MLCT and MC levels. Excitation to the MLCT (ca. 480 nm) leads to emission at ca. 650 nm. ATP addition to **(152)** in 70:30 acetonitrile:water at pH3 gave rise to a large enhancement of emission intensity, whereas chloride, sulfate and phosphate gave no significant change[106]. This was however ascribed to a physical polyanion/polycations association rather than coordination. The copper(II) complex of **(152)** showed a quenching of the emission at pH4-8 due to interaction between the two metal centres. Addition of anions such as chloride, bromide and hydroxyl lead to a recovery of emission intensity. This was due to axial coordination of the anion which would reduce the degree of interaction between the Cu(II) and Ru(tpy)$_2^{2+}$ and thereby enhance the emission. At pH 9, ATP was the only anion to produce any significant effect on the emission.

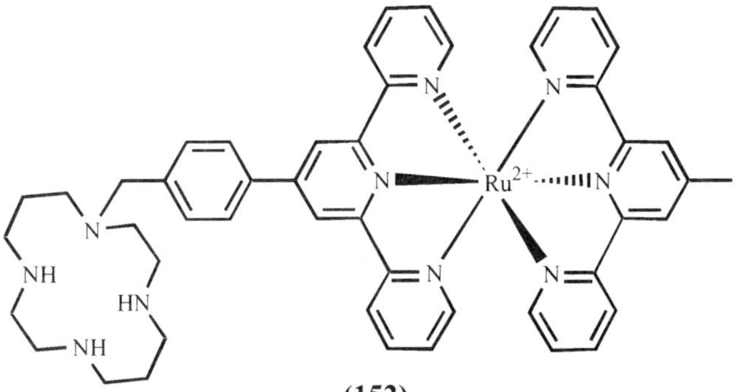

(152)

Figure 2.74. Ruthenium terpyridine-based receptor.

Williams and Goodall have made the isomeric bis-terpyridine complexes of iridium(III) shown in Figure 2.75.[107] Like ruthenium, iridium is a d^6 metal. However, iridium terpyridine is more emissive than Ru(tpy)$_2^{2+}$ and has a luminescent lifetime up to 10^5 times greater. Addition of chloride anion to aqueous solutions of receptor **(154)** quenches the luminescence, and the receptor is in fact sensitive to chloride at physiologically relevant concentrations. Other halides also quench the luminescence efficiently, but anions such as $H_2PO_4^-$, NO_3^- and SO_4^{2-} have no effect.

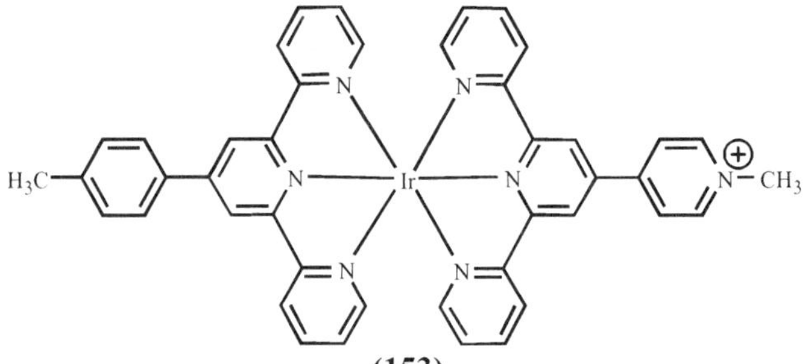

(153)

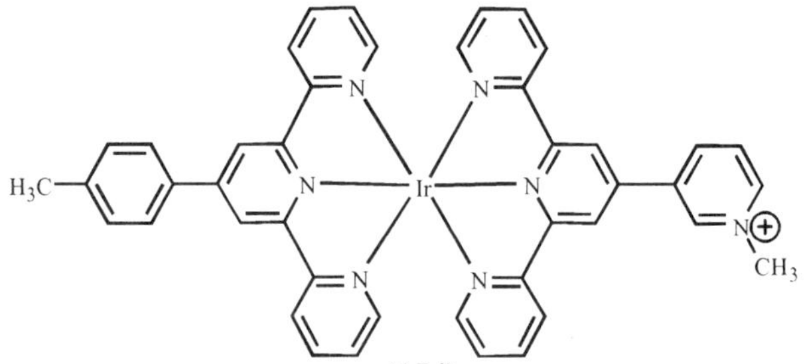

(154)

Figure 2.75. Iridium(III) based receptors.

2.3.3 Lanthanide-based receptors

Several examples of anion receptors containing lanthanide complexes have also been detailed. In general, lanthanides are not the first choice as a fluorophore, given that direct excitation is very inefficient since the f-f transitions are Laporte forbidden and very weak (ε 0.5-3 dm^{-3} mol^{-1} cm^{-1}). A strategy that circumvents these problems has been developed, which involves indirect excitation of the lanthanide by using an "antenna". The lanthanide ion is situated near a sensitising chromophore, usually organic, which absorbs UV-visible radiation. Energy transfer from the antenna to the metal centre occurs, and the emission from the long-lived lanthanide centre is obtained, as shown in Figure 2.76.

FLUORESCENT ANION COMPLEXATION AGENTS

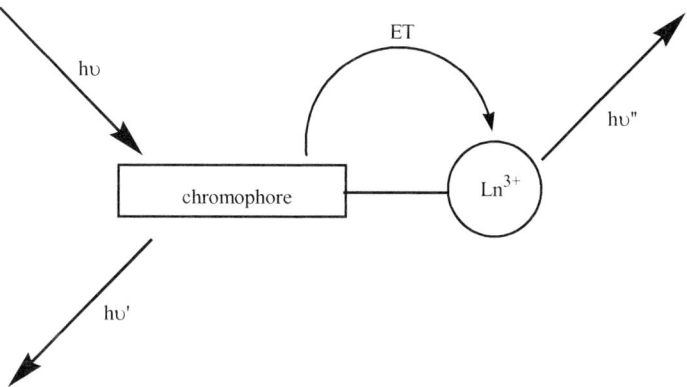

Figure 2.76. Mechanism of lanthanide antenna molecules.

It is proposed that binding of anions may perturb the excited state of the chromophore and the rate of energy transfer to the lanthanide centre, resulting in quenching of the luminescence intensity. Another proposed mechanism relies on the presence of water molecules coordinated to the lanthanide centre. Water is well known to quench luminescence via energy transfer processes between the metal centre and bound water molecules. Binding of an anion will displace the water molecules and thereby enhance the emission intensity of the complex.

The fluorescence of N-alkylated phenanthridium ions is known to be quenched by the presence of halide ions in aqueous media.[108] Quenching is believed to occur by charge transfer from the halide to the singlet excited state of the electron-poor cation. This property has been exploited in the design of biologically relevant assays for chloride.[109] This subunit has therefore been incorporated as an antenna molecule into receptors **(155)** and **(156)**[110] (Figure 2.77). The europium complexes show a decrease in phenanthridium fluorescence intensity upon addition of halides to aqueous solutions as expected. This is accompanied by a decrease in the europium luminescence, and the effect has been shown to be independent of added sodium bicarbonate, dihydrogen phosphate, lactate and citrate. Hydroxide ions have been shown to attack the N-methylphenanthridium groups at C6 in a reversible reaction,[111] resulting in a decrease in europium emission intensity. The europium emission intensity can therefore be considered to be pH independent from pH 2-9, but emission then falls in intensity by a factor of up to 200 as the pH goes beyond 13.

Figure 2.77. Cyclen/phenanthridium-based receptors.

The europium and terbium complexes of receptors **(157)** and **(158)**, shown in Figure 2.78, use the reversible displacement of water molecules coordinated to the lanthanide centre as a probe for anion signalling.[112, 113] For instance, the Tb**(157)** complex, which has two coordinated water molecules, showed emission intensity enhancement in the presence of fluoride, acetate and sulphate. No change was observed with the other halides or nitrate. However, hydrogen carbonate and carbonate, both of which are capable of displacing two water molecules from the lanthanide coordination shell, induced a larger change in measured lifetime and emission intensity. Similar results were obtained with **(158)** and the europium complexes of both ligands.

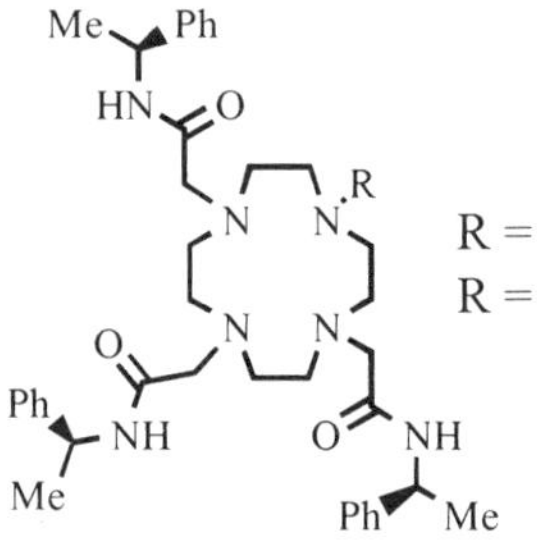

R = H, **(157)**
R = Me, **(158)**

Figure 2.78. Cyclen-based anion receptors.

The terbium complexes of receptor **(160)** (Figure 2.79) proved to be selective for p-N,N-dimethylaminobenzoate and salicylate anions in aqueous solutions.[114] Addition of these anions led to the formation of ternary complexes as the anion displaced the water molecules. In the absence of anions, none of the complexes are emissive. However, excitation at 300 nm upon titration with p-N,N-dimethylaminobenzoic acid resulted in a substantial *'switching on'* of the emission, with a luminescent enhancement factor of *ca.* 680. The primary quenching mechanism in this system involves vibrational energy transfer to bound OH oscillators. When the OH oscillators *i.e.* water molecules are displaced, quenching no longer occurs, and emission increases. The Tb**(159)** receptor resulted in similar behaviour, although with slightly less enhancement of emission (*ca.* 220 fold). A range of similar guest molecules failed to induce the same degree of increased emission intensity, and studies with p-N,N-dimethylaminobenzoic acid in the presence of other guests demonstrated the selectivity of the receptor for this particular molecule.

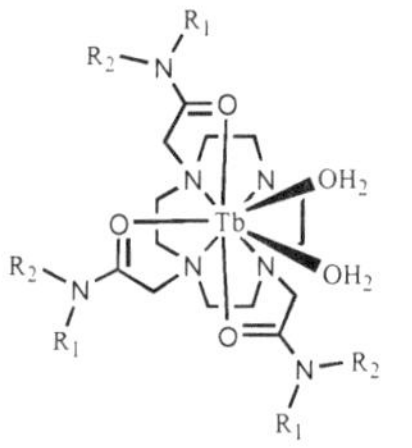

R_1 = H, R_2 = Me, **(159)**
R_1 = R_2 = Me, **(160)**

Figure 2.79. Terbium(III) complexes of cyclen-based receptors.

FLUORESCENT ANION COMPLEXATION AGENTS 111

Another class of cyclen-based receptor with a sensitising acridone chromophore attached has been prepared, an example of which is shown in Figure 2.80.[115] The other side chains have been varied to produce a series of cationic, zwitterionic and anionic europium complexes. One of the most promising molecules is receptor (**161**), containing three glutarate residues. This molecule was found to exhibit a 69 % change in the intensity ratio of the 618/588 nm europium emission bands between the intracellular HCO_3^- concentration range of 5 and 15 mM in a cell lysate medium.

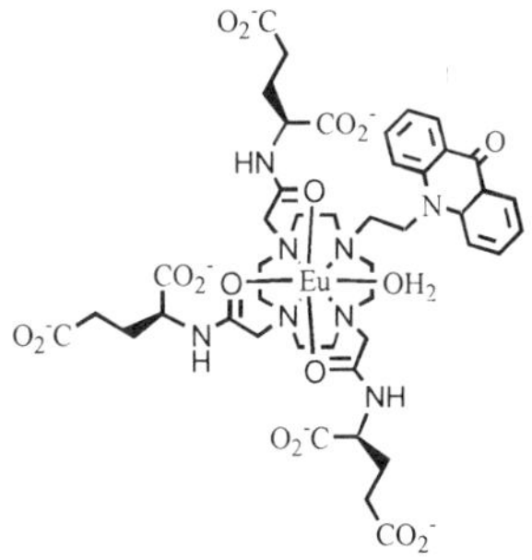

(**161**)

Figure 2.80. Cyclen-based hydrogencarbonate receptor.

Receptor (**162**) consists of a P=O subunit and two bpy subunits, which form a stable complex with both europium and terbium, with the bpy acting as a photon antenna.[116] Addition of nitrate anion to acetontrile solutions of both Eu(**162**) and Tb(**162**) induced an increase in emission intensity. Chloride produced a similar although less pronounced result, and fluoride and acetate were both far less pronounced again. The observed changes in emission were ascribed to a sequential substitution of ligands by nitrate anions. The initial increase in intensity was due to the displacement of solvent molecules from the lanthanide coordination sphere by one nitrate anion, to give the molecule shown in Figure 2.81. Coordination of a second nitrate released one bpy arm, leading to an increased efficiency of the ligand-to-metal energy transfer process between the remaining bpy coordinated to the lanthanide. This also resulted in an increase in the emission intensity. The addition of a third nitrate releases the second bpy arm, resulting in slight quenching of the emission. However, the overall effect was that of an 11 % increase in emission intensity.

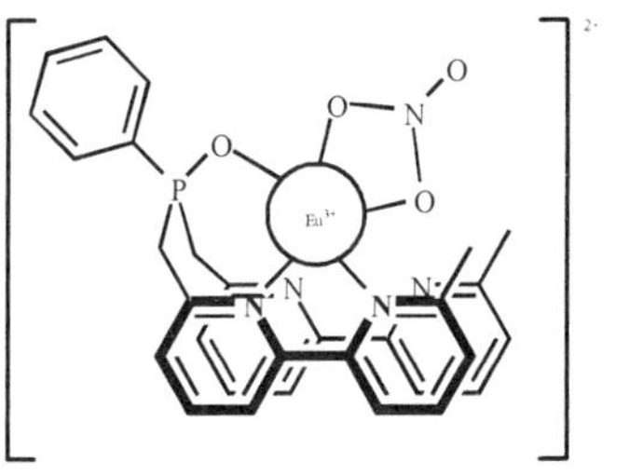

Figure 2.81. Eu(**162**) with coordinated nitrate anion.

Receptor **(163)** has been designed as a fluorescent sensor for 2,3-biphosphoglycerate (BPG),[117] which is a biologically important anionic molecule. The tetra-*N*-oxide bipyridine-europium complex combines with the two ammonium groups to complement the three anionic moieties in the guest molecule. Binding studies performed in 50 % methanol/acetonitrile resulted in a decrease in the fluorescence signal upon addition of guest. A 1:1 adduct was formed with a K_a of 6.7×10^5 mol^{-1}, and the proposed mode of binding shown in Figure 2.82. Phenyl phosphate, with only one anionic group, also binds in a 1:1 stoichiometry, although with a K_a more than three times less than that of BPG.

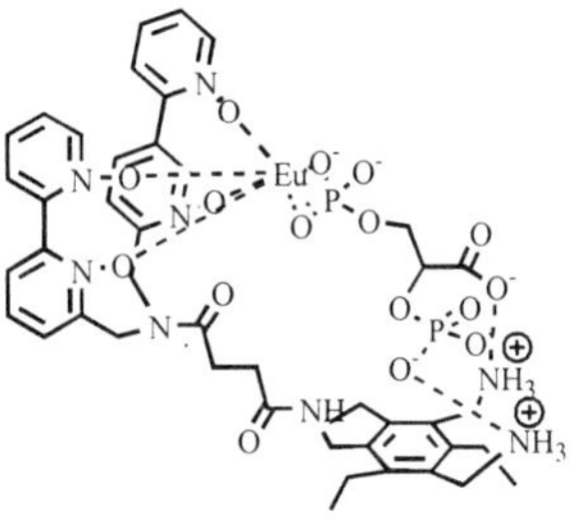

Figure 2.82. Proposed mode of binding of receptor **(163)** and BPG.

The europium and terbium complexes of both receptors in Figure 2.83 were studied in acetonitrile.[118] The luminescence spectrum was measured upon excitation of the pyridine chromophore, which served as the antenna molecule. The Eu**(164)** complex gave the most striking result with nitrate, with addition of three equivalents resulting in an enhancement in luminescence of 4.9 times, as well as giving rise to significant changes in the spectrum. Chloride produced similar but less pronounced results (1.8 times enhancement), acetate offered slight changes, and a range of other anions gave no effect at all. The Tb**(164)** complex favoured chloride anions, with emission enhancement upon addition of three equivalents of anion being increased by 5.4 times for chloride, 2.2 for nitrate and 1.1 for acetate. Experiments with the achiral ligand **(165)** produced similar results with much reduced sensitivity.

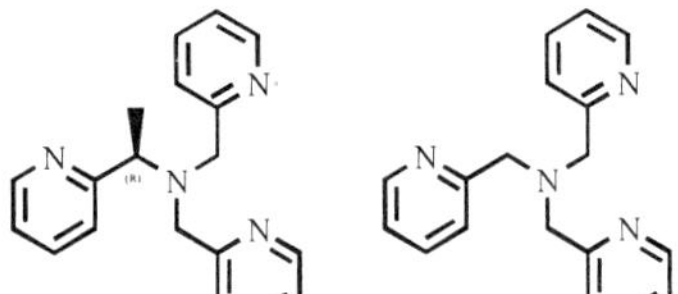

(164)   (165)

Figure 2.83. Tripodal pyridine-based anion receptors.

The simple receptors **(166)** and **(167)** in Figure 2.84 also operate by the formation of ternary complexes with the anion.[119] As in the previous example, the anion in effect acts as the antenna molecule, sensitising the lanthanide centre and switching on luminescence. Picolinic and phthalic acid were both chosen for study in buffered aqueous solutions. With picolinic acid, the emission intensity increases to a plateau, with an overall enhancement of 250-fold for the terbium complex of **(166)** and 170-fold for the europium. Similar behaviour was observed for the Eu**(167)** complex, with an enhancement of emission intensity of 120-fold upon addition of picolinate. The presence of only one sidearm in this ligand leads to a very well defined 1:2 complex:guest binding stoichiometry.

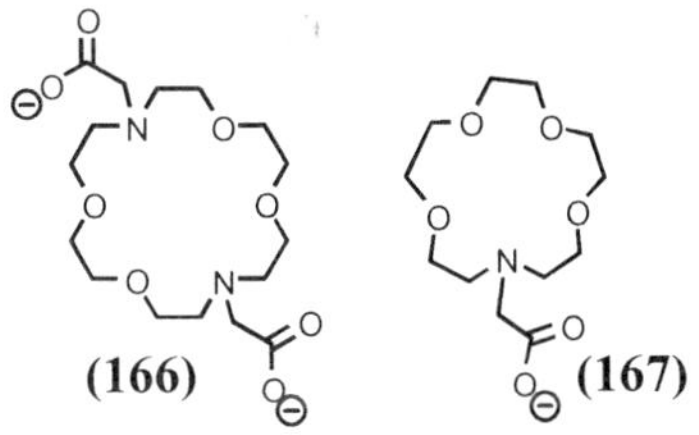

Figure 2.84. Receptors dacda **(166)** and macma **(167)**.

Lanthanide tris(fluorinated β-diketonates) have also been proven to be effective chloride receptors.[120] Upon addition of three equivalents of chloride anion to an acetonitrile solution of the europium complex **(168)** (Figure 2.85), the europium luminescence observed at 611 nm was increased by a factor of two. Addition of bromide, iodide and perchlorate anions also influenced the luminescence intensity, although to a lesser degree (< 1.3-fold). The chloride response was such that it could be observed with the naked eye. The chloride-selective receptor was also incorporated into ion-sensitive electrodes, and shown to maintain its selectivity over a host of other anions.

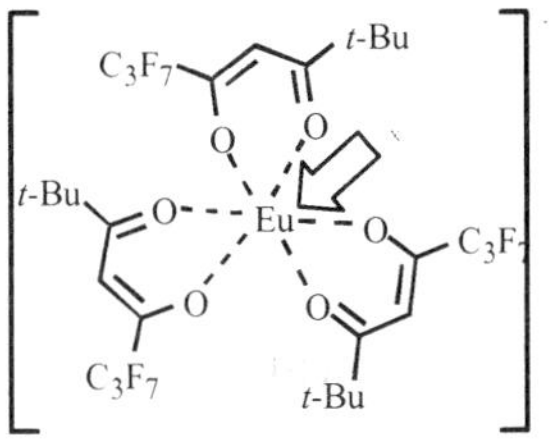

Figure 2 85. Lanthanide tris(β-diketonates) and highly coordinated anion (X) complex.

2.4. REFERENCES

1. L.G. Lange, J.F. Riordan, B.L. Vallee, Functional arginyl residues as NADH binding sites of alcohol dehydrogenases, *Biochemistry*, **13**(21), 4361-70 (1974).
2. R.M. Harrison, *Pollution: Causes, Effects and Control*, 4th ed. (RSC, Cambridge, 2001).
3. E. Holm, Radioanalytical studies of Tc in the environment: progress and problems, *Radiochimica Acta*, **63**, 57-62 (1993).
4. M.M.G. Antonisse, D.N. Reinhoudt, Neutral anion receptors: design and application, *Chem. Commun.*, (4), 443-448 (1998).
5. F.P. Schmidtchen and M. Berger, Artificial Organic Host Molecules for Anions, *Chem. Rev.*, **97**(5), 1609-1646 (1997).
6. P.D. Beer, P.A. Gale, Anion recognition and sensing: the state of the art and future perspectives, *Angew. Chem. Int. Ed. Engl.*, **40**(3), 486-516 (2001).

7. T.S. Snowden, E.V. Anslyn, Anion recognition: synthetic receptors for anions and their application in sensors, *Curr. Opinion in Chem. Biol.*, **3**(6), 740-746 (1999).
8. J.L. Sessler, J.M. Davis, Sapphyrins: Versatile Anion Binding Agents, *Acc. Chem. Res.*, **34**(12), 989-997 (2001).
9. A. Bianchi, K. Bowman-James, E. Garcia-Espana, *Supramolecular Chemistry of Anions* (Wiley VCH, New York, 1997).
10. L. Fabbrizzi, A. Poggi, Sensors and switches from supramolecular chemistry, *Chem. Soc. Rev.*, **24**(3), 197-202 (1995).
11. B. Valeur, I. Leray, Coord. Design principles of fluorescent molecular sensors for cation recognition, *Coord. Chem. Rev.*, **205**, 3-40 (2000).
12. L. Fabbrizzi, M. Licchelli, P. Pallavicini, P. Parodi and L. Taglietti, in: *Transition Metals in Supramolecular Chemistry*, edited by J.-P. Sauvage (John Wiley & Sons Ltd, Chichester, 1999), pp. 93-134.
13. A. P. de Silva, H. Q. N. Gunaratne, T. Gunnlaugsson, A. J. M. Huxley, C. P. McCoy, J. T. Rademacher, T. E. Rice, Signaling Recognition Events with Fluorescent Sensors and Switches, *Chem. Rev.*, **97**(5), 1515-1566 (1997).
14. H. Xie, S. Yi, X. Yang and S. Wu, Study on the host-guest complexation of anions based on a tripodal naphthylurea derivative, *New J. Chem.* **23**, 1105-1110 (1999).
15. G. Hennrich, H. Sonnenschein and U. Resch-Genger, Fluorescent anion receptors with iminoylthiourea binding sites-selective hydrogen bond mediated recognition of CO_3^{2-}, HCO^{3-} and HPO_4^{2-}, *Tet. Lett.* **42**, 2805-2808 (2001).
16. X. Qian and F. Liu, Promoting effects of the hydroxymethyl group on the fluorescent signalling recognition of anions by thioureas, *Tet. Lett.* **44**, 795-799 (2003).
17. E. J. Cho, J. W. Moon, S. W. Ko, J. Y. Lee, S. K. Kim, J. Yoon and K. C. Nam, A new fluoride selective fluorescent as well as chromogenic chemosensor containing a naphthalene urea derivative, *J. Am. Chem. Soc.* **125**, 12376-12377 (2003).
18. S. Amemiya, P. Bühlmann and Y. Umezawa, Fluorescence-mediated sensing of guanosine derivatives based on multitopic hydrogen bonding, *Chem. Commun.* 1027-1028 (1997).
19. Y. Kubo, M. Tsukahara, S. Ishihara and S. Tokita, A simple anion chemosensor based on naphthalene-thiuronium dyad, *Chem. Commun.* 653-654 (2000).
20. Y. Kubo, S. Ishihara, M. Tsukahara and S. Tokita, Isothiouronium-derived simple fluorescent chemosensors of anions, *J. Chem. Soc., Perkin Trans.* 2 1455-1460 (2002).
21. S. Nishizawa, Y.-Y. Cui, M. minigawa, K. Morita, Y. Kato, S. Taniguchi, R. Kato and N. Teremae, Conversion of thioureas to fluorescent isothiouronium-based photoinduced electron transfer sensors for oxoanion sensing, *J. Chem. Soc. Perkin Trans.* 2 866-870, (2002).
22. S. Pagliari, R. Corradini, G. Galaverna, S. Sforza, A. Dossena and R. Marchelli, Enantioselective sensing of amino acids by copper(II) complexes of phenylalanine-based fluorescent β-cyclodextrins, *Tet. Lett.* **41**, 3691-3695 (2000).
23. C. R. Cooper, N Spencer and T. D. James, Selective fluorescence detection of fluoride using boronic acids, *Chem. Commun.* 1365-1366 (1998).
24. H. Ikeda, Y. Lidaka and A. Ueno, Remarkably enhanced excimer formation of naphthylacetate in cation-charged γ-cyclodextrin, *Org. Lett.* **5**, 1625-1627 (2003).
25. M. E. Huston, E. U. Akkaya and A. W. Czarnik, Chelation enhanced fluorescence detection of non-metal ions, *J. Am. Chem. Soc.* **111**, 8735-8737 (1989).
26. S. A. Van Arman and A. W. Czarnik, general fluorescente assay for enzyme-catalyzed polyanion hydrolysis based on template-directed excimer formation. Application to heparin and Polyglutamate, *J. Am. Chem. Soc.* **112**, 5376-5377 (1990).
27. D. H Vance and A. W. Czarnik, Real-time assay of inorganic pyrophosphate using a high-affinity chelation-enhanced fluorescence, *J. Am. Chem. Soc.* **116**, 9397-9398 (1994).
28. M. T. Albelda, M. A. Bernardo, E. García-España, M. L. Rodino-Salido, S. V. Luis, M. J. Melo, F. Pina and C. Soriano, Thermodynamics and fluorescent emission studies on potential molecular chemosensors for ATP recognition in aqueous solution, *J. Chem. Soc., Perkin Trans.* 2 2545-2549 (1999).
29. S. Sasaki, A. Hashizume, D. Citterio, E. Fujii and K. Suzuki, Fluororeceptor for zwitterionic form amino acids in aqueous methanol solution, *Tet. Lett.* **43**, 7243-7245 (2002).
30. S. Sasaki, D. Citterio, S. Ozawa and K. Suzuki, Design and synthesis of preorganized tripodal fluororeceptors based on hydrogen bonding of thioureas groups for optical phosphate ion sensing, *J. Chem. Soc., PerkinTtrans.* 2 2309-2313 (2001).
31. T. Gunnlaugsson, A. P. Davis and M. Glynn, Fluorescent photoinduced electron transfer (PET) sensing of anions using charge neutral chemosensors, *Chem. Commun.* 2556-2557 (2001).

32. T. Gunnlaugsson, A. P. Davis, J. E. O'Brien and M. Glynn, Fluorescent sensing of pyrophosphate and bis-carboxylates with charge neutralPET chemosensors, *Org. Lett.* **4**, 2449-2452 (2002).
33. S. K. Kim and J. Yoon, A new fluorescent PET chemosensor for fluoride ions, *Chem. Commun.* 770-771 (2002).
34. H. Miyaji, P. Anzenbacher Jr, J. L. Sessler, E. R. Bleasdale and P. A. Gale, Anthracene-linked calyx[4]pyrroles: fluorescent chemosensors for anions, *Chem. Commun.* 1723-1724 (1999).
35. A. P. de Silva, G. D. McClean and S. Pagliari, Direct detection of ion pairs by fluorescence enhancement, *Chem. Commun.* 2010-2011 (2003).
36. M. Montalti and L. Prodi, A supramolecular assembly controlled by anions: threading and unthreading of a pseodorotaxane, *Chem. Commun.* 1461-1462 (1998).
37. L. Fabbrizzi, G. Francese, M. Liccheli, A. Perotti and A. Taglietti, Fluorescent sensor of imidazole and histidine, *Chem. Commun.* 581-582 (1997).
38. L. Fabbrizzi, I. Faravelli, G. Francese, M. Licchelli, A. Perotti and A. Taglietti, A fluorescent cage for anion sensing in aqueous solution, *Chem. Commun.* 971-972 (1998).
39. L. Fabbrizzi, M. Licchelli, A. Perotti, A. Poggi, G. Rabaioli, D. Sacchi and A. Taglietti, Fluorescent molecular sensing of amino acids bearing an aromatic residue, *J. Chem. Soc., Perkin Trans. 2* 2108-2113 (2001).
40. I. Bruseghini, L. Fabbrizzi, M. Licchelli and A. Taglietti, Coordinative control of photoinduced electron transfer : bulky carboxylates as molecular curtains, *Chem. Commun.* 1348-1349 (2002).
41. L. Fabbrizzi, M. Licchelli, F. Mancin, M. Pizzeghello, G. Rabaioli, A. Taglietti, P. Tecill and U. Tonellato, Fluorescence sensing of ionic analytes in water: from transition metal ions to vitamin B13, *Chem. Eur. J.* **8**, 94-101 (2002).
42. A. Ojida, S.-k Park, Y. Mito-oka and I. Hamachi, Efficient florescent ATP-sensing based on coordination chemistry under aqueous neutral conditions, *Tet. Lett.* **43**, 6193-6195 (2002).
43. F. Sancenón, A. Benito, F. J. Hernández, J. M. Lloris, R. Martínez-Mañez, T. Pardo and J. Soto, Difunctionalised chemosensors containing electroactive and fluorescent signalling subunits, *Eur. J. Inorg. Chem.* 866-875 (2002).
44. S. Patterson, B. D. Smith and R. E. Taylor, Fluorescence sensing of a ribonucleoside 5´-triphosphate, *Tet. Lett.* **38**, 6323-6326 (1997).
45. W. Yang, J. Yan, H. Fang and B. Wang, The first fluorescent sensor for D-glucarate based on the cooperative action of boronic acid and guanidinium groups, *Chem. Commun.* 792-793 (2003).
46. S. Nishizawa, Y. Kato and N. Teramae, Fluorescence sensing of anions via intramolecular excimer formation in a pyrophosphate-induced self-assembly of a pyrene-funtionalized guanidinium receptor, *J. Am. Chem. Soc.* **121**, 9463-9464 (1999).
47. S. Nishizawa, H. Kaneda, T. Uchida and N. Teramae, Anion sensing by a donor-spacer-acceptor system: an intramolecular exciplex emission enhanced by hydrogen bond-mediated complexation, *J. Chem. Soc. Perkin Trans 2* 2325-2327 (1998).
48. J.-H- Liao, C.-T. Chen and J.-M. Fang, A novel phosphate chemosensor utilizing anion-induced fluorescence change, *Org. Lett.* **4**, 561-564 (2002).
49. M. Inouye, K. Fujimoto, M. Furusyo and H. Nakazumi, Molecular recognition abilities of a new class of water-soluble cyclophanes capable of encompassing a neutral cavity, *J. Am. Chem. Soc.* **121**, 1452-1458 (1999).
50. D. H. Lee, J. H. Im, J.-H. Lee and J.-I. Hong, A new fluorescent fluoride chemosensor based on conformational restriction of a biaryl fluorophore, *Tetrahedron Lett.* **43**, 9637-9640 (2002).
51. M. W. Hosseini, A. J. Blacker and J.-M. Lehn, Multiple molecular recognition and catalysis. A multifunctional anion receptor bearing an anion binding site, an intercalating group, and a catalytic site for nucleotide binding and hydrolysis, *J. Am. Chem. Soc.* **112**, 3896-3904 (1990).
52. P. E. Kruger, P. R. Mackie and M. Nieuwenhuyzen, Optical-structural correlation in a novel quinoxaline-based anion sensor, *J. Chem. Soc., Perkin Trans. 2* 1079-1083 (2001).
53. C. B. Black, B. Andrioletti, A. C. Try, C. Ruiperez and J. L. Sessler, Dipyrrolylquinoxalines: efficient sensors for fluoride anion in organic solution, *J. Am. Chem. Soc.* **121**, 10438-10439 (1999).
54. P. Anzenbacher, Jr., A. C. Try, H. Miyaji, K. Jursikova, V. M. Lynch, M. Marquez and J. L. Sessler, Fluorinated Calix[4]pyrrole and Dipyrrolylquinoxaline: Neutral Anion Receptors with Augmented Affinities and Enhanced Selectivities, *J. Am. Chem. Soc.* **122**, 10268-10272 (2000).
55. D. Aldakov and P. Anzenbacher, Jr., Dipyrrolyl quinoxalines with extended chromophores are efficient fluorimetric sensors for pyrophosphate, *Chem. Commun.* 1394-1395 (2003).
56. C. P. Causey and W. E. Allen, Anion binding by fluorescent biimidazole diamides, *J. Org. Chem.* **67**, 5963-5968 (2002).

57. T. Mutai, Y. Abe and K. Araki, A novel bipyridine-based fluorescent host for diphenyl phosphate: affinity, photo-response and mechanism, *J. Chem. Soc., Perkin Trans. 2* 1805-1809 (1997).
58. L. A. Cabell, M. D. Best, J. J. Lavigne, S. E. Schneider, D. M. Perreault, M.-K. Monahan and E. V. Anslyn, *J. Chem. Soc., Perkin Trans. 2* 315-323 (2001).
59. M. Shionoya, T. Ikeda, E. Kimura and M. Shiro, Novel "multipoint" molecular recognition of nucleobases by a new zinc(II) complex of acridine-pendant cyclen (cyclen= 1,4,7,10-tetraazacyclododecane), *J. Am. Chem. Soc.* **116**, 3848-3859 (1994).
60. M. Takeuichi, M. Yamamoto and S. Shinkai, Fluorescent sensing of uronic acids based on a cooperative action of boronic acid and metal chelate, *Chem. Commun.* 1731-1732 (1997).
61. M. Yamamoto, M. Takeuichi and S. Shinkai, Molecular design of a PET-based chemosensor for uronic acids and sialic acids utilizing a cooperative action of boronic acid and metal chelate, *Tetrahedron* **54**, 3126-3140 (1998).
62. I. Piantanida, V. Tomisic and M. Zinic, 4,9-Diazapyrenium cations. Synthesis, physico-chemical properties and binding of nucleotides in water, *J. Chem. Soc., Perkin Trans. 2* 375-383 (2000).
63. I. Piantida, B. S. Palm, P. CEDIC, M. Zinic and H.-J. Schneider, Phenanthridinium cyclobisintercalands. Fluorescence sensing of AMP and selective binding to single-stranded nucleic acids, *Tetrahedron Lett.* **42**, 6779-6783 (2001).
64. P. Anzenbacher, Jr., K. Jursikova and J. L. Sessler, Second generation calixpyrrole anion sensors, *J. Am. Chem. Soc.* **122**, 9350-9351 (2000).
65. H. Tong, G. Zhou, L. Wang, X. Jing, F. Wang and J. Zhang, Novel highly selective anion chemosensor based on 2,5-bis(2-hydroxyphenyl)-1,3,4-oxadiazole, *Tetrahedron Lett.* **44**, 131-134 (2003).
66. K. Choi and A. D. Hamilton, A dual channel fluorescence chemosensor for anions involving intermolecular excited state proton transfer, *Angew. Chem. Int. Ed.* **40**, 3912-3915 (2001).
67. K. Kobiro and Y. Inoue, A new chiral probe for sulfate anion: UV, CD, fluorescence, and NMR spectral studies of 1:1 and 2:1 complex formation and structure of chiral guanidinium- p-dimethylaminobenzoate conjugate with sulfate anion, *J. Am. Chem. Soc.* **123**, 421-427 (2003).
68. A. Metzger and E. V. Anslyn, A chemosensor for citrate in beverages, *Ang. Chem. Int. Ed.* **37**, 649-652 (1998).
69. K. Niikura, A. Metzger and E. V. Anslyn, Chemosensor ensemble with selectivity for inositol-triphosphate, *J. Am. Chem. Soc.* **120**, 8533-8534 (1998).
70. R. Prohens, G. martorell, P. Ballester and A. Costa, A squaramide ensemble for monitoring sulphate in water, *Chem. Commun.* 1456-1457 (2001).
71. L. Fabbrizzi, A. Leone and A. Taglietti, A chemosensing ensemble for selective carbonate detection in water based on metal-ligand interactions, *Angew. Chem.Int. Ed.* **40**, 3066-3069 (2001).
72. L. Fabbrizzi, N. Marcotte, F. Stomeo and A. Taglietti, Pyrophosphate detection in water by fluorescence competition assays: inducing selectivity through the choice of the indicator, *Angew. Chem.Int. Ed.* **41**, 3811-3812 (2002).
73. M. A. Hortala, L. Fabbrizzi, N. Marcotte, F. Stomeo and A. Taglietti, Designing the selectivity of the fluorescent detection of amino acids: a chemosensing ensemble for histidine, *J. Am. Chem. Soc.* **125**, 20-21 (2003).
74. V. Balzani, F. Barigelletti, L.D. Cola, Metal complexes as light absorption and light emission sensitizers, *Top. Curr. Chem.* **158**, 31-71 (1990).
75. A. Juris, V. Balzani, F. Barigelletti, S. Campagna, P. Belser, A.von. Zelewsky, Ruthenium(II) polypyridine complexes: photophysics, photochemistry, electrochemistry, and chemiluminescence, *Coord. Chem. Rev.* **84**, 85-277 (1988).
76. F. Szemes, D. Hesek, Z. Chen, S.W. Dent, M.G.B. Drew, A.J. Goulden, et. al, Synthesis and Characterization of Novel Acyclic, Macrocyclic, and Calix[4]arene Ruthenium(II) Bipyridyl Receptor Molecules That Recognize and Sense Anions, *Inorg. Chem.*, **35**(20), 5868-5879 (1996).
77. P.D. Beer, S.W. Dent, T. Wear, Spectral and electrochemical recognition of halide anions by acyclic mononuclear ruthenium(II) bipyridyl receptor molecules, *J. Chem. Soc. Dalton Trans.*, (11), 2341-2346 (1996).
78. P.D. Beer and E.J. Hayes, Transition metal and organometallic anion complexation agents, *Coord. Chem. Rev.*, **240**(1-2), 167-189 (2003).
79. P.D. Beer, Transition-Metal Receptor Systems for the Selective Recognition and Sensing of Anionic Guest Species, *Acc. Chem. Res.* **31**(2), 71-80 (1998).
80. P.D. Beer, N.C. Fletcher and T.J. Wear, Spectral and electrochemical halide anion recognition by acyclic ruthenium(II) 5,5'-bis-amide substituted bipyridyl receptor molecules, *Polyhedron*, **15**(8), 1339-1347 (1996).

81. L.H. Uppadine, F.R. Keene, P.D. Beer, Approaches towards the enantioselective recognition of anionic guest species using chiral receptors based on rhenium(I) and ruthenium(II) with amide bipyridine ligands, *J. Chem. Soc. Dalton Trans.*, (14), 2188-2198 (2001).
82. P.D. Beer, A.R. Graydon, L.R. Sutton, Luminescent anion recognition: selective induced emission by binding of dihydrogenphosphate, *Polyhedron*, **15**(14), 2457-2461 (1996).
83. P.D. Beer, J. Cadman, Phosphate anion binding and luminescent sensing in aqueous solution by ruthenium(II) bipyridyl polyaza receptors, *New J. Chem.*, **23**(4), 347-350 (1999).
84. S. Watanabe, N. Higashi, M. Kobayashi, K. Hamanaka, Y. Takata and K. Yoshida, Stereoselective optical sensing of dicarboxylate anions by an induced-fit type Ru(II) receptor, *Tetrahedron Lett.*, **41**(23), 4583-4586 (2000).
85. T. Duff, A. Grüßing, J.-L. Thomas, M. Duati, and J.G. Vos, Luminescent anion recognition: probing the interaction between dihydrogenphosphate anions and Ru(II) polypyridyl complexes in organic and aqueous media, *Polyhedron*, **22**(5), 775-780 (2003).
86. M.J. Deetz, B.D. Smith, Heteroditopic ruthenium(II) bipyridyl receptor with adjacent saccharide and phosphate binding sites, *Tetrahedron. Lett.* **39**(38), 6841-6844 (1998).
87. V.W.-W. Yam, A.S.-F. Kai, Synthesis and optical sensing properties of a boronic acid appended rhenium(I) complex for sugar, *Chem. Commun.*, (1), 109-110 (1998).
88. S. Watanabe, O. Onogawa, Y. Komatsu, K. Yoshida, Luminescent Metalloreceptor with a Neutral Bis(Acylaminoimidazoline) Binding Site: Optical Sensing of Anionic and Neural Phosphodiesters, *J. Am. Chem. Soc.* **120**(1), 229-230 (1998).
89. P.D. Beer, S.W. Dent, G.W. Hobbs, T. Wear, Novel anion binding selectivity trends exhibited by new dinuclear rhenium(I), ruthenium(II) and osmium(II) bipyridyl cleft-type receptors, *Chem. Commun.* (1), 99-100 (1997).
90. P.D. Beer, F. Szemes, V. Balzanni, C.M. Sala, M.G.B. Drew, S.W. Dent, M. Maestri, Anion Selective Recognition and Sensing by Novel Macrocyclic Transition Metal Receptor Systems. ^{1}H NMR, Electrochemical, and Photophysical Investigations, *J. Am. Chem. Soc.* **119**(49), 11864-11875 (1997).
91. P.D. Beer, S.W. Dent, N.C. Fletcher and T.J. Wear, Anion and cation recognition by new mono-and bis-ruthenium(II) bipyridyl crown ether receptor molecules, *Polyhedron*, **15**(18), 2983-2996 (1996).
92. P.D. Beer and S.W. Dent, Potassium cation induced switch in anion selectivity exhibited by heteroditopic ruthenium(II) and rhenium(I) bipyridyl bis(benzo-15-crown-5) ion pair receptors, *Chem. Commun.*, (1), 825-826 (1998).
93. P.D. Beer, S.E. Stokes, Potassium cations allosterically switch off the halide anion recognition properties of a new cobalticinium bis benzo crown ether receptor, *Polyhedron*, **14**(19), 2631-2635 (1995).
94. J.E. Redman, P.D. Beer, S.W. Dent, M.G.B. Drew, Cooperative binding of potassium cation and chloride anion by novel rhenium(I) bipyridyl amide crown ether receptors, *Chem. Commun.*, (2), 231-232 (1998).
95. L.H. Uppadine, J.E. Redman, S.W. Dent, M.G.B. Drew, P.D. Beer, Ion Pair Cooperative Binding of Potassium Salts by New Rhenium(I) Bipyridine Crown Ether Receptors, *Inorg. Chem.*, **40**(12), 2860-2869 (2001).
96. P.D. Beer, V. Timoshenko, M. Maestri, P. Passaniti, V. Balzani, Anion recognition and luminescent sensing by new ruthenium(II) and rhenium(I) bipyridyl calix[4]diquinone receptors, *Chem. Commun.*, (17), 1755-1756 (1999).
97. J.B. Cooper, M.G.B. Drew, P.D. Beer, Heteroditopic rhenium(I) and ruthenium(II) bipyridyl calix[4]arene receptors for binding cation-anion ion pairs, *J. Chem. Soc. Dalton Trans.*, (4), 392-401 (2001).
98. I. Dumazet and P.D. Beer, Synthesis and Characterisation of Novel Ruthenium(II) Bipyridyl and Ferrocenoyl Cavitand Receptors for the Recognition of Anionic Guests, *Tetrahedron Lett.*, **40**(4), 785-788 (1999).
99. P. Anzenbacher Jr., D.S. Tyson, K. Jursiková and F.N. Castellano, Luminescence Lifetime-Based Sensor for Cyanide and Related Anions, *J. Am. Chem. Soc.*, **124**(22), 6232-6233 (2002).
100. T. Mizuno, W.-H. Wei, L.R. Eller, J.L. Sessler, Phenanthroline Complexes Bearing Fused Dipyrrolylquinoxaline Anion Recognition Sites: Efficient Fluoride Anion Receptors, *J. Am. Chem. Soc.*, **124**(7), 1134-1135 (2002).
101. R.V. Slone, D.I. Yoon, R.M. Calhoun, J.T. Hupp, Luminescent Rhenium/Palladium Square Complex Exhibiting Excited State Intramolecular Electron Transfer Reactivity and Molecular Anion Sensing Characteristics, *J. Am. Chem. Soc.*, **117**(47), 11813-11814 (1995).
102. S. Sun, J.A. Anspach, A.J. Lees, P.Y. Zavalij, Synthesis and Electrochemical, Photophysical, and Anion Binding Properties of Self-Assembly Heterometallic Cyclophanes, *Organometallics*, **21**(4), 685-693 (2002).

103. S.-S. Sun and A. J. Lees, Transition metal based supramolecular systems: synthesis, photophysics, photochemistry and their potential applications as luminescent anion chemosensors, *Coord. Chem. Rev.*, **230**(1-2), 171-192 (2002).
104. S.-S. Sun and A.J. Lees, Anion recognition through hydrogen bonding: a simple, yet highly sensitive, luminescent metal-complex receptor, *Chem. Commun.*, (17), 1687-1688 (2000).
105. S.-S. Sun, A. J. Lees and P. Y. Zavalij, Highly sensitive luminescent metal-complex receptors for anions through charge-assisted amide hydrogen bonding, *Inorg. Chem.*, **42**(11), 3445-3453 (2003).
106. M.E. Padilla-Tosta, J.M. Lloris, R. Martinez-Manez, T. Pardo, J. Soto, A. Benito and M.D. Marcos, A fluorescent chemosensor based on a ruthenium(II)-terpyridine core containing peripheral amino groups that selectively sense ATP in an aqueous environment, *Inorg. Chem. Commun.*, **3**(1), 45-48 (2000).
107. W. Goodall, J.A.G. Williams, Iridium(III) bis-terpyridine complexes incorporating pendent *N*-methylpyridinium groups: luminescent sensors for chloride ions, *J. Chem. Soc. Dalton Trans.*, (17), 2893-2895 (2000).
108. E. Wang and M. Meyerhoff, Anion selective optical sensing with metalloporphyrin-doped polymeric films, *Anal. Chim. Acta*, **283**(2), 673-682 (1993).
109. M. Vasseur, R. Frangne and F. Alvarado, Buffer-dependent pH sensitivity of the fluorescent chloride-indicator dye SPQ, *Am. J. Physiol.*, **264**(1), C27-C31, (1993).
110. D. Parker, P. K. Senanayake and J. A. G. Williams, Luminescent sensors for pH, pO$_2$, halide and hydroxide ions using phenanthridine as a photosensitiser in macrocyclic europium and terbium complexes, *J. Chem. Soc., Perkin Trans. 2*, (10), 2129-2139 (1998).
111. J. W. Bunting and W. G. Meathrel, Quaternary nitrogen heterocycles VII - Reactions of some tricyclic heteroaromatic cations in basic solutions, *Can. J. Chem.*, **52**(6), 981-987 (1974).
112. R. S. Dickins, T. Gunnlaugsson, D. Parker and R. D. Peacock, Reversible anion binding in aqueous solution at a cationic heptacoordinate lanthanide centre: selective bicarbonate sensing by time-delayed luminescence, *Chem. Commun.*, (16), 1643-1644 (1998).
113. J. I. Bruce, R. S. Dickins, L. J. Govenlock, T. Gunnlaugsson, S. Lopinski, M. P. Lowe, D. Parker, R. D. Peacock and J. J. B. Perry, S. Aime and M. Botta, The Selectivity of Reversible Oxy-Anion Binding in Aqueous Solution at a Chiral Europium and Terbium Center: Signaling of Carbonate Chelation by Changes in the Form and Circular Polarization of Luminescence Emission, *J. Am. Chem. Soc.*, **122**(40), 9674-9684 (2000).
114. T. Gunnlaugsson, A. J. Harte, J. P. Leonard and M. Nieuwenhuyzen, Delayed lanthanide luminescence sensing of aromatic carboxylates using heptadentate triamide Tb(III) cyclen complexes: the recognition of salicylic acid in water, *Chem. Commun.*, (18), 2134-2135 (2002).
115. Y. Bretonniere, M. J. Cann, D. Parker and R. Slater, Ratiometric probes for hydrogencarbonate analysis in intracellular or extracellular environments using europium luminescence, *Chem. Commun.*, (17), 1930-1931 (2002).
116. M. Montalti, L. Prodi, N. Zaccheroni, L. Charbonniere, L. Douce and R. Ziessel, A luminescent anion sensor based on a europium hybrid complex, *J. Am. Chem. Soc.*, **123**(50), 12694-12695 (2001).
117. M. D. Best and E. V. Anslyn, A Fluorescent Sensor for 2,3-Biphosphoglycerate Using a Europium Tetra-*N*-oxide Bis-bipyridine Comlex for Both Binding and Signaling Purposes, *Chem. Eur. J.*, **9**(1), 51-57 (2003).
118. T. Yamada, S. Shinoda and H. Tsukube, Anion sensing with luminescent lanthanide complexes of tris(2-pyridylmethyl)amines: Pronounced effects of lanthanide center and ligand chirality on anion selectivity and sensitivity, *Chem. Commun.*, (11), 1218-1219 (2002).
119. S. W. Magennis, J. Craig, A. Gardner, F. Fucassi, P. J. Cragg, N. Robertson, S. Parsons and Z. Pikramenou, Crown ether lanthanide complexes as building blocks for luminescent ternary complexes, *Polyhedron*, **22**(5), 745-754 (2003).
120. R. K. Mahajan, I. Kaur, R. Kaur, S. Uchida, A. Onimaru, S. Shinoda and H. Tsukube, Anion receptor functions of lanthanide tris(β-diketonate) complexes: naked eye detection and ion-selective electrode determination of Cl$^-$ anion, *Chem. Commun.*, (17), 2238-2239 (2003).

FLUORESCENT CARBON DIOXIDE INDICATORS

Andrew Mills* and Stephanie Hodgen

3.1. INTRODUCTION

There are few analytes in the world as significant as carbon dioxide, equal, as it is, in importance as oxygen and pH. Carbon dioxide is a basic chemical feedstock of life, which when coupled with green plant photosynthesis[1], i.e.

$$CO_2 + H_2O \xrightarrow{\text{sunlight}} C(H_2O) + O_2 \qquad (1)$$

where $C(H_2O)$ is a reduced form of carbon such as a sugar or starch, generates the fuel and food necessary for the continued existence of most known forms of life. The reverse of reaction (1) is the basis of most cell metabolism, releasing, as it does, the energy for life. Thus, not only is carbon dioxide usually an essential ingredient to make the prerequisite chemicals for life, it is also often used as an indicator of the existence of life and a measure of health. For example, in medicine, the key, basic analytes that are routinely monitored in the blood of hospital patients are: dissolved oxygen, pH and carbon dioxide[2]. In clinical chemistry, a whole area devoted to the monitoring of the levels of carbon dioxide in breath has emerged, i.e. capnography, in which not only the level of carbon dioxide is important, but also its temporal variation, since both provide valuable medical diagnostic information[3].

The use, presence and measurement of carbon dioxide is also important in many industries. For example, in many biotechnology industries the measurement of carbon dioxide levels forms an important part of process control. Nowhere is this more important than in the brewing industry, where the continuous monitoring of carbon dioxide levels during the fermentation process is vital for: high product yields, minimal unwanted, and not very tasty, by-products, and an optimized control strategy[4].

In the food industry, a revolution in food packaging has come about through the use of carbon dioxide in modified atmosphere packaging (MAP)[5]. In MAP the food package is flushed with an oxygen-free gas, usually carbon dioxide, before being sealed and sent off to the wholesale or retail trader. By gas flushing the food package in this way, oxygen is removed and, as a consequence, aerobic spoilage microbes cannot thrive. As a result, the food contained within a MAPed package will keep typically 3-4 times longer, without recourse to chemical preservatives, which are becoming increasingly unacceptable to the consumer. This form of packaging is now routinely used in the packaging of a wide variety of foods, including: bread, biscuits,

---

* Andrew Mills and Stephanie Hodgen, Department of Pure and Applied Chemistry, University of Strathclyde, Glasgow, G1 1XL, UK

cakes, pastries, nuts, sweets, coffee, tea, wholefat, dry foods, processed, smoked and cured meats, dairy products, fresh and pre-cooked pasta and noodles and pet food, to name but a few[5]. Carbon dioxide is commonly used as the flush gas in MAP, mainly because it is plentiful (0.03% of air is carbon dioxide), easily liquefied (critical temperature = 31°C, critical pressure = 72.9 atm) and, therefore, inexpensive. In addition, high levels of carbon dioxide have an antimicrobial action by reducing the rate of microbe metabolism even if oxygen is present[5]. As a consequence of its use in MAP the detection and measurement of carbon dioxide is very important in the food packaging industry.

Because of its low critical temperature and pressure, and chemical inert nature, carbon dioxide also features strongly in the use of supercritical fluids to dissolve and extract substances, most notably caffeine from coffee beans to generated decaffeinated coffee. Thus, in many industries, the use, or presence, of carbon dioxide is commonplace and its measurement and continuous monitoring often essential.

The measurement of carbon dioxide levels is also an important feature of environmental monitoring, providing, as it does, a rough gauge of the health of the environment under test. Thus, stagnant lakes and rivers are often characterized by high levels of dissolved carbon dioxide. The levels of carbon dioxide in our atmosphere are routinely monitored worldwide by environmentalists interested in the greenhouse effect of this gas on the Earth's delicately balanced biosphere[6]. The major source of carbon dioxide in the biosphere is combustion, generated by industry, domestic heating, burning, biomass degradation and fermentation. Knowledge of the levels of carbon dioxide in the atmosphere is obviously important, but just as important, if not more so, is that of the levels of carbon dioxide in the oceans, since these contain 60% more carbon dioxide than in the atmosphere! The oceans, by acting as a vast carbon dioxide reservoir, help reduce the effect of carbon dioxide as a greenhouse gas and so lower the potential of global warming[6]. Monitoring the levels of carbon dioxide in the atmosphere and hydrosphere is particularly important since they are not in equilibrium with each other and the rate of exchange between the two, which is so very important to the continued existence of life on this planet, depends on a wide variety of parameters including: atmospheric pressure, wind, humidity and temperature.

The quantitative and qualitative analysis of carbon dioxide in the gas phase is often routinely carried out using infra-red spectroscopy. However, the latter is prone to interference and requires long path lengths, and bulky and expensive equipment which lacks mechanical stability. Gas phase measurements of carbon dioxide are also often carried out by gas chromatography, usually using a molecular sieve column and a thermal conductivity detector. However, such instrumentation is expensive and usually requires a trained technician for its routine operation and maintenance. The measurement of dissolved carbon dioxide in an aqueous medium (e.g. riverwater, salt water or blood) *via* infrared spectroscopy or gas chromatography is more difficult and other analytical methods are preferred. Thus, the routine monitoring of carbon dioxide, especially dissolved carbon dioxide, is currently dominated by the Severinghaus electrode; an analytical device that has been in regular use for nearly 50 years without serious challenge[7,8]. This electrode utilises a pH electrode, placed in contact with a thin layer of an aqueous sodium bicarbonate solution, trapped behind a gas-permeable, ion-impermeable membrane. Carbon dioxide in the test medium diffuses through the gas-permeable membrane and causes a change in the pH of the trapped bicarbonate layer that is measured by the pH electrode. The key equilibria associated with this process are reported a little later in this article and, as we shall see, these equilibria allow the measured pH to be simply related to the partial pressure of carbon dioxide, P_{CO2}, in the test medium. Unfortunately, the Severinghaus electrode is bulky, quite expensive, prone to electrical interference, affected by acidic or basic gases, and exhibits slow response and recovery times. Other problems include: effects of osmotic pressure (caused by variable salt conditions in the test sample), reference electrode contamination and liquid junction fouling. These electrodes are also quite expensive, usually quite delicate, and require high and regular maintenance. Finally, the Severinghaus electrode does not have a disposable transducer.

In recent years there has been a growing interest in the development of optical sensors for a wide variety of analytes[9-16]. Such sensors are usually sensitive, robust, rapid-in-response, inexpensive, easily miniaturised and do have a disposable transducer. In addition, through fibre optics, optical sensors offer the possibility of remote, continuous, multianalyte analysis in low volume locations, such as the artery of a premature baby. Optical sensors are usually either colourimetric, i.e. characterised by a change in colour, or lumophoric, i.e. characterised by a change in luminescence intensity, I_L, or lifetime, τ. In this paper the basic concepts behind the major, and some minor, different luminescent optical sensors for carbon dioxide that have been reported in the literature are discussed and illustrated.

3.2. THE TWO TYPES OF CARBON DIOXIDE OPTICAL SENSOR SYSTEMS

Almost all of the optical sensors for carbon dioxide (colourimetric and lumophoric) can be classified initially into one of two major categories, namely: wet or dry sensors.

3.2.1 Wet Optical Sensors for Carbon Dioxide

The basic characteristics of any wet carbon dioxide optical sensor are: (i) a pH-sensitive dye (anionic form, D$^-$; protonated form, DH), (ii) a luminescent dye (dyes (i) and (ii) are usually one and the same), (iii) an aqueous encapsulation medium, usually containing some sodium bicarbonate, in which the dye(s), (i) and (ii), are dissolved or dispersed and (iv) a gas-permeable, ion-impermeable membrane (GPM) used to cover the wet sensor layer. A schematic illustration of such a typical wet optical sensor for carbon dioxide and its features is given in figure 3.1. Table 3.1 lists many of the wet optical sensors for carbon dioxide that have been reported to date and provides details of: (i) the luminescent dyes and encapsulating solutions used, (ii) what kind of analyses they were used for, i.e. gaseous or dissolved carbon dioxide, and (iii) the type of measurement required, i.e. luminescence intensity or lifetime[17-35]. A key to all the abbreviations used in this, and all other tables and in the text, is provided at the end of this article. Structures of some of the key luminescent dyes listed in Table 1 are illustrated in figure 3.2.

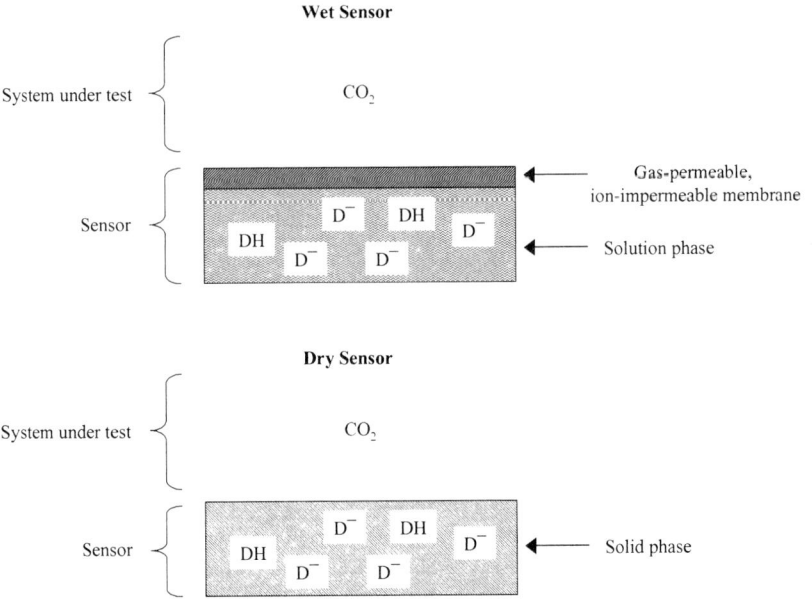

Figure 3.1. Schematic illustration of the main features of the two different optical sensors for carbon dioxide, namely wet and dry sensors.

The basic principles of operation behind all wet optical sensors for carbon dioxide are very simple and similar to those that underpin the Severinghaus carbon dioxide electrode[7]. Thus, in a typical wet optical sensor for carbon dioxide, the carbon dioxide in the test medium (gaseous or aqueous) diffuses through the GPM, see figure 3.1, and rapidly (usually within a few seconds to minutes) establishes an equilibrium with the entrapped aqueous layer of the optical sensor. In an aqueous solution in contact with carbon dioxide the following equilibria are set up[7,8]:

$$CO_2(g) \Leftrightarrow CO_2(aq) \quad (2)$$
$$K_1 = 3.3 \times 10^{-2} \text{ mol.dm}^{-3}.\text{atm}^{-1}$$

$$CO_2(aq) + H_2O \Leftrightarrow H_2CO_3 \quad (3)$$
$$K_2 = 2.6 \times 10^{-3}$$

$$H_2CO_3 \Leftrightarrow H^+ + HCO_3^- \quad (4)$$
$$K_3 = 1.72 \times 10^{-4} \text{ mol.dm}^{-3}$$

$$HCO_3^- \Leftrightarrow H^+ + CO_3^{2-} \quad (5)$$
$$K_4 = 5.59 \times 10^{-11} \text{ mol.dm}^{-3}$$

If, as is usually the case, the solution also contains sodium hydroxide or sodium bicarbonate at some known concentration [Na^+], it can be shown that at equilibrium the relationship between the partial pressure of carbon dioxide, P_{CO2}, and the proton concentration in the aqueous solution is as follows:

$$\alpha.P_{CO2} = [H_2CO_3] = \frac{[H^+]^3 + [H^+]^2[Na^+] - K_w[H^+]}{K_3([H^+] + 2K_4)} \qquad (6)$$

Where $\alpha = K_1K_2[H_2O]$, and K_w is the water dissociation constant ($K_w = [H^+][OH^-] = 10^{-14}$ mol^{2}dm^{-6}).

The Severinghaus electrode takes advantage of the above set of equilibria, i.e. the acidic nature of carbon dioxide, and monitors the value of the P_{CO2} in the test medium *via* the pH change it induces in the entrapped sodium bicarbonate layer using a pH electrode. In the Severinghaus electrode, and most wet optical sensors for carbon dioxide, the level of sodium bicarbonate in the trapped internal aqueous layer is sufficiently high (typically 10^{-2} mol.dm^{-3}) that eqn. (6) reduces to:

$$\alpha.P_{CO2} = [H_2CO_3] \approx [H^+][Na^+]/K_3 \qquad (7)$$

i.e. P_{CO2} is proportional to $[H^+]$.

Most of the wet optical sensors for carbon dioxide listed in Table 3.1 exploit the same equilibria as the Severinghaus electrode, but use a pH sensitive dye (DH/D$^-$), rather than a pH electrode, to determine the acid concentration $[H^+]$, in the thin aqueous bicarbonate layer. In an aqueous solution containing a luminescent pH-sensitive dye, the key equilibria are summarised by eqn.s (2)-(5) and, the following transduction step:

$$DH \Leftrightarrow H^+ + D^- \qquad (8)$$
$$K_a(DH)$$

Where DH and D$^-$ are the protonated and deprotonated form of the dye and K_a (DH) is the acid dissociation constant for the dye; note: by definition, $pK_a = -\log(K_a)$.

DH and D$^-$ usually have completely different absorption and emission spectral properties. The fluorescent pH-indicating dye, 1-hydroxy-pyrene-3,6,8-trisulfonate (pyranine), or HPTS for short, so often used in optical sensors for carbon dioxide (see Table 3.1), provides a good illustration of an almost ideal lumophore for use in luminescence, intensity-based sensors for carbon dioxide. The structure of D$^-$ for HPTS is illustrated in fig. 2 and the photochemical properties of both D$^-$ and DH for HPTS are summarised[36,37] in table 3.2 and their features illustrated *via* the reaction scheme in fig. 3. It can be seen from this data that the protonated and deprotonated forms of HPTS have very different λ_{max} absorption and emission values and that over a wide pH range any luminescence is due to the electronically excited state of D$^-$, i.e. HPTS$^-$, which luminesces at 510 nm, even if DH is excited. HPTS is often used as the pH indicator in carbon dioxide optical sensors because of its: (i) near-ideal pK_a for monitoring changes in P_{CO2} associated with patient monitoring, (ii) high stability and water solubility, (iii) strong visible absorption band and (iv) large Stokes shift emission band (λ_{max} (D$^-$) = 455nm; λ_{em} (D^{-*}) = 512nm).

Table 3.1: 'Wet' luminescent indicator systems for carbon dioxide

Luminescent dye	Encapsulating medium*	Gaseous (g) or dissolved (d) CO_2 measurement	Intensity (I) or lifetime (τ) measurement	Ref.
BMUB	Aqueous sodium bicarbonate in agarose	g	I	17-19
HPTS	Aqueous sodium bicarbonate in agarose	g	I	20
HPTS	Dye attached to a strong anion-exchange membrane, soaked in aqueous sodium bicarbonate	d	I	21
AcFl	A polyHEMA hydrogel, saturated with an aqueous solution of sodium chloride and sodium bicarbonate	d	I	22
HPTS	Dye covalently immobilized in cellulose granules embedded in a hydrogel soaked in aqueous sodium bicarbonate	g	I	23
Fl and other Fl derivatives	(a) dye adsorbed onto porous glass bead with hygroscopic lithium nitrate to keep moist and (b) an aqueous solution. In both cases, the sensor element was sealed in a capillary tube, with a gas bubble as the GPM. Photobleaching reduced using an antioxidant additive..	g	I	24
Fl and HPTS	Fl copolymersised with HEMA or HPTS-adsorbed-onto acrylamide. Both sensors soaked in aqueous sodium bicarbonate	g	I	25

Table 3.1: (Continued)

Luminescent dye	Encapsulating medium*	Gaseous (g) or dissolved (d) CO_2 measurement	Intensity (I) or lifetime (τ) measurement	Ref.
HPTS	Aqueous sodium bicarbonate entrapped in expanded PTFE (Gore Tex™)	d	I	26
HPTS	Dye with or without cross-linked polyacrylamide beads, soaked in sodium bicarbonate and embedded in silicone rubber	g	I	27
HPTS	Dye plus cross-linked with aminoethylcellulose fibres, embedded in an ion-permeable polyurethane based hydrogel, attached to a polyester foil.	d	I	28
c-SNARF	Dye in aqueous sodium bicarbonate micelles, stabilized with PVP, dispersed in siloxane polymer. No additional GPM.	g	I	29
c-SNAFLc	Dye in aqueous sodium bicarbonate soaked in an NVP hydrogel	d	I	30
HCA	MCP added to compress pH response range in an aqueous sodium bicarbonate solution with sodium chloride added to match the osmotic pressure of the salt water.	d	I	31

* unless stated otherwise all sensors covered by a GPM (silicone rubber or PTFE usually)

Table 3.1: (Continued)

Luminescent dye	Encapsulating medium*	Gaseous (g) or dissolved (d) CO_2 measurement	Intensity (I) or lifetime (τ) measurement	Ref.
HPTS	Non-fluorescent NR added to compress pH response range in an aqueous sodium bicarbonate solution with sodium chloride added to match the osmotic pressure of the salt water. NR precipitates at high ionic strength, so limiting it use.	d	I	31
Eosin, R6G or THR	Dyes encapsulated in a poly(HEMA) hydrogel, soaked with sodium bicarbonate and with FRET acceptor dye PR or BTB	g	τ	32
$Ru(pzth)_2^{2+}$	Dye attached to sephadex beads that are then made to gel.	g	τ	33-35

* unless stated otherwise all sensors covered by a GPM (silicone rubber or PTFE usually)

Table 3.2: Photochemical and Chemical Characteristics of HPTS in aqueous solution[36,37] and in a plastic film*

Property	HD	D⁻
Aqueous solution		
Absorption λ_{max}/nm	403	455
Emission λ_{max}/nm	435	512
Excited state lifetime/ns	5.3	4.8
pK_a of HPTS	7.3-8.0	
pK_a of excited HPTS	1.4	
Plastic film*		
Absorption λ_{max}/nm	394	467
Emission λ_{max}/nm	440	517

*Plastic film comprises[37]: TAOH as the phase transfer agent, EC as the polymer and TBP as the plasticiser

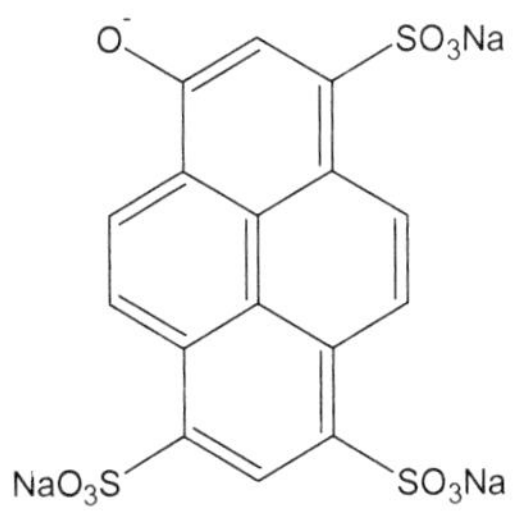

FLUORESCENT CARBON DIOXIDE INDICATORS

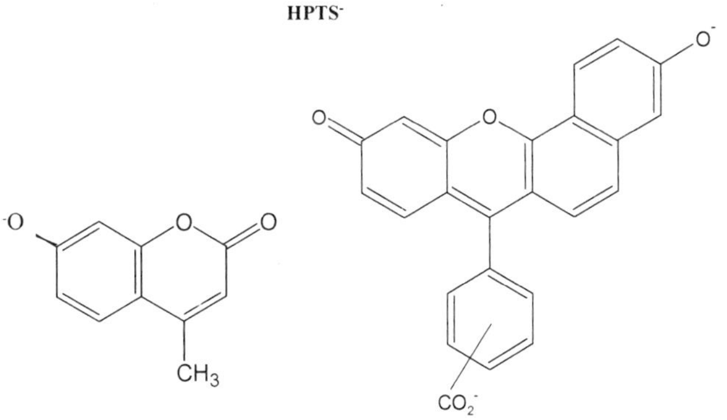

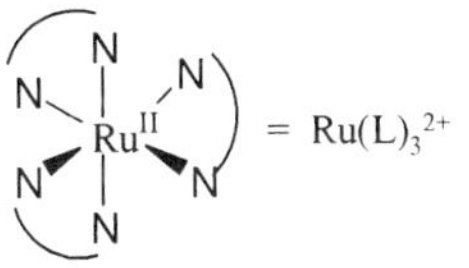

Figure 3.2. Structures of selected lumophores listed in table 3.1 and used in wet optical sensors for carbon dioxide.

From the data given in table 3.2 and the schematic illustration of the major processes associated with HD and D⁻, in fig. 3, it would appear that by simply monitoring the emission intensity at 512 nm, due to D^{-*}, of an internal aqueous HPTS solution containing sodium bicarbonate, covered by a GPM, generated using λ_{excit} = 455 nm, it should be possible to determine the proton concentration of the internal aqueous solution and, via eqns. (7) and (8), the level of P_{CO2} in the external medium under test. If HPTS is used under such conditions, provided the concentration of dye used is very small compared to that of the buffer, sodium bicarbonate, then eqns. (7) and (8) can be combined to yield the following expression:

$$\alpha \cdot P_{CO2} = [H_2CO_3] \approx K_a[DH][Na^+]/(K_3 \cdot [D^-]) \quad (9)$$

Provided the pH sensing dye is optically dilute, i.e. absorbance at λ_{excit} is typically < 0.1, then I_L, the sensor film luminescence due to D^-*, in the presence of carbon dioxide at a level of P_{CO2}, will be proportional to $[D^-]$. Thus, eqn. (9) becomes:

$$\alpha \cdot P_{CO2} = [H_2CO_3] \approx K_a(I_L^\circ - I_L)[Na^+]/(K_3 \cdot I_L) \quad (10)$$

where I_L° is the measured luminescence intensity for the system in the absence of carbon dioxide, i.e. when all the dye, HPTS in this case, is in its deprotonated form, D^-. Note that as a consequence of eqn. (10), the smaller the concentration of sodium bicarbonate, the narrower and lower the dynamic range of the carbon dioxide optical sensor.

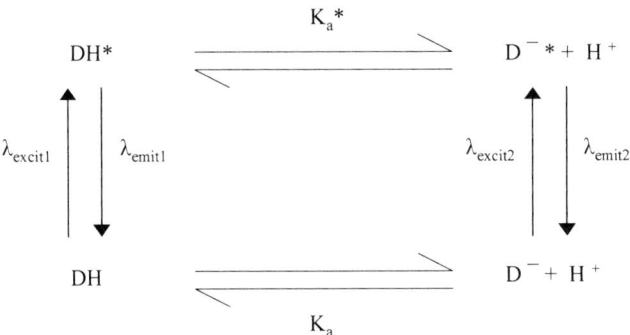

Figure 3.3. Schematic illustration of the major processes[36,37] associated with the electronic excitation of either the protonated, DH, or deprotonated, D^-, forms of a luminescent, pH-sensitive dye, such as HPTS. Usually, the pK_a of the electronically excited state of DH is much lower than that of its ground state, i.e. $pK_a^* \ll pK_a$. As a consequence over a wide range of pH the only luminescence that will be observed is due to D^-* even if DH is excited. This is certainly the case for D = HPTS.

Other factors that will affect the above equilibria, both with respect to the equilibria associated with carbon dioxide and the deprotonation of the dye include: temperature and ionic strength. With either temperature or ionic strength, an increase invariably produces a decrease in the sensitivity of the carbon dioxide wet, luminescent optical sensor, or, as others[26] have more positively put it, 'an increase in the dynamic range'. Either way, optical sensors for carbon dioxide are notoriously sensitive to changes in temperature, ionic strength (relevant to dissolved carbon dioxide measurements) and humidity (relevant to gaseous carbon dioxide measurements), and this fact should never be forgotten.

One of the first luminescence intensity-based wet optical sensors for carbon dioxide, featured in table 3.1, was reported by Lübbers and Opitz[17-19] in 1975, using the pH sensitive dye, β-methyl umbelliferon, BMUB, ($\lambda_{excit}(D^-)$ 357 nm; $\lambda_{em.}(D^-*)$ 445 nm), dissolved in a thin aqueous layer containing: 5 mmol dm^{-3} sodium bicarbonate with 1% agarose added for improved mechanical stability and covered by a gas-permeable, ion-impermeable membrane, such as PTFE. Eight years later these same authors reported the first 'wet' optical HPTS sensor for carbon dioxide, using otherwise the same formulation as their PTFE/BMUB/agarose/sodium bicarbonate sensor[20]. Figure 3.4(a) illustrates a typical set of excitation spectra ($\lambda_{em.}$ = 510nm) reported by these workers for their PTFE/HPTS/agarose/sodium bicarbonate sensor as a function of P_{CO2}. Figure 3.4(b) illustrates the observed variations in I_L as a function of P_{CO2} for the Lübbers and Opitz HPTS carbon dioxide sensor for 3 different concentrations of sodium bicarbonate in the thin aqueous layer. From eqn. (10) a plot of the sensor's luminescence intensity at 510nm, I_L, in the following form: $(I_L^0 - I_L)/I_L$ as a function of P_{CO2} should generate a straight line with a zero intercept. This prediction is confirmed for the Lübbers and Opitz HPTS carbon dioxide sensor by the plot of the results in fig. 4(b) in this latter format, illustrated in the insert diagram of fig. 4(b) for a range (1-6 mM) of different concentrations of sodium bicarbonate in the thin aqueous layer. Note the decrease in sensitivity of the optical sensor with increasing sodium bicarbonate concentration, as predicted by eqn. (10) and highlighted earlier.

Most optical sensors for carbon dioxide give a response curve, i.e. I_L vs. P_{CO2} curve, similar in nature to those illustrated by the main diagram in fig. 4(b). Given the nature of eqn (10), it is not surprising to note that such sensors are said[19,38] to exhibit a hyperbolic response, or to be hyperbolic-type sensors. One of the common features of a hyperbolic response sensor is a shorter response time (say for 90% of the overall signal change) than the corresponding recovery time, with the difference increasing with increasing magnitude in the change in P_{CO2}. This feature is readily explained if the response and recovery process are both controlled by the same process, such as the diffusion of carbon dioxide into and out of the sensor[38].

The pioneering work of Lübbers and Opitz[17-20] in the area of optical sensors for carbon dioxide helped encourage the subsequent generation of many other luminescence intensity-based optical sensors for carbon dioxide, as indicated by the extensive list in table 3.1. Subsequent variations on the Lübbers and Opitz wet luminescence intensity-based carbon dioxide optical sensor include using the dye encapsulated in, or bound to, a sponge like material, soaked in an aqueous sodium bicarbonate solution[26], or the dye attached to an inert support in a 'sea' or gel of bicarbonate solution[23,27,28]. In all cases listed in table 3.1 the sensors were protected by a gas-permeable, ion-impermeable layer, mainly to prevent water loss through evaporation to the usual gaseous medium under test. When such sensors were used to monitor dissolved levels of carbon dioxide, the gas-permeable membrane prevents the dissolution of the thin aqueous layer, containing the luminescent, pH-sensitive dye and sodium bicarbonate.

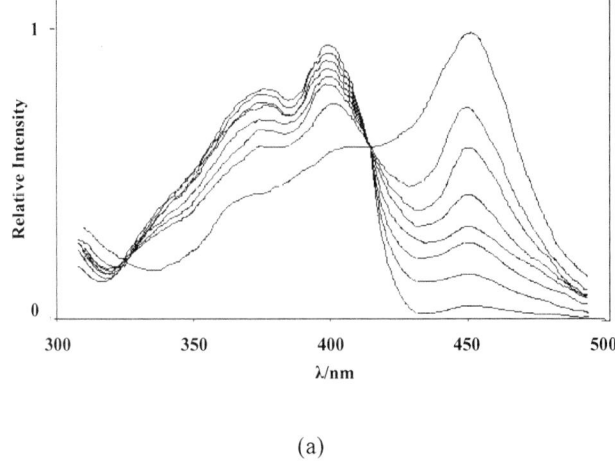

(a)

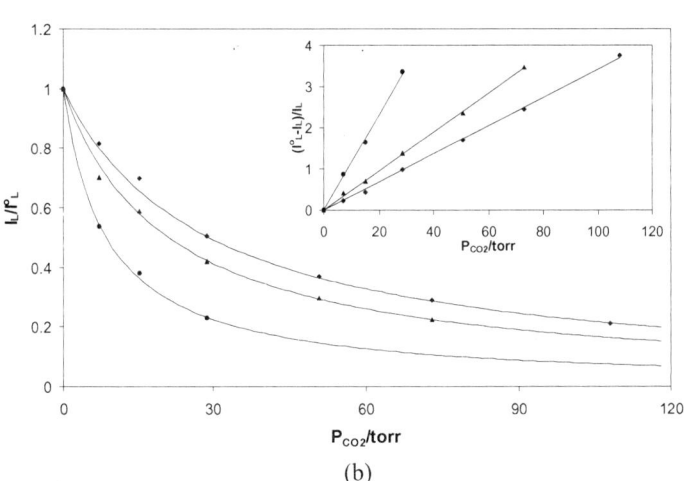

(b)

Figure 3.4. (a) Typical corrected excitation spectra ($\lambda_{em.}$ = 510 nm) for a wet optical sensor for carbon dioxide comprising: PTFE/HPTS/agarose/sodium bicarbonate solution, when exposed to the following levels of % CO_2 in the gas phase under test (from top to bottom): 0, 1, 2, 4, 7, 10, 20 and 100, respectively[20]. (b) Plot of the relative intensity of emission (I_L/I_L^0) exhibited by a PTFE/HPTS/agarose/sodium bicarbonate solution optical sensor ($\lambda_{excit.}$ = blue LED (λ_{max} = 490 nm); $\lambda_{em.}$ = 510 nm) as a function of P_{CO2}, with the concentration of the sodium bicarbonate used set at: 6 (♦), 3 (▲) and 1 (•) mM, respectively. The insert diagram illustrates the good linear relationship between ($I_L^0 - I_L$)/ I_L and P_{CO2}, for the three different films as predicted by eqn. (10)[20].

As noted above, most of the wet luminescence indicator systems reported to date and listed in table 3.1 are based on the measurement of the intensity of luminescence of a pH-sensitive dye, such as HPTS as a function of P_{CO2}. However, in at least one case listed in table 3.1, luminescence lifetime, τ, measurements are used instead to ascertain the pH of the internal aqueous sodium bicarbonate solution and so, via eqn. (10), the partial pressure of carbon dioxide in the external test medium[32]. This example uses a very different method of transduction, pioneered by Lakowicz and his co-workers, which involves the use of long-range, non-radiative Fluorescence Resonance Energy Transfer (FRET)[32]. In this and all FRET sensing systems for carbon dioxide reported here the colour change exhibited by a pH-insensitive colourimetric dye (the acceptor, A^-) is used to affect the decay time of a, usually, pH-sensitive lumophore (the donor, D). In such sensor systems, the electronically excited state of the donor lumophore, i.e. D^*, is usually quenched by the deprotonated form of the acceptor pH-sensitive dye, A^- i.e.

$$D^* + A^- \xrightarrow{FRET} D + A^{-*} \qquad (11)$$

Where A^{-*} is the electronically excited state of the deprotonated acceptor, A^{-*}. Typically A^{-*} undergoes rapid non-radiative decay to return to the ground-state of the dye, A^-, i.e. A^{-*} does not usually luminesce and, if it does, not at the same wavelengths as D^*. The protonated form of the quencher, AH, is assumed either not to quench D^* or, if it does, at a rate that is significantly lower than that for A^-.

In a FRET-based carbon dioxide sensor, in the absence of carbon dioxide the pH of the encapsulation medium, in this case aqueous sodium bicarbonate, is initially set sufficiently high that all, or most of the pH-sensitive, colourimetric acceptor dye is present in its anionic form, A^-, and thus should, ideally, quench the pH-insensitive, electronically-excited luminescent donor dye molecules, D^*, very effectively and efficiently via reaction (11). As a consequence, the lifetime of the luminescent donor will be very short and at a minimum value, τ_0, in the absence of any carbon dioxide in the medium under test. However, if carbon dioxide is present, then the pH of the sensor's encapsulation medium will be decreased and the concentration of the acceptor quencher, in reaction (11), A^- will decrease, due to the following equilibrium:

$$A^- + H^+ \Leftrightarrow AH \qquad (12)$$
$$1/K_a(AH)$$

where $K_a(AH)$ is the acid dissociation constant for the protonated form of the acceptor, AH. As a consequence, D^* will be less quenched via reaction (11) the higher the level of carbon dioxide in the test medium. Consequently, the measure lifetime of D^*, τ, will increase with increasing P_{CO_2}, i.e. $\tau > \tau_0$. As noted above, this transduction process presumes that AH doesn't itself quench D^*, and this assumption appears fairly reasonable given the stringent conditions required to observe efficient quenching of an electronically excited donor by FRET in the first place.

Thus, in order for FRET to be most effective the following conditions need to be satisfied: (i) the absorption spectrum of the colourimetric acceptor should overlap strongly with the emission spectrum of the lumophoric donor, (ii) both transitions, $D^* \rightarrow D$ and $A^- \rightarrow A^{-*}$, should be highly allowed and, finally, (iii) (for unlinked D-A pairs) A^- should be ca. 1-10 mmol.dm^{-3} (so that D and A^- are within the Förster distance)[32].

In the pioneering work of Lakowicz and his co-workers on FRET optical sensors for carbon dioxide, the donors used were: Texas Red Hydrazine (TRH), Eosin and Rhodamine 6G (R6G) and the acceptors used were: Bromothymol Blue (BTB) and Phenol Red (PR)[32]. In this work and most, if not all, lifetime-based optical carbon dioxide sensor studies, the lifetime of the luminescent donor at different levels of P_{CO_2} were determined by a phase-modulation technique[39]. In this method the increase in lifetime of the donor is detected in the frequency, rather than the time, domain, as a change in the phase angle, between the modulated excitation light and the forced oscillation of the emission from the electronically excited donor molecules. The experimental observables are the shift in phase angle of the emission, ϕ, and its modulation, m, in intensity, both relative to the phase and modulation of the source. An average decay time, τ, can be calculated from the phase angle shift or modulation through either one of the following equations[39]:

$$\tan \phi = \omega\tau \qquad (13)$$

and

$$m = (1 + \omega^2\tau^2)^{-1/2} \qquad (14)$$

where ω is the angular modulation frequency ($= 2\pi f$, where $f =$ frequency in Hz of the modulated excitation light). A simple illustration of the output of a phase-modulated system is illustrated in figure 3.5. The broken line depicts the sinusoidal variation in excitation light, in this case set at a typical value $f = 100$ MHz, as a function of time. The solid line shows how the emitted light intensity generated by a fluorophore, with a lifetime, τ, (in this case τ is set at 10ns), comparable to $1/f$ would vary as a function of time if it absorbed some of the excitation light. Due to the finite lifetime of the excited state, the emission will be delayed in time relative to the excitation, which is measured as a phase shift, ϕ, which can be used to calculate τ via eqn. (13). The relative amplitude of the emission is also reduced compared to that of the excitation. This process of demodulation allows the parameter, m, to be calculated via the following equation:

$$m = (A/B)/(a/b) \qquad (15)$$

using the data in Figure 3.5. For any value of m a value of τ can be calculated via eqn. (14). It should be briefly noted that the above equations hold only if the luminescence decay is described by a single-exponential decay. Interestingly, for FRET-based carbon dioxide sensors, and, for that matter, most FRET-based sensors, e.g. for oxygen, invariably the assumption of mono-exponential excited state decay kinetics is not valid, especially for luminescent dyes dispersed in a heterogeneous medium such as a polymer[40,41]. Fortunately, as we shall see, the variation in the apparent average lifetime, as measured using phase modulation spectroscopy and eqn. (13), as a function of P_{CO_2} still appears to fit the simple equations, based on the assumption of a single exponential decay, that allow the data to be linearised.

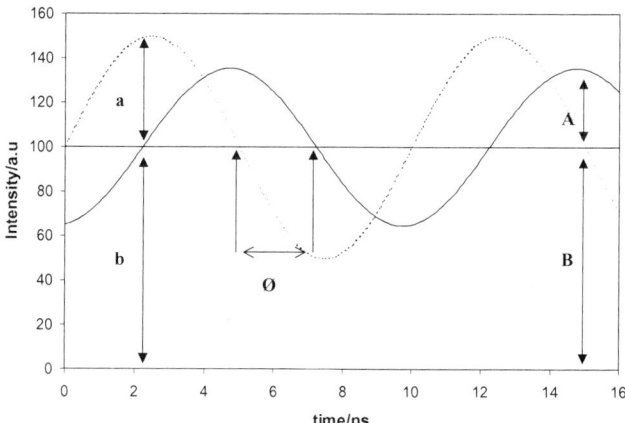

Figure 3.5. Schematic illustration of the variation in intensity of a modulated excitation light source (broken line) operating at a frequency = 100 MHz and the resulting phase-shifted luminescence (solid line) generated by a lumophore, which absorbs some of the excitation light, with a lifetime of 10 ns. Under such conditions, from eqn. (13), ϕ = 86.96° (≡ 2.25 ns). "A" and "B" and "a" and "b" in this diagram are the measured "amplitude" and "background" light intensity levels for the lumophore and excitation source, respectively, from which a value for m can be calculated using eqn. (15).

Typically, for any FRET-based sensor system, an optimum single modulation frequency is chosen, typically $f \cong 1/\tau$, and the values of ϕ exhibited by the FRET-based sensor are then measured as a function of P_{CO_2}. Thus, Lakowicz and his co-workers used $f = 133$ or 155 MHz in their early studies of FRET-based wet optical sensors for carbon dioxide[32] using the donor-acceptor pairs: TRH-BTB, Eosin-PR and R6G-PR in a poly(HEMA) hydrogel noted earlier. The observed variations of the relative phase angle shifts as a function of P_{CO_2} for these three different sensors are illustrated in fig. 6 and show that the TRH-BTB and Eosin-PR sensors are particularly responsive to variations in the level of carbon dioxide using FRET[32].

The use of lifetime measurements, via time-resolved or phase-modulated techniques, has the advantages that the measured decay times are generally not dependent upon the macroscopic optical properties of the sample and are not sensitive to the fluctuations in the exciting light intensity. In contrast, the more traditional steady-state intensity measurements, which are used for the intensity-based optical sensors for carbon dioxide that dominate table 3.1, are sensitive to these parameters and thus measurements will be sensitive to light loss, lamp drift and dye bleaching, amongst other things. These drawbacks, which necessitate regular sensor calibration in intensity-based luminescence systems, are mitigated only by the simplicity of the measurements and the inexpensive nature of the associated necessary equipment.

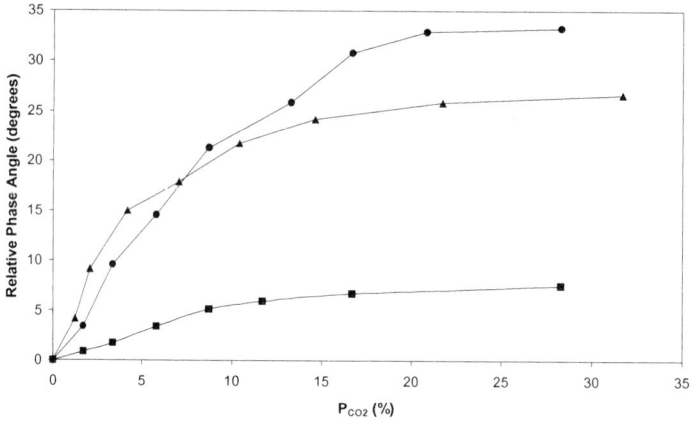

Figure 3.6. Dependence of the donor relative phase angle on P_{CO_2} for a series of different donor-acceptor pairs in a poly(HEMA)-hydrogel at 25°C. The donor-acceptor pairs used were: THR-BTB ($\lambda_{em.}$ = 600 nm; f = 133 MHz) (●), Eosin-PR ($\lambda_{em.}$ = 580 nm; f = 155 MHz) (▲); and R6G-PR ($\lambda_{em.}$ = 600 nm; f = 133 MHz) (■). In all cases $\lambda_{excit.}$ = 543 nm[32].

Some of the difficulties associated with luminescence intensity measurements can also be obviated using a simple wavelength-ratiometric method. For example, Uttamlal and Walt developed a fibre-optic carbon dioxide sensor for fermentation monitoring[26], in which the lumophore, HPTS, was dissolved in a sodium bicarbonate solution and entrapped in an expanded PTFE support held at the distal end of the optical fibre by a gas permeable membrane. In this work, the luminescence emission of the optical sensor at 515 nm (due to D⁻*) was measured for two different excitation wavelengths, namely: 470 nm (which excites D⁻ only) and 405 (which excites mostly DH). The ratios of the two measured luminescence intensities, i.e. $I_L(470)/I_L(405)$, were used[26] to calibrate the sensor, both in 0.45 M sodium chloride solution and distilled water, as illustrated by the data in fig. 7. By ratioing the luminescence intensities measured at the two different excitation wavelengths, it is possible to eliminate problems due to errors arising from lamp intensity fluctuations and drifts in the photodetection system. If the lumophore is present at optically dilute concentrations, (i.e. its absorbance at λ_{excit} is <0.1 then this ratiometric method can also compensate for changes in dye concentration. This wavelength – ratiometric compensation method[26] obviously assumes that any change in optical artefact affects both wavelengths in a similar manner and this and other underlying assumptions need to be recognised. However, the above wavelength-ratiometric technique has proved to be a popular method for reducing errors and providing reliable calibration plots for intensity-based carbon dioxide, and pH, optical sensors.

In contrast to intensity-base measurements, lifetime measurements, whether they be time resolved or phase modulated, often require expensive, bulky and sophisticated equipment. Thus, in their early work on FRET-based optical sensors for carbon dioxide, Lakowicz and his co-workers used several expensive lasers, (such as a 543nm-He-Ne laser and a cavity dumped dye laser)[32,39], which had their intensities modulated with an acoustic-optic modulator, working at, typically, 155 MHz. Subsequent to this work, these and other workers have attempted to address the issue of the prohibitive expense of lifetime measurements for optical sensor interrogation by studying longer-lived luminescent donor dyes in FRET-based systems that can be probed using cheap diode light sources, modulated at considerably lower frequencies, i.e. $f < 100$ MHz, than those used previously. This move to more affordable lifetime measurement systems for carbon dioxide optical sensors will become more apparent in the next section.

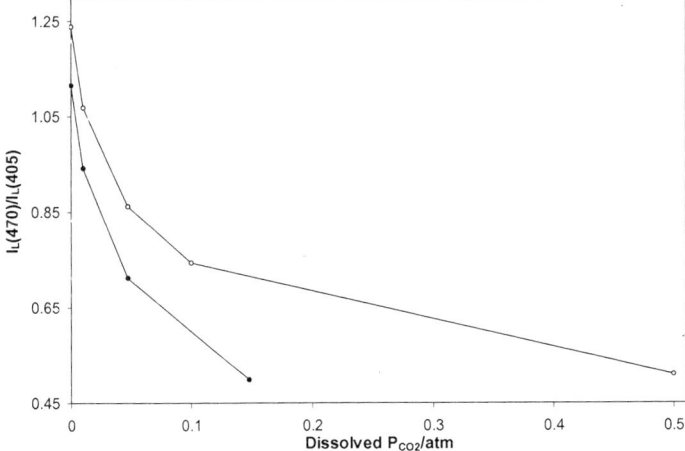

Figure 3.7. Observed[26] variation is the ratio of the intensity of luminescence due to HPTS when excited at 470 and 405 nm, i.e. $I_L(470)/I_L(405)$ as a function of the level of dissolved carbon dioxide, P_{CO_2}. In this work the sensor used comprised: PTFE/HPTS/sodium bicarbonate solution, with the latter two entrapped in an expanded PTFE support. The measurements were performed using 0.45 M sodium chloride (○) and distilled water (●).

In the area of optical sensors for carbon dioxide, the use of FRET has the added advantage over intensity measurements that the lumophore itself no longer needs to be pH-sensitive, which is very liberating since the number of dyes that are both pH-sensitive and luminescent are very limited. Thus, a FRET-carbon dioxide sensor can use a wide range of pH-insensitive donor lumophores, provided a suitable colourimetric, pH-sensitive acceptor dye can be found. Fortunately, there are many such donor-acceptor dye combinations and, as a consequence, there is the potential for a wider variety of FRET-based carbon dioxide sensors than there are of luminescence-based ones.

Finally, under the heading of wet optical sensors for carbon dioxide, it is appropriate to mention the work of Orellana and his co-workers[33-35] and their development of a carbon dioxide sensor based on the quenching of a novel ruthenium (II) complex by proton transfer. In this work, the novel lumophore employed was the tris(2-(2-pyrazinyl)thiazole)-ruthenium (II) cation, $Ru(pzth)_3^{2+}$, the structure of which is illustrated in fig. 2. This complex is water soluble and absorbs strongly at 331 nm ($\varepsilon_{max} = 47000$ dm^3 mol^{-1} cm^{-1}) and 460 nm ($\varepsilon_{max} = 17000$ dm^3 mol^{-1} cm^{-1}) and emits at 652 nm. The proton transfer agent used to quench the ruthenium (II) complex is a Brönsted acid (HB), and was typically dihydrogen phosphate or dihydrogenphthalate. Fig. 8 illustrates the essential features of the various processes involved in the quenching of $Ru(pzth)_3^{2+}$ (equivalent to D⁻ in fig. 8) by HB and shows that with an aqueous solution of $Ru(pzth)_3^{2+}$, with $\lambda_{excit} = 337$ nm or 460 nm, the observed emission intensity, at $\lambda_{em.} = 652$ nm will decrease with increasing concentration of HB through an irreversible proton transfer quenching reaction[33-35].

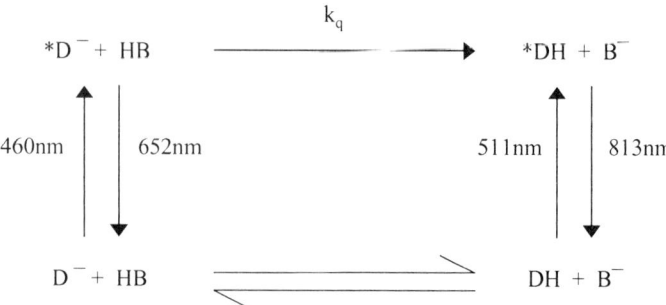

Figure 3.8. Sensing scheme[33-35] for a carbon dioxide sensor based on the quenching of the luminescence of a dye, D⁻, where D⁻ ≡ $Ru(pzth)_3^{2+}$, by a Brönsted acid, HB, where HB = hydrogen pthalate. The vertical up and down arrows refer to the absorption and emission processes associated with D⁻ and DH, the λ_{max} values of which are given next to the arrows and refer to the deprotonated and protonated lumophore $Ru(pzth)_3^{2+}$.

As a consequence of the photochemical properties of $Ru(pzth)_3^{2+}$, illustrated in fig. 8, Orellana and his co-workers were able to create a novel optical sensor for carbon dioxide comprising: $Ru(pzth)_3^{2+}$(D⁻) immobilised onto CM-Sephadex particles, saturated with 0.1 mol.dm⁻³ hydrogen phthalate (HB) buffer and covered with a silicone membrane[33-35]. This wet optical sensor for carbon dioxide allows the evaluation of P_{CO2} through either lifetime or intensity measurements, as illustrated by the normalised luminescence intensity or emission lifetimes results recorded for this sensor as a function of P_{CO2} profiles in fig. 9, by Orellana and his co-workers[35].

The sensor is temperature sensitive, as most carbon dioxide optical sensors are, and exhibits a slight sensitivity towards oxygen. Orellana et al also showed[34] that the addition of the enzyme carbonic anhydrase to the film formulation produced a marked improvement (typically a factor of 3) in the response time of the optical sensor by increasing the rate of the forward step in the aquation of carbon dioxide equilibrium step, eqn. (3). Similar observations of the beneficial effects of carbonic anhydrase on the response characteristics of optical sensors for carbon dioxide have also been made by others[19] working on luminescence-based wet optical sensors for carbon dioxide.

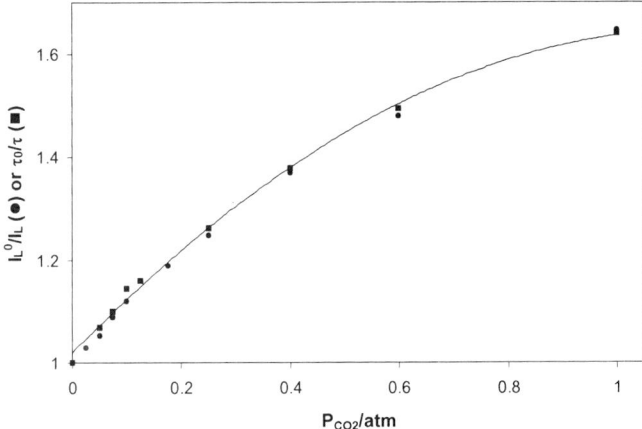

Figure 3.9. Stern-Volmer plot of the relative luminescence intensity (I_L^0/I_L) or emission lifetime (τ_0/τ) as a function of P_{CO_2} as reported by Orellana et al for their fibre-optic sensor for carbon dioxide. The sensor comprised: Ru(pzth)$_3^{2+}$ immobilised onto CM-Sephadex soaked in a pH 7.25 hydrogen pthalate solution and covered with a silicone gas permeable membrane[35].

The above sensor represents a very novel approach to the luminescence optical sensing of carbon dioxide, although the number of lumophores that have easily measurable lifetimes and are quenched by a proton transfer mechanism still appear very limited, even today.

Interestingly, others have recently promoted[42] the possibility of generating optical-sensors for carbon dioxide based on the quenching of an excited state by a pH-sensitive quencher. In this case, the quenching process is photoinduced electron transfer in non-aqueous solution between a fluorophore, such as naphthalene and anthracene, and an unprotonated quencher, such as an amine, which is unable, or less efficient, in its quenching ability when protonated. By combining these two features in one molecule, as in 1-naphthylmetylamine (NMA), Herman and his co-workers showed[43] that the fluorescence of NMA dissolved in dioxane increased when exposed to carbon dioxide. So far only the initial results of experiments conducted in non-aqueous solution have been reported and an optical sensor for carbon dioxide based on this type of process has yet to be generated, despite its apparent promise.

3.2.2 Dry Optical Sensors for Carbon Dioxide

The response characteristics of wet optical sensors for carbon dioxide are altered if the water vapour pressure (for gas phase measurements) or osmotic pressure (for dissolved CO_2 measurements) of the system under test is significantly different from that of the sensor. Under the latter circumstances, either some hydration or dehydration of the sensor film will occur upon exposure to the test sample and, as a consequence, the sensor will need recalibrating. The latter drawback has long been recognised and effectively restricts the use of such sensors to test and calibrate systems with osmotic, or water vapour, pressure values similar to the wet sensor itself. That is why, for example, Uttamlal and Walt calibrated[26] the PTFE/HPTS/sodium bicarbonate luminescence based sensor reported earlier in 0.45M NaCl solution, so that the calibration medium matched the ionic strength of the fermentation test medium, see fig. 7. In addition, the same group [31,25] have noted that such wet optical sensors permanently lose intensity upon prolonged exposure at high P_{CO_2} levels and are very slow to respond (>30 min) at very low P_{CO_2} levels. These drawbacks make such wet carbon dioxide sensors far from ideal. The ideal solution to these problems is the Holy Grail of carbon dioxide sensors, namely a solid state device comprising an indicator encapsulated in a gas-permeable, ion-impermeable membrane, with no apparent aqueous solution bicarbonate layer. The key features of such a dry sensor for carbon dioxide are illustrated in fig. 2.

In 1991, Raemer and his co-workers published a patent[43] on colourimetric carbon dioxide indicators for placement of tracheal tubes, in which it was noted that tetra t-butyl ammonium hydroxide (TBAH), a phase transport agent, 'enhanced the response of the dye'. The agent was thought to accelerate the exchange of carbon dioxide to and from a liquid phase sensor. In this work, the pH-sensitive dye was usually bound to controlled pore glass particles and, most of the sensors that utilised TBAH, gave a reversible response to carbon dioxide. These sensors clearly fall short of the ideal of a 'solid sensor' for carbon dioxide, since they are not encapsulated in a GPM and, therefore, not appropriate for dissolved carbon dioxide work, unless a gas-permeable membrane cover is added. In addition, they are likely to exhibit response features that are markedly affected by changes in humidity (for gas phase work). However, despite these caveats, this patent is notable as one of the first examples of phase transfer agent use in the context of optical sensors for carbon dioxide.

A real advance in solid dry sensor design came with the discovery by others[44,45] that a phase transfer agent, (PTA), such tetraoctyl ammonium hydroxide (TOAH) or, in its more general form, Q^+OH^-, could be used to solubilise the anionic form of a colourimetric pH indicator dye, such as *m*-cresol purple (MCP), in a hydrophobic solvent, such as toluene, that is mutually compatible with that of a water-insoluble polymer, such as ethyl cellulose (EC) or poly(vinyl butyral), (PVB). Thus, it was found that many PTA's, i.e. Q^+OH^-'s, when mixed with a pH sensitive, hydrophilic indicator dye anion, D, form ion pairs, Q^+D^-, that can be dissolved in non-aqueous, and often hydrophobic, solvents in which a variety of different hydrophobic polymers could also be dissolved. The product of such work was effectively a series of coloured inks, comprising: pH-sensitive dye/PTA/ polymer/solvent, which could be cast, by printing, doctor-blade or spin-coating, to generate a range of thin, coloured plastic films containing a pH-sensitive dye in its highly coloured, deprotonated anionic form, i.e. as Q^+D^-. Although this formulation goes some way towards generating a solid dry sensor, it is not obvious why the encapsulated ion-paired dye, Q^+D^-, would respond to the presence of carbon dioxide. However, it is known that associated with most ion pairs are usually a few molecules of water; thus, the PTA cation, usually a quaternary ammonium cation, pH- indicator dye anion ion-pair entrapped in a polymer film combination is more appropriately formulated as $Q^+D^-.xH_2O$. As a consequence, in the final dried ink films, the encapsulated pH-sensitive dye anion, D^-, can interact with carbon dioxide as if there is water nearby; and the overall process can be summarised by the following equilibrium process[44,45]:

$$Q^+D^-.xH_2O + CO_2 \overset{\alpha}{\Leftrightarrow} Q^+HCO_3^-.(x-1)H_2O.HD \qquad (16)$$

Where α the equilibrium is constant associated with the process. In order to aid the diffusion of carbon dioxide through the polymer often a plasticiser, such as tributyl phosphate (TBP), is included in the film formulation. Based on this simple formulation, i.e. dye/PTA/plasticiser/polymer/solvent, a number of different colourimetric plastic films were generated by Mills et al[37,44-46] and others[47] and all proved very effective as solid-state optical sensors for carbon dioxide.

As you would expect of such solid dry sensors, research reveals that they can be used for the determination of carbon dioxide in dry and humid gases[44-47] and when dissolved in aqueous solution[48]. These dry films typically exhibit response and recovery times of the order of a few seconds, and very thin films based on this technology, and which respond even faster, i.e. typically <0.1s, have been used for capnography[49]. In a dry gas environment it might be expected that such solid film sensors would eventually lose their water of hydration and cease to function. However, the water of hydration associated with the ion-pair combination: PTA cation - pH-sensitive dye anion, appears very tightly bound and as a consequence it appears that these films can be used and stored under very dry conditions without any appreciable loss in performance.

When used for dissolved carbon dioxide measurements these dry optical sensors for carbon dioxide are less sensitive to changes in the osmotic pressures of the samples under test than the wet, i.e. Severinghaus, sensor types discussed so far. Silicone rubber, with its high hydrophobicity and permeability towards carbon dioxide appears a good encapsulation material for such work, although plasticised ethyl cellulose has been used to great effect to create optical sensors for carbon dioxide that can operate over the pH range 3-10. However, when such dry, ion-pair type carbon dioxide sensors are used for the continuous and prolonged measurement of P_{CO_2} in solution then a gas-permeable, ion-impermeable membrane cover is usually required.

Unfortunately, the presence of acidic vapours, such as the dioxides of nitrogen or sulfur, which are often found in the laboratory, have a marked and irreversible deleterious effect on the response features of dry sensors for carbon dioxide. Thus, on the open bench in a laboratory, the dye in the sensor films can change from its deprotonated (D^-) form to its protonated (DH) form, and as consequence become unresponsive, within 24h. Fortunately, in most environments, the levels of these acidic, oxidising gases are very low and so the shelf life of most ion-pair sensors are usually long (> 1 year), especially if placed in sealed bags and stored in the dark; refrigerated conditions also helps to preserve the films.

It has been suggested[45] that one mode of loss of sensitivity of these ion pair dry sensors for carbon dioxide is the thermal degradation of the phase transfer base through a Hofmann β-hydrogen elimination reaction. The rather elegant work of Chang et al[50] appears to lend support to this proposal, in their study of steam sterilisable fluorescence lifetime-based sensing films for carbon dioxide. Thus, by comparing films containing the same quantity but different quaternary ammonium hydroxides, these workers found that the order of sensor film stability (with number of β-hydrogens in parenthesis) was: TOAH (8) < CTAH (2) < TMAH (0). The Q^+D^- $.xH_2O$ ion pairs are so stable when TMAH is used as the PTA that the final films hardly respond at all to carbon dioxide, i.e. the increased ion-pair stability can ultimately affect the key equilibrium process, reaction (16). Thus, nowadays CTAH appears the preferred PTA for ion-pair dry sensor films. In order to increase sensor film longevity, most researchers used a vast excess of base, i.e. $Q^+D^-.xH_2O$, in the film formulation to produce sensors with long operational and storage lifetimes. The excess base in the thin dry film carbon dioxide sensors acts as a lipophilic bicarbonate buffer system and exists in the form of $Q^+HCO_3^-.xH_2O$. Increasing the background concentration of the excess base also appears to decrease the sensitivity of the final optical sensor, as it does in its Severinghaus type wet counterpart, see eqn. (9). Table 3.3 provides details of many of the dry luminescent optical films for carbon dioxide that have been reported in the literature[51-70] and, as with the wet indicators, reviewed earlier, see table 3.1, most are intensity- rather than lifetime-, based sensors. Almost all these sensors utilise the ion-pair technology developed by Mills et al[44,45]. Figure 3.10 gives some of the structures of the key lumophores that are listed in table 3.3.

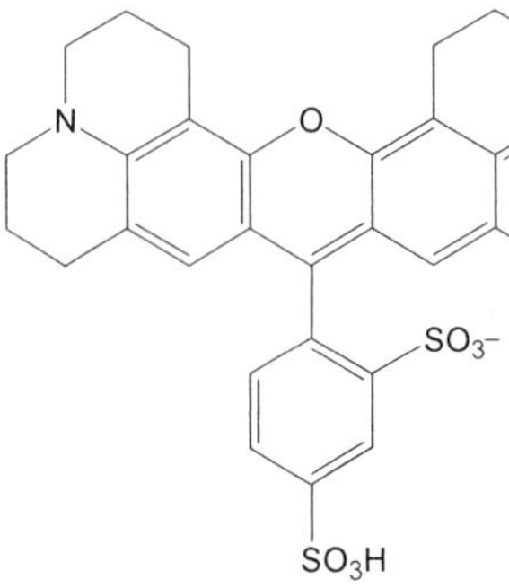

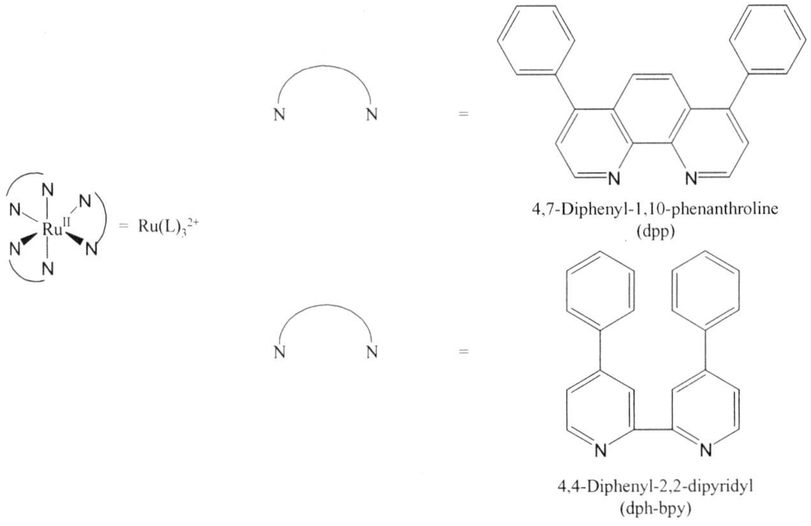

Figure 3.10. Structures of some of the key lumophores listed in table 3.3 and used in dry, ion-pair optical sensors for carbon dioxide.

Table 3.3: 'Dry' Luminescent indicator systems for carbon dioxide

Luminescent dye	Encapsulating medium	Gaseous (g) or dissolved (d)* CO_2 measurement	Intensity (I) or lifetime (τ) measurement	Ref.
Fl	Polyethylene glycol, pretrated with sodium hydroxide	g	I	51
HPTS	Dye plus phase transfer agent, TOAH, in EC polymer with TBP as a plasticiser	g	I	52,37
HPTS	Dye plus phase transfer agent (TBAH, MAPTAC and CTAH) in EC and polystyrene polymer	g	I	53
HPTS	Dye plus phase transfer agent, TOAH, in EC with TBP as a plasticiser	d	I	54
HPTS	Dye plus phase transfer agent, generates HPTS$^-$CTA$^+$, with excess TOAOH, electrostatically bound to aminocellulose granules then encapsulated in silicone rubber	d	I	55
HPTS	Dye plus phase transfer agent (TOAH) encapsulated in an organically modified silica glass	g	I	56,57
[Eu(tta)$_3$]	Luminescence of europium (II) complex in polystyrene reduced, by absorption screening, using a pH indicator dye (TB, PR or CR) – TOAH ion pair, encapsulated in EC, with TBP as plasticiser. The two films were on opposite sides of the same glass slide.	g	I	58-60

Table 3.3: (Continued)

Luminescent dye	Encapsulating medium	Gaseous (g) or dissolved (d)* CO_2 measurement	Intensity (I) or lifetime (τ) measurement	Ref.
HPTS	Dye plus phase transfer agent, TOAH, in EC dissolved in ethanol/toluene solvent. Viscous solution used in resevoir-type capillary optical senor with a PTFE GPM.	g	I	61
HPTS	Dye plus phase transfer agent, generates HPTS$^-$CTA$^+$ ion-pair that is then encapsulated in a silicone rubber film with an excess of base, CTAH	d	I	62
HPTS	Dye immobilized in a base-catalysed silica sol-gel by electrostatic attraction. Sensor film then covered with a hydrophobic sol-gel to reduce pH cross reactivity.	d	I	63
SRh	Luminescent donor dye, SR, mixed with a non-fluorescent acceptor dye (eg. TB or MCP) with TOAH, TBP in EC to produce a plastic film suitable for FRET	g	τ	64,65
SRh	SR, plus MCP-CTAH in silicone rubber creates a steam-sterilizable sensor. Detection by FRET.	d	τ	50
Ru(dph-bpy)$_3^{2+}$:	Ru(dph-bpy)$_3^{2+}$-TMS ion pair plus TB anion – TMDA$^+$ in an excess of TOAH encapsulated in EC was used to prepare FRET films	g	τ	66

Table 3.3: (Continued)

Lumines-cent dye	Encapsulating medium	Gaseous (g) or dissolved (d)* CO_2 measurement	Intensity (I) or lifetime (τ) measurement	Ref
Ru(dph-bpy)$_3^{2+}$:	Ru(dph-bpy)$_3^{2+}$-TMS ion pair plus MCP anion – TMDA$^+$ in an excess of TOAH encapsulated in EC was used to prepare FRET films	g	τ	67
Ru(dpp)$_3^{2+}$	Ru(dpp)$_3^{2+}$-TMS ion pair plus Sudan III in an excess of TOAH encapsulated in a hydrophobic silica sol-gel/EC hybrid produced films for FRET.	g	τ	68
Ru(dpp)$_3^{2+}$	Ru(dpp)$_3^{2+}$ dye doped nanobeads provide a luminescent reference for the luminescence of HPTS$^-$CTA$^+$ ion pairs, all encapsulated in a hydrophobic organically modified silica film. Basis of sensing via DLR.	g	τ	69

A luminescence intensity-based dry optical sensor film for carbon dioxide which used the phase transfer technology discussed above was reported[37] as early as 1993. In this work, the dye used was HPTS, the phase transfer agent, TOAH, the polymer, EC, and the plasticiser, TBP. In such a system, assuming eqn (16) is the key equilibrium process, then a new parameter, R, can be defined as follows:

$$R = (I_L^\circ - I_L)/I_L = \alpha.P_{CO_2} \qquad (17)$$

Where, I_L° and I_L are the luminescence intensities of $Q^+D^-.xH_2O$ in the absence and presence of carbon dioxide, respectively. Given the similar natures of the two processes, it is not surprising to note that the wet and dry luminescence intensity-based optical sensors have very similar key equations, i.e. eqns. (10) and (17), respectively. As with eqn. (10), eqn. (17) assumes that an excitation wavelength is selected at which only $Q^+D^-.xH_2O$ absorbs, and that in the absence of carbon dioxide all the dye will be in its anionic form. The spectral properties of the first reported HPTS/TOAH/EC/TBP sensor film for carbon dioxide are given[37] in table 3.2. Thus, the absorption and emission spectra of the protonated and deprotonated forms of HPTS in the plastic film are very similar to those of the dye in aqueous solution, although shifted slightly in a bathochromic direction. This shift is due mainly to the formation of the quaternary ammonium cation-dye anion ion pairs, and, to some extent, to the change in the polarity of the surrounding environment.

A typical example of the observed variation in luminescence intensity, measured at 517 nm and due to the deprotonated form of HPTS in a HPTS/TOAH/EC/TBP film, as a function of the partial pressure of carbon dioxide in the test medium, is illustrated[37] in fig. 11. The subsequent plot of this data in the form of R $(=(I_L^° - I_L)/I_L)$ vs. P_{CO2} is illustrated in the insert diagram in fig. 11 and reveals a reasonable straight line as predicted by eqn (17). As noted earlier, in aqueous solution the excited state of the protonated form of HPTS, i.e. DH*, emits the green light (λ_{max} = 512 nm) associated with the excited state of the deprotonated form of the dye, i.e. Q^+D^{-*} (see fig. 2). This phenomenon is due to the very rapid equilibrium between DH* and D^{-*} and H$^+$ in solution. However, in a solid film this equilibrium does not appear to be so well established[37] and emission from DH*, or more precisely $Q^+HCO_3^-$ $(x-1).H_2O.HD$ at 440 nm is observed in a HPTS/TOAH/EC/TBP film at high P_{CO2} levels.

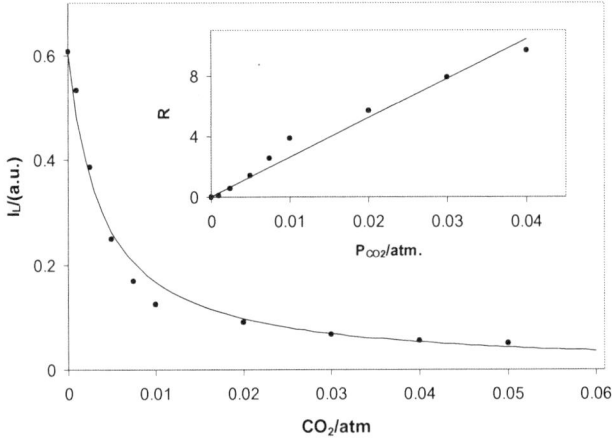

Figure 3.11. Reported[37] variation in the observed luminescence intensity of a HPTS/TOAH/EC/TBP dry film sensor as a fuction of P_{CO2}, recorded using $\lambda_{excit.}$ = 430 nm and $\lambda_{em.}$ = 517 nm. The insert diagram is a replot of the data in the main diagram, where R is defined by eqn. (17).

A number of other researchers, see table 3.3, have subsequently used HPTS, coupled with a PTA, encapsulated in a polymer, such as EC or silicone rubber, to create similar luminescence intensity-based optical sensors for carbon dioxide. Probably the most notable of these is the high-stability, non-invasive auctoclavable naked optical sensor for dissolved carbon dioxide measurement reported by Rao et al[62]. The sensor film comprises: HPTS/CTAH/RTV-silicone rubber; the latter being chosen because of its highly hydrophobic nature. This sensor has a shelf life of several months and exhibits no cross-sensitivity to salt in the range 0-0.2 M, or to pH, in the range 5.6-8.0, despite the fact that no gas-permeable membrane cover layer was used. The latter feature explains the use of the term 'naked' in the paper's title[62]. The robustness of this sensor film towards ion-exchange, and heat treatment (it's autoclavable) is due to the robust and hydrophobic nature of the encapsulation polymer employed. Typical plots of the normalised emission, and excitation, spectra exhibited by this HPTS/CTAH/silicone rubber film recorded[62] for this film for dissolved carbon dioxide concentrations ranging from 0-18.15% are illustrated in fig. 3.12. Interestingly, the optical characteristics of HPTS⁻ in this film, and its response toward carbon dioxide, are much like they are for the dye in aqueous solution, i.e. there is no evidence that DH* is stable and not readily dissociated in this film, unlike the HPTS/TOAH/EC films discussed earlier.

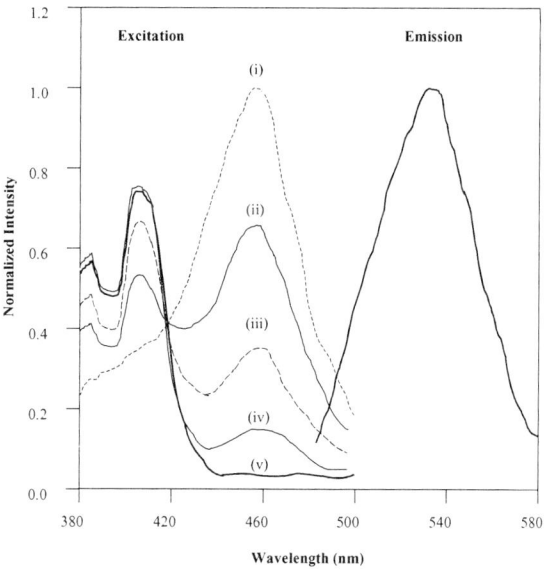

Figure 3.12. Normalised reported[62] excitation and emission spectra of a naked, i.e. no GPM cover, HPTS/silicone rubber film when exposed to water purged with the following levels of carbon dioxide: (i) 0, (ii) 0.13, (iii) 0.37, (iv) 0.58 and 18.15%.

Continuing the theme of novel encapsulating media, a dry optical sensor for carbon dioxide, employing the HPTS-tetraoctyl ammonium cation ion pair has been developed[56] which uses an organically modified silica glass (ormosil), rather than an organic polymer, as an encapsulation medium. In this study a relatively high humidity was maintained in all test and sample gases, since work showed the dry luminescent sensor film to be quite sensitive towards variations in humidity. The latter feature, largely absent from plastic film based sensors, appears a possible drawback to the use of ormosil in dry optical sensors for carbon dioxide.

To summarise at this point, most of the luminescence intensity-based dry optical sensors for carbon dioxide listed in table 3.3 work on similar principles as their wet optical sensor counterparts. Thus, these sensors utilise a pH-sensitive, luminescent dye, such as HTPS, encapsulated in a polymer in its anionic form using a phase transfer agent. The key equations that describe the interaction of the dye with carbon dioxide in the test medium are (16) and (17), in which it is assumed that a wavelength of excitation, λ_{excit}, is chosen so that only the anionic form of the dye absorbs and that the measured intensity of luminescence, I_L, is due only to D^{-*}.

A different type of luminescence intensity-based dry optical sensor for carbon dioxide has been reported recently by Nakamura and Amao[58-60], using a combination of a luminescent europium (II) complex, [Eu(tta)$_3$], and a pH-sensitive colourimetric dye, such as TB, PR or CR. As with almost all dry indicator systems for carbon dioxide, the pH-sensitive dyes used are always in the form of an ion pair, i.e. Q^+D^- xH_2O. In this work, TOAH was used as the PTA, EC as the encapsulation polymer and TBP as the plasticiser. This indicator is, in effect, a simple, standard colourimetric dry indicator for carbon dioxide, with the added exception to its formulation of a fluorophore, [Eu(tta)$_3$], which is not quenched by FRET and has an absorption spectrum that overlaps with that of the anionic form of the non-luminescent, pH-sensitive dye, $Q^+D^-.xH_2O$. As a consequence, as the level of P_{CO2} in the test medium is increased, the intensity of luminescence due to [Eu(tta)$_3$] will increase, because the concentration of $Q^+D^-.xH_2O$ decreases and so, therefore, does its absorbance at λ_{excit}. Such a decrease in absorbance allows more of the excitation light to be absorbed by the lumophore, [Eu(tta)$_3$], and, therefore, the luminescence of the latter will increase intensity, the higher the level of P_{CO2}. Typical results generated[58] by three such pH-intensive lumophore/pH-sensitive colourimetric dye type sensors, namely: [Eu(tta)$_3$]/TB/TOAH/EC/TBP, [Eu(tta)$_3$]/PR/TOAH/EC/TBP and [Eu(tta)$_3$]/CR/TOAH/EC/TBP, are illustrated in figure 3.13. For this system it can be shown, via eqn.(16), that the concentration of the protonated pH-sensitive dye, i.e. [$Q^+HCO_3^-.(x-1) H_2O.DH$], is related to P_{CO2} via the expression:

$$[Q^+HCO_3^-.(x-1) H_2O.DH] = \frac{[D]_{TOTAL} \alpha P_{CO2}}{1 + \alpha P_{CO2}} \qquad (18)$$

where [D]$_{TOTAL}$ is the total concentration of the colourimetric, ion-paired pH-sensitive dye in the film, i.e.[D]$_{TOTAL}$ = [$Q^+HCO_3^-.(x-1) H_2O.DH$] + [$Q+D^-.x H_2O$]. In the work of Nakamura et al, on these [Eu(tta)$_3$]/pH-sensitive dye films, the intensity of luminescence of [Eu(tta)$_3$] appears proportional to [$Q^+HCO_3^-.(x-1) H_2O.DH$], although this is only expected for optically dilute systems, which these are probably not. However, given this apparent relationship, it follows that, I_L is related to P_{CO2} via the following expression:

$$\frac{I_L - I_L^\circ}{I_L^{100} - I_L^\circ} = \frac{KP_{CO2}}{1 + \alpha P_{CO2}} \qquad (19)$$

I_L° and I_L^{100} are the intensities of luminescence of this type of film at 0 and 100% CO_2, respectively, and where K is a constant. It follows from eqn. (19) that the data illustrated in fig. 13 for this type of optical sensor can be linearised by plotting (I_L^{100} − I_L°)/(I_L − I_L°) versus P_{CO2}^{-1}, as reported by Nakamura et al[58]. Like the wet FRET sensors described earlier, this type of optical sensor, described as 'colourimetric with a reference luminescent dye', overcomes the problem of the limited availability of fluorescent pH-sensitive dyes with pK_a values that span a wide pH range. Instead, unlike the FRET-based carbon dioxide sensors, this type of sensor requires only a luminescent dye which has an absorption spectrum that's similar to that of one of the forms of the pH-sensitive, colourimetric dye. Usually, this absorption spectral overlap is between the pH-insensitive fluorophore and the anionic form of the pH colourimetric dye. As with the FRET-type sensors, the possible combinations of pH-insensitive fluorophore and pH-colourimetric dye are many. However, it is not clear what benefits this luminescence intensity-based sensor offers over its simpler, colourimetric counterpart, which has the same formulation, with the exception of the lumophore, other than it allows the measurement of P_{CO2} via luminescence intensity, rather than absorbance, measurements.

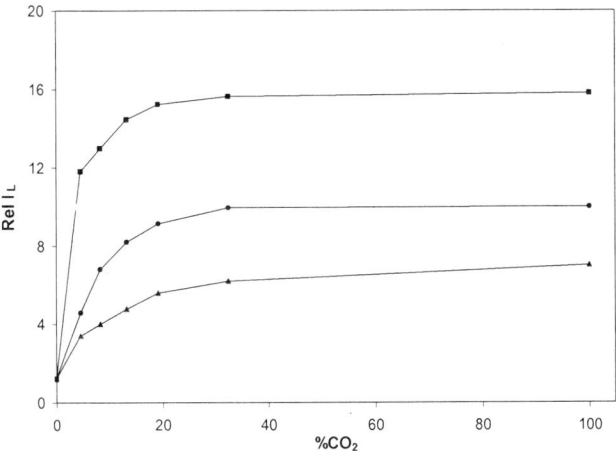

Figure 3.13. Observed[58] variation in the relative intensity of luminescence as a function of P_{CO_2} for three different dry 'colourimetric with reference luminescent dye' type sensors. The three different sensors used comprised: the lumophore [Eu(tta)$_3$], EC/TBP and the following three different pH-sensitive colourimetric dyes: TB (■), PR (●) or CR (▲). The results of this work can be linearised using eqn. (19).

As early as 1995, researchers[64] began to report FRET-based dry optical sensors for carbon dioxide. The features of such sensors, namely a fluorescent, pH-insensitive donor (D) and a colourimetric pH-sensitive acceptor (A), with a great deal of overlap between the emission spectrum of D and the absorption spectrum of A, were much like those noted earlier for wet type, FRET–based sensors, with the exception that the pH-sensitive dye was combined with a PTA to render it soluble in a hydrophobic medium, such as a polymer. Thus, for example, Sipior and co-workers[64] in their 1995 study of dry, FRET-based optical sensors for carbon dioxide used, amongst others, the luminescent dye sulforhodamine (SR) as the donor and a thymol blue (TB)–TOAH ion-pair as the acceptor; all entrapped in an ethyl cellulose polymer matrix. The frequency responses of this SR/TB-TOAH/EC film as a function of the level of carbon dioxide in a test humid

nitrogen gas-phase medium are illustrated in fig. 14(a), for which only the reported phase shift data for this system are illustrated. A plot of this data, recorded at a frequency of 138.14 MHz in the form of phase angle, φ, versus %CO$_2$ is illustrated in fig. 14(b). In such sensors the lifetime of the lumophore is at a maximum, τ_{max}, when there is no FRET, i.e. when all the pH-sensitive dye is protonated due to a high level of carbon dioxide in the test medium.

It follows that:

$$\frac{\tau_{max}}{\tau} = \frac{k_{other} + k_{FRET}}{k_{other}} \tag{20}$$

where k_{FRET} is the rate of fluorescence resonance energy transfer and k_{other} is the rate constant for all other deactivations, including luminescence. Since k_{FRET} will be proportional to the concentration of the deprotonated pH-sensitive dye, i.e. $[Q^+D^-.xH_2O]$, then,

$$\frac{\tau_{max}}{\tau} - 1 = \frac{k_{FRET}}{k_{other}} = K'[Q^+D^-.xH_2O] \qquad (21)$$

where K' is a proportionality constant. However, from eqn. (16) it can be shown that:

$$\alpha P_{CO2} = \frac{[Q^+D^-.xH_2O]_o - [Q^+D^-.xH_2O]}{[Q^+D^-.xH_2O]} \qquad (22)$$

where $[Q^+D^-.xH_2O]_o$ is the concentration of $[Q^+D^-.xH_2O]$ when no carbon dioxide is present and all the pH-sensitive dye is in its deprotonated, ion-paired form. Combining eqns. (21) and (22) together the following expression can be derived:

$$\frac{(\tau_{max}/\tau_o - 1) - (\tau_{max}/\tau - 1)}{(\tau_{max}/\tau - 1)} = \alpha P_{CO2} \qquad (23)$$

where τ_o and τ are the lifetimes of the donor lumophore in the absence and presence of carbon dioxide.

or, more simply:

$$\frac{(1/\tau_o - 1/\tau)}{(1/\tau - 1/\tau_{max})} = R = \alpha P_{CO2} \qquad (24)$$

FLUORESCENT CARBON DIOXIDE INDICATORS

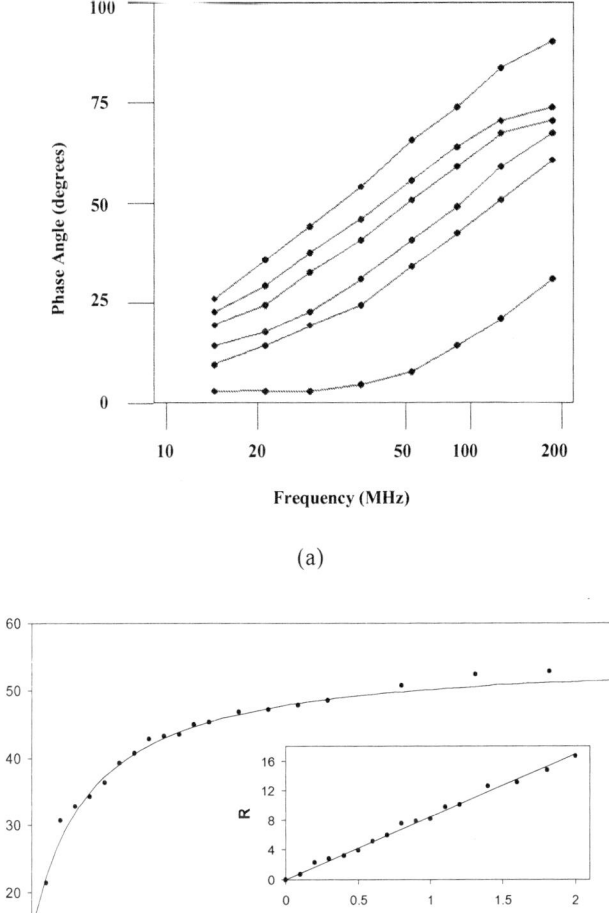

Figure 3.14. (a) Observed variation in phase angle, ϕ, as a function of modulation frequency for an SR/TB-TOAH/EC dry sensor film. The different profiles are the film responses when exposed to the following different levels of carbon dioxide (from top to bottom): ∞, 10, 2, 0.5, 0.25 and 0%. The '∞' CO_2 response curve was recorded for a sensor film without any TB-TOAH ion pair, i.e. for SR alone[64].(b) Observed variation in phase angle as a function of %CO_2, for the SR/TB-TOAH/EC film sensor used in (a), using 442 nm light modulated at a frequency of 138.14 MHz. The insert diagram is the linearised plot of the data in the main diagram, where the function of R is defined by eqn. (25), assuming ϕ_0 and ϕ_{max} are equal to 14.3 and 55°, respectively[64].

In the work of Sipior et al[64], and all FRET-based dry optical sensors for carbon dioxide, a phase modulation technique was used to measure the lifetime of the pH-insensitive donor lumophore. As a consequence, it is more relevant and useful to rewrite eqn. (24) with respect to the measured phase angle, given eqn. (13). Thus, eqn. (24) can be more usefully written as:

$$\frac{(\cot\phi_o - \cot\phi)}{(\cot\phi - \cot\phi_{max})} = R = \alpha P_{CO2} \qquad (25)$$

where R is a parameter that can be easily calculated, using eqn. (25) and ϕ_o, ϕ and ϕ_{max} which are the measured phase angles for the sensing system in the absence, presence and overwhelming presence (so that all the pH-sensitive dye is protonated) of carbon dioxide. The insert diagram in fig. 14(b) illustrates the plot of the data in the main diagram, taken from the work of Sipior et al[64] for their SR/TB-TOAH/EC films, in the form of R vs. %CO_2, where R is defined by eqn. (25). The linearity of this plot helps confirm the validity of eqn. (25) and the underlying assumption that the degree of FRET quenching is proportional to $[Q^+D^-.xH_2O]$.

As noted earlier, a major problem with FRET, as a method for interrogating optical sensors, such as FRET-based wet, or dry, optical sensors for carbon dioxide, is the need for expensive equipment. It is the short lifetime, typically 0.5 – 5ns, of the donor lumophores that are usually used in such sensors that make it necessary to use expensive optical sources, such as lasers, frequency generators, optics and electronics. In an attempt to address this problem, Klimant and his co-workers[66] developed a Ru(dph-bpy)$_3^{2+}$- (TMS)$_2$/MCP$^-$TMDA$^+$/EC film sensor for carbon dioxide that required a much less expensive lifetime measuring system. These workers used the ion pair technique to both solubilise the anionic form of the pH-sensitive colourimetric acceptor dye, and the cationic donor lumophore Ru(dph-bpy)$_3^{2+}$. Thus, the pH-sensitive acceptor MCP$^-$ and the pH-insensitive, long-lived donor Ru(dph-bpy)$_3^{2+}$ were rendered lipophilic, and therefore soluble in the EC supporting polymer, using the lipophilic cation and anions, TMDA$^+$ and TMS$^-$, respectively. An excess of TOAH was also added to the film formulation in order to ensure prolonged functionality. Klimant and his co-workers[66] chose to use a donor with a long lifetime, >1µs, since it enabled lifetime measurements to be performed on this FRET-based optical sensor for carbon dioxide using just a bright blue light emitting diode (LED), λ_{max} (emission) = 470 nm, operating at a frequency of only 75 kHz; making it much cheaper than the high frequency, laser-based phase modulation lifetime measuring systems used previously in such work. Using this inexpensive frequency domain fluorescence spectroscopy system to determine the lifetime of the lumophore in the Ru(dph-bpy)$_3^{2+}$-(TMS$^-$)$_2$/MCP$^-$-TMDA$^+$/TOAH/EC dry film sensor, as a function of P_{CO2}, Klimant et al found τ to vary as a function of P_{CO2} as illustrated[66] in fig. 15. Once again this data can be linearised using eqn. (24) for a FRET-based optical sensor for carbon dioxide as illustrated by the insert diagram in fig. 15. More recently, McCraith and his co-workers[68] have reported a Ru(dpp)$_3^{2+}$-(TMS$^-$)2 /Sudan (III) – TOAH/(silica sol-gel/EC) dry FRET-based optical sensor for carbon dioxide that used the same bright blue LED as used by Klimant et al but operated at only 20 kH. Because of the high pK_a of Sudan (III), McCraith et al were able to operate their dry, FRET-based carbon dioxide sensors up to P_{CO2} levels of 1 atm. Thus, from these two examples, it is clear that it is possible to devised FRET-based sensors for carbon dioxide that require relatively low frequency, i.e. <<100 MHz, modulated excitation light generation and detection, which are much more affordable.

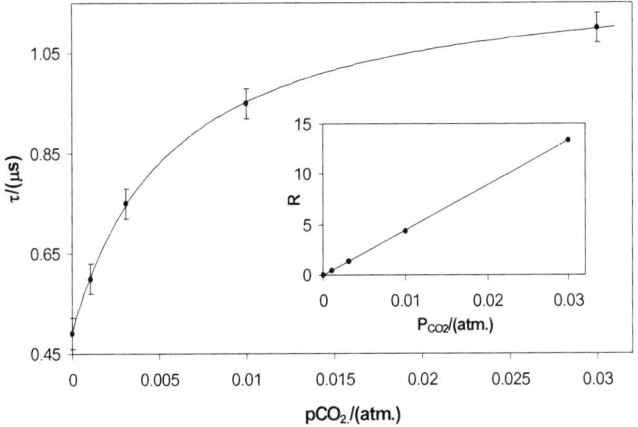

Figure 3.15. Observed[66] variation in luminescence lifetime, for a Ru(dph-bpy)$_3^{2+}$-(TMS)$_2$/MCP$^-$-TMDA$^+$/TOAH/EC FRET-based film sensor as a function of P$_{CO2}$. The lifetime data were determined using a phase-modulated blue LED excitation source (λ_{ex} = 470 nm) and a red-sensitive PMT ($\lambda_{em(max)}$ = 630 nm). The insert diagram is a linearised plot of the lifetime data in the main diagram, where the parameter R has been calculated for each value of P$_{CO2}$ using eqn. (24) and τ_0 and τ_{max} = 0.49 and 1.21 µs, respectively.

A brief inspection of the many examples of dry luminescent indicator systems for carbon dioxide reveals that most of those reported to date are intensity-based and, amongst the few other, and admittedly more recent, examples, the FRET method of detection dominates. One other sensing method that doesn't fall into any of the categories considered so far, is the measurement of carbon dioxide levels using phase-modulated fluorimetry coupled with a Dual Lumophore Referencing (DLR) proposed by MacCraith et al[69]. In DLR, two different luminescence signals are generated in the sensing membrane by the modulated excitating light source; in this case a bright blue 470 nm LED. In DLR, the two lumophores must have a significant overlap in their excitation and emission spectra, so that both luminescence signals can be excited by one light source (the modulated LED) and detected by one filter-detector combination. As a consequence of this combination, it follows that the total signal is a superposition of the two signals. However, one of the lumophores is selected so that it is pH, and therefore P$_{CO2}$, insensitive and has a lifetime that is similar to that of $1/f$ where f = frequency of the modulated excitation light; this is referred to as the reference lumophore. The other lumophore is selected on the basis that it is pH, and therefore P$_{CO2}$, sensitive and has a lifetime that is so short that the phase shift is zero; this is referred to as the analyte lumophore. In the work of McCraith et al[69] the reference lumophore was Ru(dpp)$_3^{2+}$ doped nano-beads and the analyte lumophore was HPTS ion-paired with CTMAH to render it soluble in the hydrophobic organically-modified silica encapsulation matrix used in this work. In such a system the intensity of luminescence due to the reference lumophore, $I_{L(REF)}$, is given by:

$$I_{L\,(REF)} = I_{B1} + I_{REF} \sin(\omega t + \phi_{REF}) \qquad (26)$$

where I_{B1} and I_{REF} are the background and amplitude luminescence intensities due to the reference lumophore in the optical film subjected to an excitation light of angular frequency ω and producing a fixed phase shift, ϕ_{REF}. For the analyte lumophore, HPTS in this case, which has a very short lifetime, (i.e. $\tau_{analyte} \ll 2\pi/\omega$, i.e. $\phi_{analyte} = 0$, the variation in the intensity of luminescence due to the analyte lumophore, is, as a consequence, given by the following expression:

$$I_{L\,(HPTS)} = I_{B2} + I_{HPTS} \sin(\omega t) \qquad (27)$$

It follows that:

$$I_L(total) = I_B(total) + I_{REF} \sin(\omega t + \phi_{REF}) + I_{HPTS} \sin(\omega t) \qquad (28)$$

where $I_L(total)$ is the total background luminescence light intensity $= I_{B1} + I_{B2}$. Using the mathematic expression: $\sin(v + u) = \sin(v)\cos(u) + \cos(v)\sin(u)$, it follows that:

$$\sin(\omega t + \phi_{REF}) = \sin(\omega t)\cos(\phi_{REF}) + \cos\omega t\,\sin(\phi_{REF}) \qquad (29)$$

thus

$$I_{REF} \sin(\omega t + \phi_{REF}) + I_{HPTS} \sin(\omega t) = \sin(\omega t)(I_{HPTS} + I_{REF} \cos(\phi_{REF})) + \cos(\omega t)(I_{REF} \sin(\phi_{REF})) \qquad (30)$$

But, using the previous mathematical expression it can be shown that:

$$C \sin(\omega t) + D \cos(\omega t) = M \sin(\omega t + \phi_M) \qquad (31)$$

where $C = M \cos(\phi_M)$ and $D = M \sin(\phi_M)$ and $M = (C^2 + D^2)^{1/2}$. Comparing eqns. (31) with (30) it follows therefore:

$$I_{REF} \sin(\omega t + \phi_{REF}) + I_{HPTS} \sin(\omega t) = M \sin(\omega t + \phi_M) \qquad (32)$$

where

$$M = ((I_{HPTS} + I_{REF} \cos(\phi_{REF}))^2 + ((I_{REF} \sin(\phi_{REF}))^2)^{1/2} \qquad (33)$$

and

$$\cot(\phi_M) = \frac{I_{HPTS} + I_{REF} \cos(\phi_{REF})}{I_{REF} \sin(\phi_{REF})} \qquad (34)$$

i.e.
$$\cot(\phi_M) = \cot(\phi_{REF}) + \frac{I_{HPTS}}{I_{REF}} \times \mathrm{cosec}(\phi_{REF}) \qquad (35)$$

It also follows that

$$I_L \text{ (total)} = I_B \text{ (total)} + M \sin(\omega t + \phi_M) \tag{36}$$

i.e. the effect of adding the two varying luminescence intensities due to the reference and analyte lumophores is that the overall intensity variation with time appear phase shifted by ϕ_M with respect to the frequency of the exciting light. In addition, the value of ϕ_M, which is one of the measurable quantities in phase modulation spectroscopy, is simply related to the amplitudes of the analyte (HPTS) and reference ($Ru(dpp)_3^{2+}$), i.e. I_{HPTS} and I_{REF}, via eqn. (34). In this work the wavelength of the excitation light was selected so that only the anionic form of HPTS was excited. As a consequence, I_{HPTS} will be proportional to $[Q^+D^-xH_2O]$ and, given eqn. (22), it follows that:

$$\frac{I_{HPTS}^o - I_{HPTS}}{I_{HPTS}} = \alpha P_{CO2} \tag{37}$$

where I_{HPTS}^o is the value of the amplitude in the absence of carbon dioxide. Combining eqns. (35) and (37) it follows:

$$\frac{\cot(\phi_M^o) - \cot(\phi_M)}{\cot(\phi_M) - \cot(\phi_{REF})} = \alpha P_{CO2} \tag{38}$$

where ϕ_M^0 is the observed phase angle when all the analyte dye is in its deprotonated form, i.e. no carbon dioxide present, and ϕ_{REF} is the observed phase shift due to the reference alone, i.e. when all the analyte dye is in its protonated form, i.e. at infinite P_{CO2}, at which point ϕ_M will be at a minimum value.

Figure 3.16 illustrates the single and combined sine waves generated by the reference and analyte lumophores in (a) the absence and (b) the presence of carbon dioxide in the test phase for a DLR optical sensor for carbon dioxide, under interrogation by phase modulated fluorimetry. From this diagram it can be seen, as predicted by eqns. (36) and (38) that in the absence of carbon dioxide the signal due to the analyte dominates and the phase angle of the overall signal is nearer to that of the HPTS signal for which $\phi = 0$, rather than that of the reference. In contrast, in the presence of a high level of carbon dioxide, the signal due to the reference dominates and the overall signal is nearer to that of the reference, for which $\phi = \phi_{REF}$, than that of the analyte signal. Figure 3.17 illustrates a typical set of data reported by McCraith et al[69] for their DLR carbon dioxide film comprising: a $Ru(dpp)_3^{2+}$ reference lumophore, encapsulated in nano-beads, with an HPTS$^-$- CMTA$^+$ analyte lumophore; all encapsulated in an organically modified silica. An analysis of the data illustrated in fig. 17 reveals a good fit to eqn.(38), using values for cot (ϕM0) and cot (ϕ_{REF}) of 3.84 and 1.45, respectively.

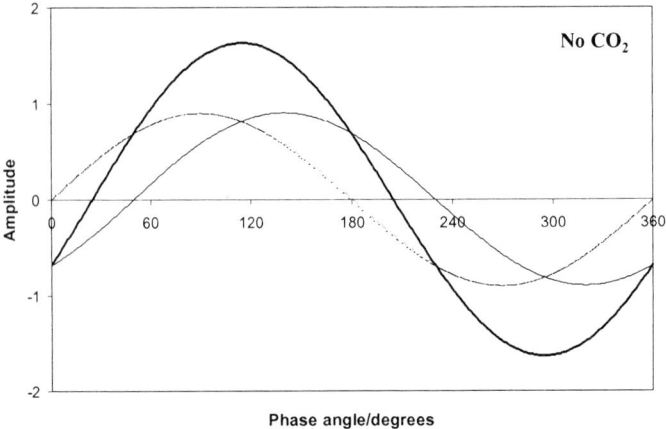

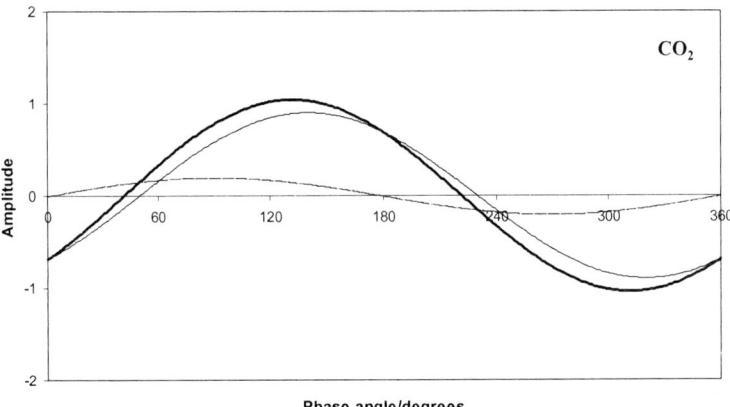

Figure 3.15. Single and combined sine waves of the luminescent intensity generated by a DLR sensor system subjected to a modulated excitation light in the absence and presence of carbon dioxide[69]. The analyte and reference single sine waves are depicted by the broken and solid lines. The overall, combined signal sine waves are denoted by the thickest solid lines.

FLUORESCENT CARBON DIOXIDE INDICATORS

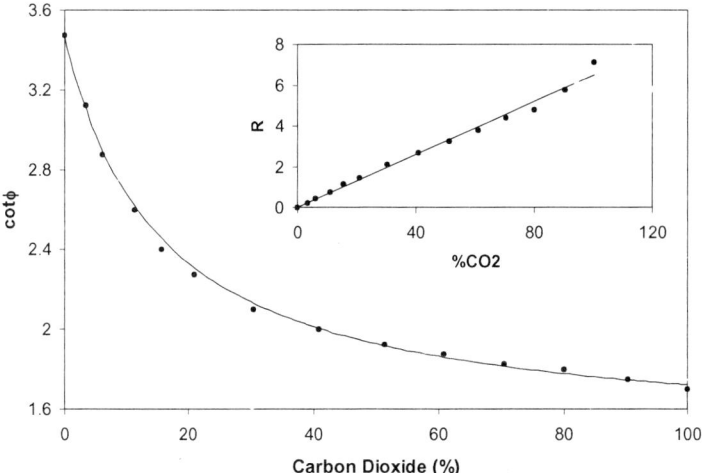

Figure 3.16. Plot of the observed variation in the contingent of the overall phase angle, cot(ϕ_M), for a Ru(dpp)$_3^{2+}$ nanobeads/HPTS-CMTA/silica DLR-based sensor as a function of P_{CO_2}. The insert diagram is a linearised form of the data in the main diagram, calculated using eqn. (38), using cot (ϕ_M) and cot (ϕ_{REF}) = 3.84 and 1.45, respectively[69].

3.3. APPLICATIONS AND PRACTICAL SYSTEMS

This article began with a brief look at the possible areas of application of optical sensors for carbon dioxide. A number of studies, carried out using dry and wet luminescent optical sensors for carbon dioxide, have illustrated the efficacy of these indicators for the detection and measurement of carbon dioxide in: blood[24], food packages[69], bioreactors[26,50] and seawater[30]. However, despite this success and promise, these sensors have been slow to take off as commercial products and the detection and analysis of carbon dioxide is still dominated by the Severinghaus electrode (dissolved work) and infrared spectroscopy (gaseous work). The reasons for this poor transition from research bench to market place are numerous but include: consumer resistance to new, and largely still unproven technology and, more seriously, basic concerns regarding the technology itself. Thus, as we have seen intensity-based measurements are fraught with several niggling and undermining problems, such as signal drift and dye bleaching, although many can be eliminated using a wavelength ratiometric method[26] as we have seen. Lifetime-based measurements were very expensive and are still considered so for many who might otherwise readily adopt this technology, such as research laboratories, despite the recent notable inroads made by workers using long-lived donor lumophores[66,68]. All carbon dioxide optical sensors are also temperature sensitive, most markedly so, and many will exhibit some sensitivity to changes in humidity or osmotic pressure. Most optical sensors for dissolved carbon dioxide measurements require a gas-permeable membrane cover, to prevent dye-leaching and ion-exchange taking place, both of which can cause such sensors to fail. Despite the above technological concerns, there are one or two examples of apparent commercial success in the transfer of optical sensors for carbon dioxide technology from the laboratory to the market place. Thus,

Yellow Springs Instrument (YSI) currently promote a wet, luminescence intensity-based carbon dioxide monitor that allows the precise, real-time measurement of dissolved carbon dioxide in situ[70]. Their YSI-8500 instrument has a range of 1-25% CO_2 and an accuracy of typically ± 5% of reading. Their sensor system exhibits only a 2% drift per week but, as you might expect given its wet nature, has a long 90% response time of <7 minutes with the recovery even longer, due to the hyperbolic response characteristics of such sensors. The sensor is not very bulky (12mm diameter with a 70-320 mm insertion depth) and utilises HPTS as the pH-sensitive lumophore. The YSI systems employs a ratiometric[26] analysis of the dye's fluorescence, exciting at two different wavelengths and monitoring the luminescence intensity at one, in order to minimise the various instrumental effects associated with such intensity-based measurements. Fig. 18 provides a schematic illustration of the sensor head for the YSI-8500, highlighting its major features. The system uses replaceable, disposable sensor capsules, thus promoting one of the attractive features of most optical sensors; the disposable nature of the transducer element.

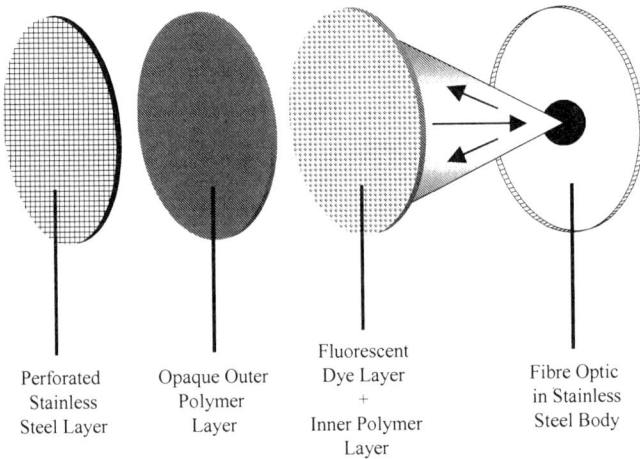

Perforated Stainless Steel Layer | Opaque Outer Polymer Layer | Fluorescent Dye Layer + Inner Polymer Layer | Fibre Optic in Stainless Steel Body

Figure 3.18. Schematic illustration[70] of the sensor head of a YSI 8500, wet optical sensor for carbon dioxide that is currently on the market.

Whereas YSI have opted for a wet, intensity-based optical sensor for carbon dioxide, OceanOptics, with their FCO2-R fibre optical sensor, have opted for a dry sensor which encapsulates the pH-sensitive dye, HPTS, in a sol-gel medium, and is covered by a black silicone GPM coating[71]. The sensor has a dynamic range of 0-25% CO_2 and a resolution of at least 0.03%. The instrumentation does not use a ratiometric technique (i.e. no two excitation sources) to interrogate the sensor film as employed by YSI, but instead monitors the whole luminescence spectral output using a miniature spectrophotometer. Like the YSI-8500, the FCO2-R uses a blue LED as the excitation source. The response time of the FCO2-R is typically 10 minutes and the probe itself is quite small (1.6 mm) diameter. However, as might be expected for a sol-gel based sensor, it must be stored in water at least 2 days before use and it is recommended to store it subsequently in water at all times after.

It is possibly surprising that there are no major lifetime-based optical sensors in the market, despite the fact that lifetime based optical sensors for oxygen already exist (e.g. OxySense™). However, it is most likely that the higher cost of such technologically advanced systems is still proving a major barrier to its market transition. The recent move towards cheaper, lower frequency, diode-based phase modulated systems, whether they be FRET- or DLR-based, may eventually lead to a relatively inexpensive, commercially viable product that challenges and widens the market for small intensity-based optical sensors for carbon dioxide. However, for this to happen a great deal more research is necessary. It remains to be seen, therefore, what the commercial future of optical sensors for carbon dioxide is which a slightly disappointing conclusion is given its history of initial rapid development and the importance of the area of analysis to life, industry and the environment.

3.4. LIST OF TERMS AND SYMBOLS

AcFl: Acryloly florescein
BMUB: β-methyl umbelliferon
BTB: Bromothymol blue
c-SNAFL: 5'and 6'-Carboxyseminaphtholfluorescein
c-SNARF: 5'and 6'-Carboxyseminaphtholrhodamine
CR: Cresol Red
CTA^+: Hexadecyl trimethyl ammonium cation
CTAH: Hexadecyl trimethyl ammonium hydroxide
DLR: Dual lumophore referencing
DNPA: 2-(2,4-Dinitrophenylaxo)-1-naphthol-3,6disulphonic acid
EC: Ethyl cellulose
$[Eu(tta)_3]$: tris(thenoyltrifluoroacetonato) europium (III)
Fl: Fluorescein
FRET: Fluorescence energy resonance transfer
GPM: gas-peremeable membrane
HCA: 7-hydroxycoumarin-4-acetic acid
HEMA: 2-Hydroxy methylmethacrylate
HPTS: 1, Hydroxypyrene-3,6,8-trisulphonic acid
MAPTAC: methacrylamidopropyl trimethyl ammonium chloride
MCP: *meta*-Cresol purple
NMA: 1-naphthylmetylamine
NR: Neutral red
NVP: Poly(N-vinylpyrrolidone)
PolyHEMA: poly(hydroxyethyl methacrylate)
PR: Phenol Red
PTA: Phase Transfer Agent
PVP: Poly(vinylpyrrolidone)
R6G: Rhodamine 6G
$Ru(dph-bpy)_3^{2+}$: Ruthenium(II) tris[4,4'-diphenyl-2,2'-bipyridyl] cation
$Ru(dpp)_3^{2+}$: Ruthenium (II) tris(4,7-diphenyl-1,10-phenanthroline) cation
$Ru(pzth)_2^{2+}$: Ruthenium(II) tris[2-(2-pyrazinyl)thiazole] cation
SRh: Sulforhodamine 101
TB: Thymol blue
TBAH: tetra butyl ammolnium hydroxide
TBP: Tributyl phosphate
$TDMA^+$: tridodecylmethylammonium cation
TMAH: Tetramethyl ammonium hydroxide
TMS^-: 3-Trimethylsilyl-1-propane sulphonate anion
THR: Texas Red hydrazine
TOAH: tetra octyl ammonium hydroxide

3.5. REFERENCES

1. E. Rabinowitch. and Govindjee, *Photosynthesis* (Wiley, New York, 1969).
2. C.L. Lake, *Clinical Monitoring* (W.B. Saunders Co., Philadelphia, 1990).
3. J.S. Gravenstein, *Gas Monitoring and Pulse Oximetry* (Butterworth-Heinemann, Boston, 1990).
4. D.G. Mou, Process dynamics: instrumentation and control, *Biotech. Adv.*, **1**, 229-245 (1983).
5. M.L. Rooney, *Active Food Packaging* (Blackie Academic & Professional, London, 1995)
6. R.P. Wayne, *Chemistry of Atmospheres*, 3rd Edition, (Oxford University Press, Oxford, 2000).
7. J.W. Severinghaus and A.F. Bradley, Electrodes for blood P_{O2} and P_{CO2} determination, *J. Appld. Physiol.*, **13**, 515-520 (1958).
8. M.A. Jensen and G.A. Rechnitz, Reponse characteristics of the pCO_2 electrode, *Anal. Chem.*, **51**, 1972-1977 (1979).
9. W.R. Seitz, Chemical sensors based on fibre-optics, *Anal. Chem.* **56**, 16A-34A (1984).
10. Fiber Optical Chemical Sensors and Biosensors, volume 1, edited by O.S. Wolfbeis (CRC Press, Boca Raton, Florida, 1991),
11. Fiber Optical Chemical Sensors and Biosensors, volume 2, edited by O.S. Wolfbeis (CRC Press, Boca Raton, Florida, 1991),
12. G. Rao, S. B. Bambot, C.W. Kwong, H. Szmacinski, J. Sipior, R. Holavanahali and G. Carter, Application of fluorescence sensing to bioreactors, in:*Topics in Fluorescence Spectroscopy. Volume 4: Probe Design and Chemical* Sensing, edited by J.R. Lakowicz (Plenum,, New York, 1994), pp 417-448.
13. H.N. McMurray and J. Albadran, Colorimetric and fluorimetric polymer membrane gas-sensing materials, *MRS Bulletin*, 55-59 (1999).
14. O.S. Wolfbeis, Fibre-optic chemical sensors and biosensors, *Anal. Chem.*, **72**, 81R-89R (2000).
15. A. Mills and K. Eaton, Optical sensors for carbon dioxide: an overview of sensing strategies past and present, *Quim. Anal.*, **19**, 75-86 (2000).
16. O.S. Wolfbeis, Fibre-optic chemical sensors and biosensors, *Anal. Chem.*, **74**, 2663-2678 (2002).
17. D.W. Lübbers and N. Opitz, Die pCO_2/pO_2-Optode: Eine neue pCO_2/bzw pO_2 meßsonde zur messung des pCO_2 oder pO_2 von gasen und flüssigkeiten, *Naturforsch.*, **30c**, 532-533 (1975)
18. D.W. Lübbers and N. Opitz, Blood gas analysis with fluorescent dyes as an example of their usefulness as quantitative chemical sensors, *Anal. Chem. Symp. Ser.*, **17**, 609-619 (1983).
19. D.W. Lübbers and N. Opitz, Quantitative fluorescence photometry with biological fluids and gases, *Adv. Exp. Med. Biol.*, **75**, 65-68 (1976).
20. N. Opitz and D.W. Lübbers, Compact CO_2 gas analyser with favourable signal-to-noise ratio and resolution using special fluorescent sensors (optodes) illuminated by blue LED's, *Adv. Exp. Med. Biol.*, **180**, 757-762 (1983).
21. Z. Zhujun and W.R. Seitz, A carbon dioxide sensor based on fluorescence, *Anal. Chim. Acta*, **160**, 305-309 (1984).
22. J.A. Ferguson, B.G. Healey, K.S. Bronk, S.N. Barnard and D.R. Walt, Simultaneous monitoring of pH, CO_2 and O_2 using an optical imaging fibre, *Anal. Chim. Acta*, **340**, 123-131 (1997).
23. O.S. Wolfbeis, L.J. Weis, M.J.P. Leiner and W.E. Ziegler, Fibre-optic fluorosensor for oxygen and carbon dioxide, *Anal. Chem.* **60**, 2028-2030 (1988).
24. T. Hirschfeld, F. Miller, S. Thomas, H. Miller, F. Milanovich and R.W. Gaber, Laser-fibre-optic "optrode" for realtime *in vivo* blood carbon dioxide level monitoring, *J. Lightwave Technol.*, L-**5**, 1027-1033 (1987).
25. C. Munkholm, D.R.Walt and F.P. Milanovich, A fibre-optic sensor for CO_2 measurement, *Talanta*, **35**, 109-112 (1988).
26. M. Uttamlal and D.R. Walt, A fibre-optic carbon dioxide sensor for fermentation monitoring, *Biotechnol.*, **13**, 597-601 (1995).
27. M.J.P. Leiner, Optical sensors for *in vitro* blood-gas analysis, *Sensors and Actuators B*, **29**, 169-173 (1995).
28. M.J.P. Leiner, Luminescence chemical sensors for biomedical applications: scope and limitations, *Anal. Chim. Acta*, **255**, 209-222 (1991)
29. J.W. Parker, O. Laksin, C. Yu, M-L. Lau, S. Klima, R. Fisher, I. Scott, and B.W. Atwater, Fibre-optic sensors for pH and carbon dioxide using a self-referencing dye, *Anal. Chem.*, **65**, 2329-2334 (1995).
30. M.B.Tabacco, M. Uttamlal, M. McAllister and D.R. Walt, An autonomous sensor and telemetry system for low-level pCO_2 measurements in seawater, *Anal. Chem.*, **71**, 154-161, (1999).
31. D.R. Walt, G. Gabor, and C. Goyet, Multiple-indicator fibre-optic sensor for high-resolution pCO_2 seawater measurements, *Anal. Chim. Acta*, **274**, 47-52 (1993).
32. J.R. Lakowicz, H. Szmacinski and M. Karakelle, High-stability non-invasive autoclavable naked optical CO_2 sensor, *Anal. Chim. Acta*, **272**, 179-186 (1993).

33. G. Orellana, N. C. Morino-Bondi, B. Segovia, and M.D. Marazuela, Fibre-optic sensing of carbon dioxide based on excited-state proton transfer to a luminescent ruthenium (II) complex, *Anal. Chem.* **64**, 2210-2215 (1992).
34. M.D. Marazuela, N.C. Moreno-Bondi and G. Orellana, Enhanced performance of a fibre-optic luminescent CO_2 sensor using carbonic anhydrase, *Sensors and Actuators B*, **29**, 126-131 (1995).
35. M.D. Marazuela, M.C. Moreno-Bondi and G. Orellana, Luminescence lifetime quenching of ruthenium (II) polypyridyl dye for optical sensing of carbon dioxide, *Applied Spectroscopy*, **52**, 1314-1320 (1998).
36. O.S. Wolfbeis, E. Fürlinger, H. Kroneis and H. Marsoner, Fluorimetric analysis: a study on fluorescent indicators for measuring near-neutral ("physiological") pH-values, *Fresenius Z Anal. Chem.*, **314**, 119-124 (1983).
37. A. Mills and Q. Chang, Fluorescence plastic thin-film sensor for carbon dioxide, *Analyst*, **118**, 839-843 (1993).
38. A.Mills and Q. Chang, Modelled diffusion-controlled response and recovery behaviour of a naked optical film sensor with a hyperbolic-type response to analyte concentration, *Analyst*, **117**, 1461-1466 (1992).
39. H. Szmacinski and J.R. Lakowicz, Lifetime-based sensing, in:*Topics in Fluorescence Spectroscopy. Volume 4: Probe Design and Chemical Sensing*, edited by J.R. Lakowicz (Plenum,, New York, 1994), pp 295-334.
40. A. Mills, Response characteristics of optical sensors for oxygen: models based on a distribution in τ_o or k_q, *Analyst*, **124**, 1301-1308 (1999).
41. A. Mills, Response characteristics of optical sensors for oxygen: model based on a distribution in τ_o and k_q, *Analyst*, **124**, 1309-1314 (1999).
42. P. Herman, Z. Murtaza and J.R. Lakowicz, Sensing of carbon dioxide by a decrease in photoinduced electron transfer quenching, *Anal. Biochem.*, **272**, 87-93 (1999).
43. D. B. Raemer, D.R. Walt and C. Munkholm, CO_2 indicator for placement of tracheal tubes, US Patent No. 5,005,572 (1991).
44. A. Mills and Q. Chang, Carbon dioxide detector, US patent No. 5,480,611 (1996).
45. A.Mills, Q. Chang, and N. McMurray, Equilibrium studies on colorimetric plastic film sensors for carbon dioxide, *Anal. Chem.* **64**, 1383-1389 (1992).
46. A.Mills, G. Chang, and N. McMurray, Equilibrium studies on colorimetric plastic film sensors for carbon dioxide, *Anal. Chem.* **64**, 1383-1389 (1992).
47. B.H. Weigl and O.S. Wolfbeis, Sensitivity studies on optical carbon dioxide sensors based on ion pairing, *Sensors and Actuators B*, **28**, 151-156 (1995).
48. A. Mills and L. Wild, Measurement of dissolved carbon dioxide using colourimetric polymer films, in: *Proceedings of Medical Sensors and Fibre Optic Sensors and Delivery Systems volume 2631*, edited by G. Orellana and M. A. Scheggi (SPIE , Barcelona, 1995), pp. 100-109.
49. A. Mills, A. Lepre and L. Wild, Breath-by-breath Measurement of Carbon Dioxide Using a Plastic Film Optical Sensor, *Sensors and Actuators B*, **38-39**, 419-425 (1997).
50. Q. Chang, L. Randers-Eichhorn, J.R. Lakowicz and G. Rao, Steam-sterilisable fluorescence lifetime-based sensing film for dissolved carbon dioxide, *Biotechnol. Prog.*, **14**, 326-331 (1998).
51. Y. Kawabata, T. Kamachika, T. Imasaka and N. Ishibashi, Fibre-optic sensor for carbon dioxide with a pH indicator dispersed in a poly(ethylene glycol) membrane, *Anal. Chim. Acta*, **219**, 223-229 (1989).
52. A. Mills and Q. Chang, Carbon dioxide detector, US patent No. 5,480,611 (1996).
53. C. Munkholm, Method for activation of polyanionic fluorescent dyes in low dielectric media with quaternary onium compounds, U.S. Patent No. 5,387,525 (1995).
54. P. Müller and P.C. Hauser, Fluorescence optical sensor for low concentrations of dissolved carbon dioxide, *Analyst*, **121**, 339-343 (1996).
55. O.S. Wolfbeis, B. Kovacs, K. Goswami and S.N. Klainer, Fibre-optic fluorescence carbon dioxide sensor for environmental monitoring, *Mikrochim. Acta*, **129**, 181-188 (1998).
56. C. Malins and B. D. MacCraith, Dye-doped organically modified silica glass for fluorescence-based carbon dioxide gas detection, *Analyst*, **123**, 23373-2376 (1998).
57. C. Malins, M. Niggermann and B.M. MacCraith, Multi- and light-optical chemical sensor employing a plastic substrate, *Meas. Sci. Technol.*, **11**, 1105-1110 (2000).
58. N. Nakamura and Y. Amao, Optical sensor for carbon dioxide combining colorimetric change of a pH indicator and a reference luminous dye, *Anal. Bioanal. Chem.*, **376**, 642-646 (2003).
59. N. Nakamura and Y. Amao, An optical sensor for CO_2 using thymol blue and europium (III) complex composite film, *Sensors and Actuators B*, **92**, 98-101 (2003).
60. N. Nakamura and Y. Amao, Optical CO_2 sensor for the combination of colorimetric change of pH indicator and internal reference luminous dye, *Bull. Chem. Soc. Jpn.*, **76**, 1459-1462 (2003).
61. K. Ertekin, I. Klimant, G. Neurauter and O.S. Wolfbeis, Characterisation of a reservoir-type capillary optical microsensor for pCO_2 measurements, *Talanta*, **59**, 261-267 (2003).

62. X. Ge, Y. Kostov and G. Rao, High-stability non-invasive autoclavable naked optical CO_2 sensor, *Biosensors and Bioelectronics*, **18**, 857-865 (2003).
63. D.A. Nivens, M.V. Schiza, S.N. Angel, Multilayer sol-gel membranes for optical sensing applications: single layer pH and dual layer CO_2 and NH_3 sensors, *Talanta*, **58**, 543-550 (2002).
64. J. Sipior, S. Bambot, R.M. Smith, G.N. Carter, J.R. Lakowicz and G. Rao, A lifetime-based optical CO_2 gas sensor with blue or red excitation and Stokes or anti-Stokes detection, *Anal. Biochem.*, **227**, 309-318 (1995).
65. J. Sipior, L. Randers-Eichhorn, J.R. Lakowicz, G.M. Carter and G. Rao, Phase fluorimetric optical carbon dioxide gas sensor for fermentation off-gas monitoring, *Biotechnol. Prog.*, **12**, 266-271 (1996).
66. G. Neurauter, I. Klimant and O.S. Wolfbeis, Microsecond lifetime-based optical carbon dioxide sensor using luminescence resonance energy transfer, *Anal. Chim. Acta*, **382**, 67-75 (1999).
67. G. Liebsch, I. Klimant, B. Frank, G. Holst, and O.S. Wolfbeis, Luminescence lifetime imaging of oxygen, pH, and carbon dioxide distribution using optical sensing, *Applied Spectroscopy*, **54**, 548-559 (2000).
68. C. Von Bültzingslöwen, A. K. McEvoy, C. McDonagh, and B.D. MacCraith, Lifetime-based optical sensor for high-level pCO_2 detection employing fluorescence resonance energy transfer, *Anal. Chim. Acta*, **480**, 275-283 (2003).
69. C. Von Bültzingslöwen, A.K. McEvoy, C. McDonagh, B.D. MacCraith, I. Klimat, C. Krause and O. Wolfbeis, Sol-gel based optical carbon dioxide sensor employing dual lumiphore referencing for applications in food packaging technology, *Analyst*, **127**, 1478-1483 (2002).
70. www.ysi.com
71. www.OceanOpticsBV.com

FLUORESCENCE-BASED NITRIC OXIDE DETECTION

Scott A. Hilderbrand, Mi Hee Lim, and Stephen J. Lippard[*]

4.1. INTRODUCTION

Nitric oxide (NO) is a neutral free radical gas. The NO molecule has long been recognized as an environmental contaminant and a potential health hazard in the atmosphere. It was not until recently that beneficial roles for NO were discovered in biological systems.[1-3] In the years since the groundbreaking discovery of NO signaling in biology, further work has shown that NO is a ubiquitous messenger in the cardiovascular, immune, and nervous systems.[1, 4-12] Even though NO is a relatively stable species with reported half-lives under physiological conditions of up to five seconds,[5, 13, 14] it readily reacts with a variety of species commonly found in living organisms. The targets of NO include dioxygen, oxygen radicals, thiols, amines, and transition metal ions.[15] The reactions of NO with dioxygen and superoxide result in formation of reactive nitrogen oxide species (RNOS) NO_2 and $ONOO^-$, respectively. Both products are more reactive than NO itself. In aqueous environments, the reactions of NO with dioxygen can also yield NO_2^-.

In vivo, NO is synthesized by the enzyme nitric oxide synthase (NOS), which catalyzes the transformation of L-arginine to L-citrulline in the presence of dioxygen, releasing NO (Scheme 4.1).[16, 17] There are three isoforms of NOS found in humans. Nitric oxide produced by the first two forms, neuronal NOS (nNOS) and endothelial NOS (eNOS), is believed to be involved in a variety of regulatory functions in the central nervous system (CNS) and vasculature. The other isoform, inducible NOS (iNOS), is activated only under certain conditions and produces higher concentrations of NO for a short period of time. The micromolar levels of NO produced by iNOS, which result in

---

Scott A. Hilderbrand, Mi Hee Lim, and Stephen J. Lippard · Department of Chemistry, Massachusetts Institute of Technology, 77 Massachusetts Avenue, Cambridge, MA 02139. *Author to whom correspondence should be addressed; Tel (617) 253-1892, Fax: (617) 258-8150, Email: lippard@mit.edu

formation of RNOS, have been implicated in carcinogenesis and several neurodegenerative disorders such as Alzheimer's disease, Parkinson's disease, multiple sclerosis, and Huntington's disease.[4, 18, 19] In lower concentrations, where formation of RNOS is minimal, NO has beneficial roles in vasodilation, protection against bacterial invaders, and possibly in long term potentiation (LTP).[1, 9-12, 15, 20] The concentration-dependent lifetime of NO, and its ability to diffuse passively through cellular membranes, make it difficult to decipher the biological functions of nitric oxide. Under certain conditions NO has a lifetime of up to 10 minutes in solution.[21] During this time it has an estimated diffusion range of 100-200 μm from its point of origin.[22] Such a diffusion range corresponds to an area that covers approximately two million synapses.[23] Thus NO produced by one cell may have a variety of effects on both neighboring and more distant cells.

In the central nervous system, NO may play a role in LTP, the process by which neuronal connections are re-enforced and the basis for memory.[20] LTP requires a neurotransmitter that is capable of diffusing from the postsynaptic neuron to the presynaptic neuron (Figure 4.1). Studies have indirectly identified NO as a possible candidate for this retrograde neurotransmitter.[20, 24, 25] A submaximal presynaptic stimulus results in the activation of nNOS. The NO produced does not react in the postsynaptic neuron but is believed to diffuse back into the presynaptic neuron where it activates guanylate cyclase. The cGMP produced by guanylate cyclase starts a signal cascade that ultimately results in the release of more neurotransmitter by the presynaptic neuron. The net effect of this feedback loop is to strengthen the neuronal connection. However, there is currently no method to probe directly the involvement of NO in LTP.

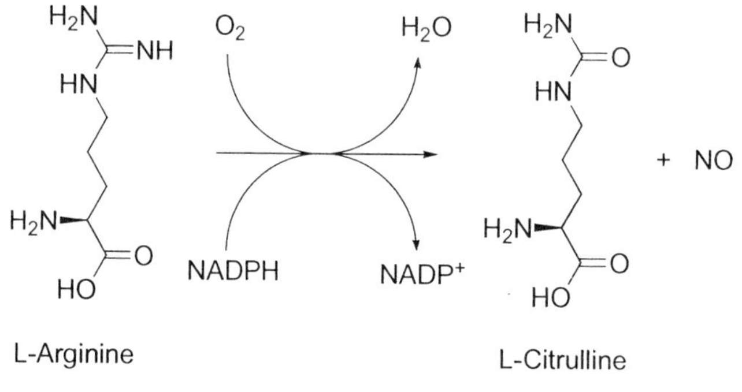

Scheme 4.1. Synthesis of NO from L-arginine.

4.2. CURRENT NON-FLUORESCENT DETECTION METHODS

To investigate adequately the role of NO in LTP, a suitable sensor is necessary to detect the presence of NO directly. Over the past two decades, several NO sensors have been developed, but all have their limitations. The Griess assay, one of the earliest NO detection methods, is not suited for in vivo NO detection. This assay provides an estimate of total NO production by measuring nitrite and is incapable of giving real time measurements of NO.[26] Although they are more sensitive than the Griess assay, chemiluminescence methods require purging of aqueous samples with an inert gas to transfer NO to an analyzer and are therefore not easily adaptable to in vivo detection.[27] Electrochemical-based sensors provide real time direct detection, but they only give information on NO concentrations directly at the electrode tip.[28-30] With these and any other invasive detection methods, it is necessary to puncture the cell to obtain measurements. EPR-based methodologies that use iron dithiocarbamate complexes as spin traps have been developed, but they only have μM sensitivity for NO.[31-34] A more detailed review of these non-fluorescence-based NO imaging techniques has recently been published.[35]

Although these methods have been valuable for elucidating NO-related biochemistry, new methods with improved characteristics are necessary. None of the current NO detection methodologies are well suited for investigating the roles that NO may play in the central nervous system. In particular, we require non-invasive techniques that are capable of selective and direct detection of NO and which provide both temporal and spatial information without requiring complicated instrumentation not commonly available in biochemical research labs. One approach to provide these properties is through the use of fluorescence methodologies.

The suitability of fluorescence-based sensors for in vivo work is exemplified by the Ca^{2+} sensors developed by Tsien.[36, 37] and by the Znpyr family of Zn^{2+} sensors prepared in our lab.[38-41]

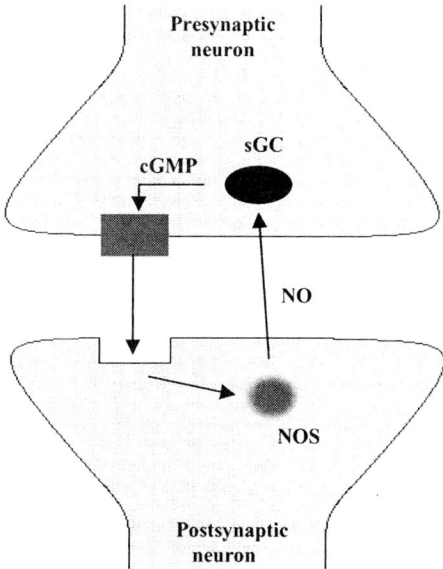

Figure 4.1. Illustration of the postulated function of NO as a retrograde neurotransmitter, where it is proposed to diffuse from the postsynaptic neuron back to the presynaptic neuron.

4.3. EARLY FLUOROMETRIC IMAGING OF NO

The first fluorescence-based NO sensors were originally prepared for the measurement of nitrite ion in solution. The sensor molecule, 2,3-diaminonaphthalene (DAN) undergoes diazotization at one of the amines under acidic conditions in the presence of nitrite. Following the diazotization, 2,3-naphthotriazole is formed (Scheme 4.2).[42] After an aqueous workup, the triazole species can be detected by fluorometric analysis. In the work of Damiani and Burini, the optimal pH for triazole formation was determined to be 1.6 and the maximal fluorescence emission was obtained at a pH of 11.65.[43] Nitrate ion can also be detected by this methodology if it is first reduced to nitrite.[44] Since NO in aqueous solution can lead to the formation of nitrite, DAN was also used to monitor nitrosative stress in *E. coli*.[45] Later, it was determined that nitrosation of DAN and ultimately triazole formation in *E. coli* proceeds by formation of strong nitrosylating agents such as N_2O_3 or N_2O_4, which arise from the rapid reaction of NO with O_2.[46] Research in the Tannenbaum laboratory with immortalized macrophages stimulated with *E. coli* lipopolysaccharide (LPS) and γ-interferon is also consistent with oxidation of NO to form N_2O_3 and N_2O_4.[47] The connection between 2,3-naphthotriazole formation and NO synthesis was further confirmed by research at Monsanto. Application of the NOS inhibitor, aminoguanidine, to LPS activated macrophages resulted in decreased nitrite formation as monitored by DAN using a protocol adapted from Damiani and Burini.[48, 49] In 1993, a procedure was developed by Grisham and coworkers in which NO could be monitored via RNOS and subsequent triazole formation with DAN at pH 7.4 followed by fluorometric analysis of the basified reaction mixture.[50] It was also observed that formation of 2,3-naphthotriazole is pH dependent, presumably a result of the pH dependence for formation of the nitrosamine intermediate.[50] DAN, however, is not suitable for detection of NO produced by stimulated porcine pulmonary artery endothelial cells, which contain eNOS in contrast to the iNOS in the macrophage cell lines.[51] DAN was unable to detect NO production from the stimulated endothelial cells, whereas other methodologies such as chemiluminescence analysis, which reports NO, NO_2^-, and NO_3^-, collectively known as NO_x, indicated significant increases in NO_x levels.[51] In 1995 Hirobe proposed mechanisms for triazole formation with N_2O_3 and N_2O_4, formed by the reaction of NO with O_2 (Scheme 4.3).[52] The study showed that NO_2^- does not react with amines in organic solvents or in neutral aqueous solutions.[52] Although DAN can be used to monitor NO production by means of reactive products formed by its reaction with O_2, it is not suited for observation of NO in living cells. Because of its non-polar nature, DAN leaks out of cells after loading.[53]

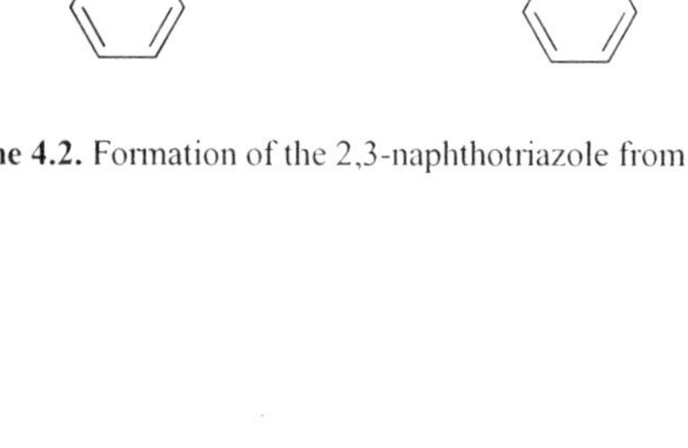

Scheme 4.2. Formation of the 2,3-naphthotriazole from NO_2^-.

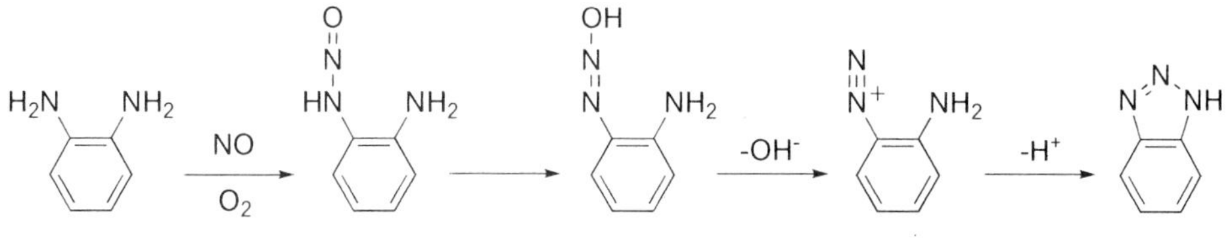

Scheme 4.3. Formation of the 1,2-triazole from NO and O_2.

The first steps towards modification of the diamine-based compounds for intracellular NO visualization were undertaken by Nagano and coworkers. To overcome the issue of sensor leakage from the cells after loading, DAN was derivatized to add an ester moiety (Scheme 4.4).[53] Ester derivatization is a well known approach to afford trappable molecules that do not diffuse out of cells after loading.[54] Cell permeable DAN-1-EE is hydrolyzed into the non-permeable DAN-1 by intracellular esterases, thus trapping the probe inside the cell.[53] Once hydrolyzed inside the cell, DAN-1 was used to image NO production in activated vascular smooth muscle cells isolated from male Wistar rats. The formation of the DAN-1 triazole species was monitored by fluorescence microscopy with excitation of the 360 nm absorption band of the triazole, and its emission at 447 nm was monitored by using a 420 nm long-pass emission filter. The absence of a proton on the DAN-1 triazole functionality eliminates the pH dependence observed with 2,3-naphthotriazole, where the fluorescence emission was maximized at alkaline pH.[53] Although DAN-1 EE can be used indirectly to visualize NO production inside living cells, it suffers from high background resulting from autofluorescence of intracellular components when excited with UV light.[53]

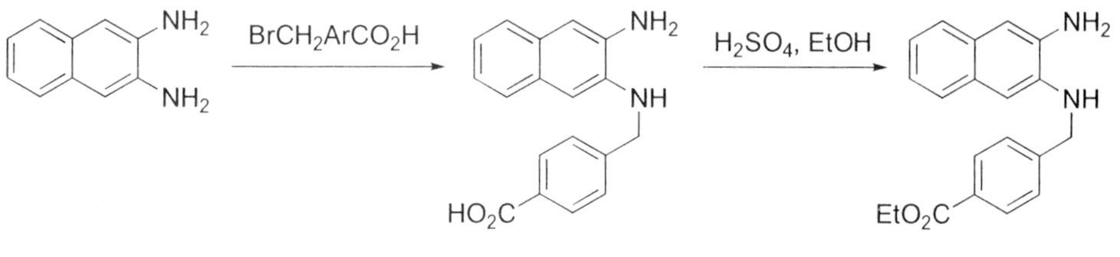

Scheme 4.4. Preparation of the cell-permeable DAN-1 EE.

4.4. THE DIAMINOFLUORESCEIN PLATFORM

To improve the fluorescence properties of the *o*-diamine-based compounds for intracellular NO detection, fluorescein was introduced as a scaffold. It was previously reported that conversion of the analinic amine on fluoresceinamine to an amide resulted in increased fluorescence emission.[55] This result implies that conversion of the electron-releasing amine to a less donating group will restore the fluorescein fluorescence. To investigate this possibility, a series of diaminofluoresceins (Figure 4.2) were prepared and their reactivity with NO in the presence of O_2 was investigated.[56] Presumably the low quantum yields ($\Phi = 0.002$-0.007) of the diaminofluorescein-based sensors are the result of photoinduced electron transfer (PET) quenching of the fluorophore excited state by the free amine groups. Conversion of the amines on a diaminofluorescein to a triazole results in enhanced fluorescence emission (Φ = 0.53-0.92). Of the diamines investigated, DAF-2 has the most desirable fluorescence properties with a quantum efficiency of 0.005 for the diamine and 0.92 for the triazole derivative.[56] The excitation band at 486 nm does not require UV light, reducing possible interference caused by autofluorescence.

The fluorescence emission from the triazole forms of the DAF sensors are, however, dependent upon pH. Fluorescence emission from the triazole, DAF-2 T, decreases rapidly as the pH falls below 7.[56] This property is in part due to the pK_a of 6.27±0.02 for the phenolic proton on DAF-2 T.[56] More recently, a new derivative DAF-FM, which contains fluoro substituents at the 2' and 7' positions of the xanthene ring and is N-methylated, was prepared to address the issue of pH dependence.[57] Fluorination lowers the pK_a responsible for the decrease in fluorescence and improves the overall photostability of the fluorophore.[58] As observed with 2,3-naphthotriazole, the triazole proton may contribute to the observed pH instability of the diaminofluorescein dyes. Introduction of a N-methyl group will result in formation of a triazole that is not protonated after reaction with NO and O_2.[57] With these modifications, the pK_a of the phenolic OH of DAF-FM T is lowered to 4.38±0.05 and the fluorescence intensity is stable above pH 5.8.[57] The DAF sensors can be converted to their diacetate derivatives, which results in formation of a lactone ring, giving a nonpolar dye that can penetrate cell membranes and enter the cytosol (Figure 4.3). Subsequent hydrolysis of the acetates by intracellular esterases traps the sensor inside the cell in a manner similar to DAN-1 EE. Sensors based on the o-diamine functionality have become the most widely used fluorescence-based methodology for imaging intracellular NO, their chief liability being the requirement for O_2, which itself reacts with NO.

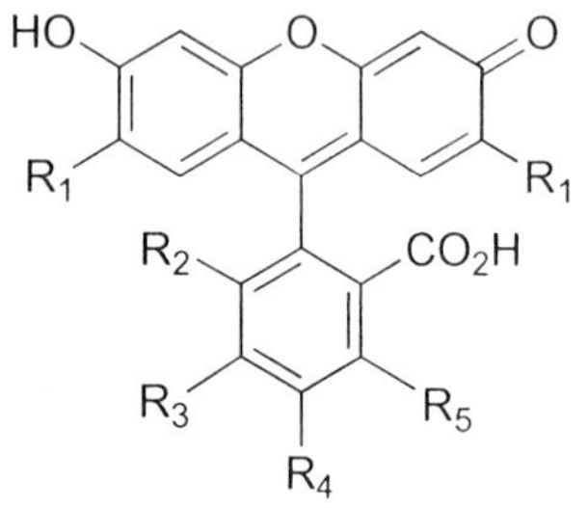

	R_1	R_2	R_3	R_4	R_5
DAF-1	H	H	H	NH_2	NH_2
DAF-2	H	H	NH_2	NH_2	H
DAF-3	H	NH_2	NH_2	H	H
DAF-4	Cl	H	H	NH_2	NH_2
DAF-5	Cl	H	NH_2	NH_2	H
DAF-6	Cl	NH_2	NH_2	H	H

Figure 4.2. A family of diaminofluorescein derivatives.

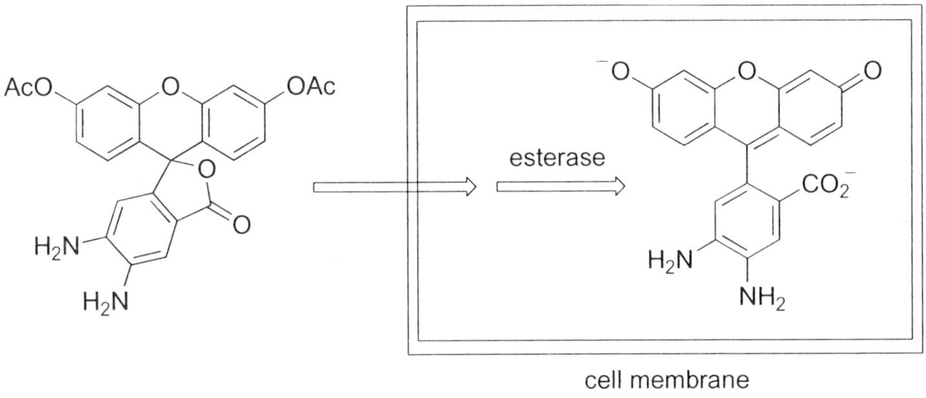

Figure 4.3. Schematic representation of DAF-2 DA, its diffusion across the cell membrane, and subsequent hydrolysis by intracellular esterases to afford DAF-2.

4.5. OTHER o-DIAMINE NO SENSORS

Several related o-diamine-based sensors have been prepared in recent years. Nagano and coworkers have developed a series of diaminorhodamine analogs to the DAF sensors.[59, 60] These rhodamine-based sensors do not display the same pH-dependent fluorescence and have stable emission above pH 4. The detection limit of DAR-4M is 7 nM NO, which is slightly higher than the 3 nM detection limit observed for DAF-FM.[57, 59] Probes based on the anthracene, coumarin, acridine, and BODIPY fluorophores have also been synthesized (Figure 4.4).[61, 62] These sensors are synthetically more accessible and their fluorescence emission should not be substantially affected by pH variations. Additional derivatives including 5,6-diamino-1,3-naphthalene disulfonic acid[63] and 1,2-diaminoanthraquinone[64] have been investigated, but offer no significant improvement over the existing o-diamine based sensors. Even though the o-diamine-based sensors react with NO^+ equivalents such as N_2O_4 or N_2O_3, they can still be used to monitor intracellular NO, although it may be difficult to draw accurate conclusions due to the lack of direct reaction with NO. The following section gives details on the various cellular systems in which o-diamine-based NO sensors and, in particular, DAF-2 have been used for the detection of biological NO.

4.6. BIOLOGICAL NO DETECTION WITH o-DIAMINE-BASED SENSORS

Initial work by Nagano and coworkers demonstrated the utility of DAF-2 for detecting intracellular NO. In non-activated macrophages, substantial fluorescence response due to triazole formation was observed 4.2 min after treatment with DAF-2, this result reflects the elevated NO production in these cells even without stimulation from LPS.[56, 65] It was also possible to image intracellular NO in activated aortic smooth muscle cells following incubation with DAF-2 diacetate, as monitored by confocal laser scanning microscopy. In these studies, there was no indication of any cytotoxicity as a result of administration of the probes.[56, 65]

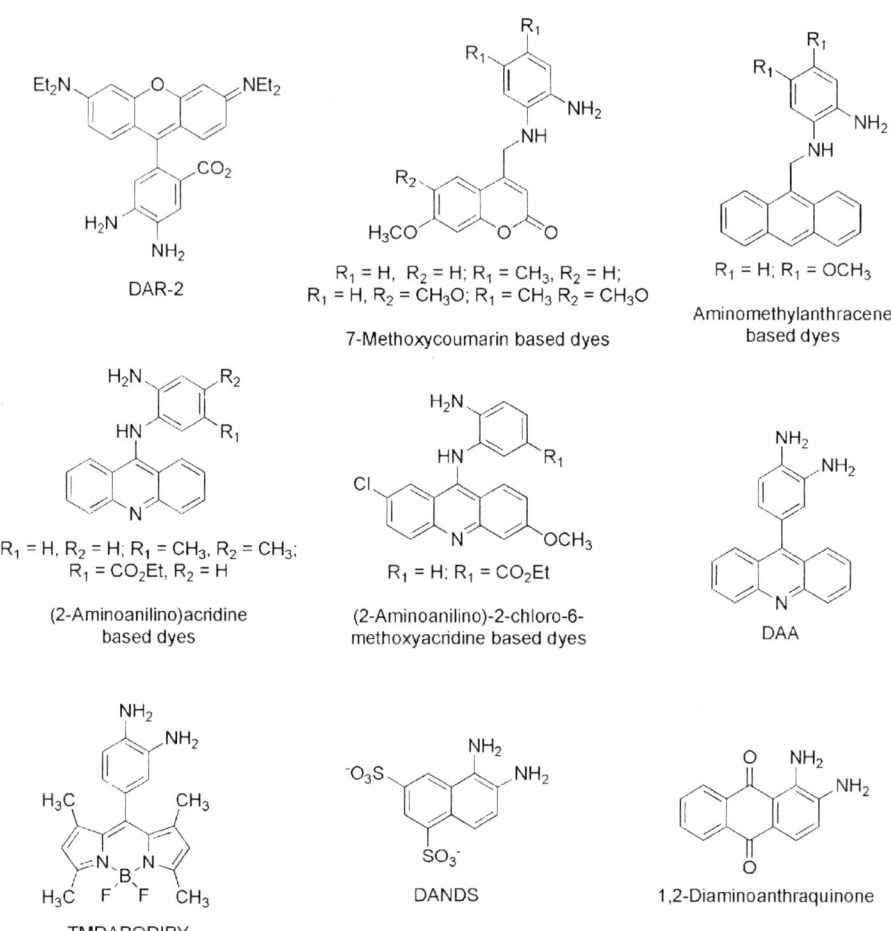

Figure 4.4. Fluorophore systems functionalized with an *o*-diamine moiety to act as NO sensors.

Methodologies for the use of DAF-2 diacetate to detect intracellular NO in non-activated endothelial cells have also been developed. The non-activated endothelial cells show low levels of NO synthesis from eNOS in contrast to the elevated NO synthesis by iNOS in macrophages. Dirsch and coworkers discovered that, by incubating endothelial cells with low (0.1 μM) concentrations of DAF-2 coupled with subtraction of the initial DAF-2 autofluorescence from the measured total fluorescence, it was possible to observe basal NO production.[66]

The validity of using o-diamine-based sensors such as DAF-2 for quantitative intracellular NO measurements has been questioned. Since it does not react directly with NO, but rather with a RNOS, the kinetic and thermodynamic properties underlying the oxidation of NO by O_2 under physiological conditions may have a significant impact on the observed spatial and temporal intracellular fluorescence response. DAF-2 only directly observes the local distribution of the highly reactive species, N_2O_3 and N_2O_4, which are not necessarily the same as the intracellular distribution of NO. Additionally, other intracellular species such as superoxide (O_2^-) may have a dramatic effect on the observed fluorescence response of the DAF sensors. In the presence of either horseradish peroxidase and H_2O_2 or peroxynitrite, DAF-2 is oxidized to an unstable intermediate that combines directly with NO, bypassing the NO/O_2 chemistry.[67] By pre-forming this activated DAF species, addition of anaerobic NO results in a 7- to 10-fold increase in the amount of recovered DAF-2 T as compared to the analogous experiment with NO and O_2.[67] These data indicate that results from DAF-2 assays may be difficult to interpret in environments where oxidants such as O_2^- are present in addition to NO. It is possible that intracellular oxidation of DAF-2 may result in increased NO-dependent fluorescence that could be confused with an increase in NO levels.[67]

4.7. OTHER ORGANIC SENSORS

Dichlorofluorescin has been used as a fluorometric assay for NO in a variety of intracellular applications.[68, 69] It is difficult, however, to draw any concrete conclusions about the distribution and production of NO using dichlorofluorescin. Dichlorofluorescin is a more general fluorometric assay for intracellular oxidants showing significant fluorescence response in the presence of oxidants such as H_2O_2, O_2^-, ONOO$^-$, NO, HOCl, and ˙OH (Scheme 4.5).[70-72]

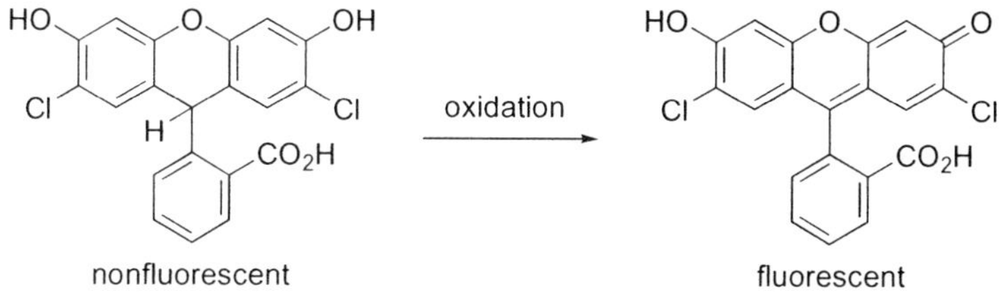

Scheme 4.5. The oxidation of nonfluorescent dichlorofluorescin to the highly fluorescent dichlorofluorescein.

Another organic-based compound developed for detecting NO is rhodamine B hydrazide (RBH). The reaction of RBH with NO_2^- under acidic conditions (pH < 5), which favor formation of NO$^+$ equivalents, results in ultimate conversion of the nonfluorescent RBH to the fluorescent rhodamine B with the shortest reaction time of 1 h observed in pH = 2 aqueous solution at 60 °C (Scheme 4.6).[73] Exposure of RBH to NO in the presence of O_2 also elicits an increase in fluorescence emission. however, presumably via a NO_2 or N_2O_3 reactive intermediate. Because of its indirect reactivity with NO, RBH has shortcomings similar to those observed with the *o*-diamine-based sensors.

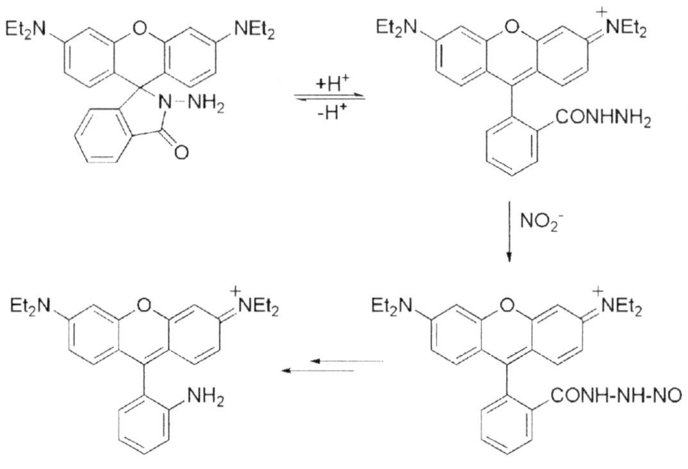

Scheme 4.6. The proposed mechanism for the conversion of RBH into a fluorescent species on reaction with NO_2^-.

One novel approach capable of direct detection of NO is based on modification of cheletropic traps, which are employed for EPR-based NO detection. The fluorescent nitric oxide cheletropic traps (FNOCTs) start as nonfluorescent dyes and add NO across a diene moiety giving a nitroxide radical species which then must react with an external reductant such as ascorbate to give the corresponding hydroxylamine (Scheme 4.7).[74, 75] The reaction of NO with FNOCT in the presence of ascorbate gives a substantial increase in fluorescence, where the hydroxylamine product has a pH-dependent emission with a maximal intensity between pH 8-9.[74] The source of the pH-based fluorescence dependence was identified as protonation of the dimethylamino group that was introduced to red-shift the fluorescence excitation wavelength of 315 nm in **1** and make it more amenable for use in living cells. Despite the pH dependence, the optical properties of **1** are sufficient for intracellular studies (λ_{ex} = 380 nm, λ_{em} = 460 nm) and it can detect sub-μM concentrations of NO. In addition to reactivity with NO, FNOCTs are also sensitive to peroxynitrite. Addition of a small excess of peroxynitrite results in an increase in fluorescence, whereas addition of a large excess leads to diminution of fluorescence emission.[74] The details of the peroxynitrite reactivity are unclear. The ability of FNOCTs to detect NO in living cells was demonstrated by using alveolar macrophages induced with LPS to produce large amounts of NO.[74] The requirement of an external reductant and the potential issues involving peroxynitrite reactivity may limit the use of FNOCTs in certain biological applications, such as NO quantification.

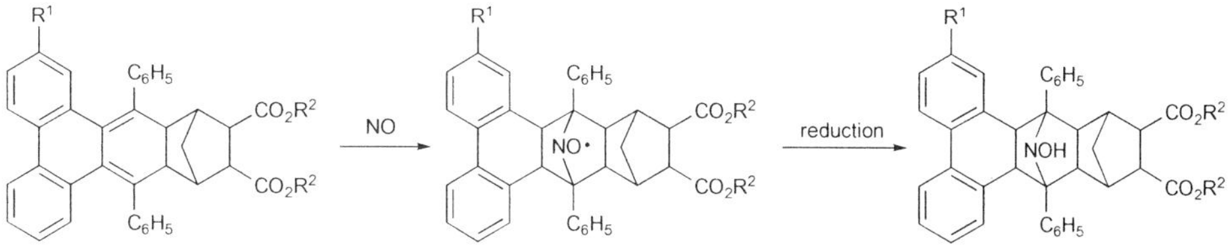

FLUORESCENCE-BASED NITRIC OXIDE DETECTION

	R^1	R^2
1	H	H
2	H	CH$_2$CO$_2$CH$_3$
3	N(CH$_3$)$_2$	H
4	N(CH$_3$)$_2$	CH$_2$CO$_2$CH$_3$

Scheme 4.7. Reaction of the FNOCTs to form fluorescent species.

4.8. TRANSITION METAL SYSTEMS AND FIBER-OPTIC NO DETECTION

All of the purely organic fluorescence-based sensors developed so far rely upon destruction or formation of covalent bonds and are non-reversible. One way to engineer a reversible sensor is to employ the use of transition metals, which are well known to bind NO in a reversible fashion. This approach to developing NO sensors for biological imaging is growing in popularity.

One strategy is the use of fluorophores attached to optical fibers coated in colloidal gold. Gold films have previously been used to detect gaseous NO by monitoring changes in the electrical resistance of the film in response to NO adsorption.[76] For modification to fabricate a fiber optic sensor, gold colloid coated optical fibers were treated with a solution of 4-carboxy-2',7'-difluorofluorescein succinimidyl ester followed by immersion of the sensor in a suspension of fluorescent carboxylate-modified polystyrene microspheres. When the sensor is exposed to NO, the gas adsorbs onto the surface of the gold colloid, which presumably causes the difluorofluorescein dye molecules to reorient relative to the gold film, resulting in a change in the transition dipole of the dye, which gives a decrease in fluorescence emission.[77] The fluorescent labeled polystyrene microspheres, which are unaffected by the presence of NO, are then used to determine the local NO concentration by ratiometric analysis. The sensor is reversible with a response time of 0.25 s or less for fluorescence quenching in the presence of NO and has a detection limit of 20 μM. Using this probe, a concentration of 190±70 μM NO was measured when released by macrophages activated with mouse interferon-γ and LPS.[77]

Related fiber optic sensors prepared using the heme domain of soluble guanylate cyclase (sGC) or cytochrome c' have also been investigated. These probes were prepared by a method similar to the 4-carboxy-2',7'-difluorofluorescein succinimidyl ester based colloidal gold sensor. Instead of soaking the gold coated optical fiber in the fluorophore, however, the fiber was immersed in either the heme domain of sGC or cytochrome c', both labeled with 4-carboxy-2',7'-difluorofluorescein.[78,79] In both cases, a decrease in fluorescence intensity from the dye conjugate is observed on binding of NO to the heme active site of the proteins. The origin of the fluorescence response of the dye is presumably due to protein conformational changes which, in turn, alter the fluorescence emission of the conjugated fluorophore.[80] Through the use of the fluorophore-protein adducts, detection limits of 1 and 8 μM NO were achieved with the sGC and cytochrome c' based probes, respectively.[78, 79] In addition to their reversibility, they are not affected by the potential interferants NO_2^-, NO_3^{2-}, O_2, O_2^-, H_2O_2, and $ONOO^-$.[79, 80] The probes were prepared using 100 μm optical fibers, which are too large for intracellular measurements but could be miniaturized by use of sub-μm tips for intracellular NO detection. Although innovative, the low NO sensitivity and inability to give spatial information, disadvantages common to all fiber optic and electrochemical based sensors, limit the utility of this approach for detecting biologically generated NO.

4.9. IRON COMPLEXES

One of the first small molecule fluorescent NO probes to employ a transition metal was designed by analogy to the active site of sGC. In the resting state of sGC, the heme moiety contains a coordinated histidine. Upon reaction with NO, a heme nitrosyl is formed, which results in dissociation of the coordinated histidine.[81] The conformational change elicited by this reaction activates the enzyme. The iron(II) quinoline pendant cyclam complex (Figure 4.5) contains two distinct fragments, an iron(II) cyclam, which can be viewed as a simple model for the heme center in sGC, and a pendant quinoline, which is analogous to the axial histidine in the active site of sGC. When excited at 366 nm, the fluorescence emission from the Fe(II) complex at 460 nm is significantly enhanced in comparison to that of the free ligand. Although no detailed studies were conducted, it was proposed that the observed fluorescence enhancement was the result of electronic interactions between the lone pair on the quinoline and the Fe(II) center.[81] Such coordination-induced fluorescence enhancement differs from the more commonly observed quenching properties of transition metals with partially filled d-shells.[82] When exposed to NO in pH 7.4 buffer, the 460 nm emission decreases markedly, the lower detection limit being 1 μM NO.[81] No interference was observed in the presence of NO_2^- or CN^-;[81] however, because Fe(II) cyclam is oxygen sensitive, the sensor may not function properly in an aerobic environment. In addition, an increase in fluorescence emission after analyte binding is generally preferred over a decrease for biological imaging.

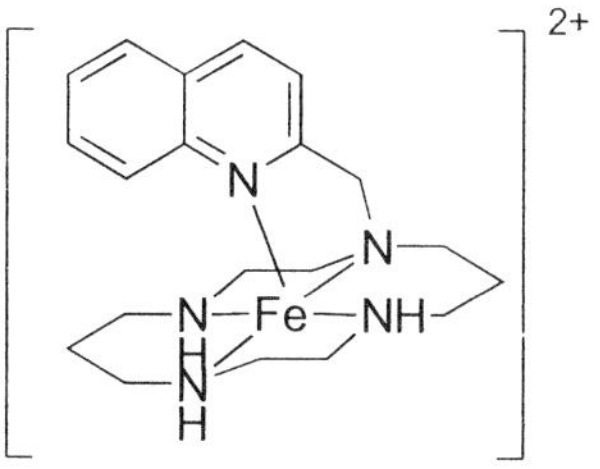

Figure 4.5. An Fe(II) quinoline pendant cyclam sensor, the design of which was based upon the active site of sGC.

A different mechanism has recently been employed to develop a transition-metal-based fluorescent NO sensor. In this system, a nitroxyl radical coordinated to an Fe(II) dithiocarbamate complex is exchanged in the presence of NO with concomitant formation of an iron nitrosyl species. EPR spectroscopy had been used to detect the liberated nitroxyl.[83] Modification of the nitroxyl ligand with a fluorophore, such as acridine, allows the use of fluorescence detection. Fluorescence emission from the acridine moiety in acridine-TEMPO is quenched by the nitroxyl radical and, when the nitroxyl coordinates to the Fe(II) dithiocarbamate, the fluorescence is restored. Exposure of the Fe(II) dithiocarbamate complex of acridine-TEMPO to NO affords the iron nitrosyl, releasing the acridine-TEMPO. The result is a diminution of the acridine fluorescence emission (Scheme 4.8).[84] With this system it was possible to detect 100 nM NO. Because it displays the less desirable decrease in fluorescence upon analyte binding, however, this strategy is not ideal. Moreover, as a consequence of using an Fe(II) dithiocarbamate complex, the system may be susceptible to competition from dioxygen.

A similar approach was applied to prepare a ratiometric NO sensor containing fluorescamine and coumarin dyes (Scheme 4.9).[85] In this case, an Fe(II) methoxycoumarinomethyl-cyclam (Mmc-cyclam) scaffold was used to bind either the nitroxyl, fluorescamine-PROXYL, or NO. In an aqueous pH 7.4 buffered solution of 40 μM Fe(Mmc-cyclam) (λ_{ex} 360 nm, λ_{em} 410 nm) and 40 μM fluorescamine-PROXYL (λ_{ex} 385 nm, λ_{em} 470 nm), excitation of the coumarin at 360 nm resulted in fluorescence resonance energy transfer (FRET) to the fluorescamine moiety and emission at 470 nm. When this system was exposed to the NO-releasing agent NOC-7, over time a diminution in fluorescamine emission along with an increase in coumarin emission was observed.[85] A detection limit of less than 100 nM NO was reported. Due to the use of an Fe(II) cyclam, the sensor may be susceptible to interference from O_2, and the reaction of NO with the metal center is apparently too slow to trap NO released in biological environments.[86]

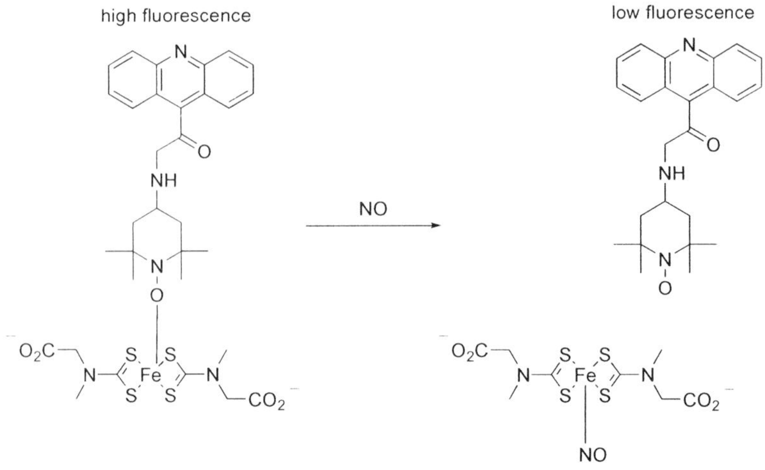

Scheme 4.8. Reaction of an Fe(dtc)$_2$-(acridine-TEMPO) conjugate with NO.

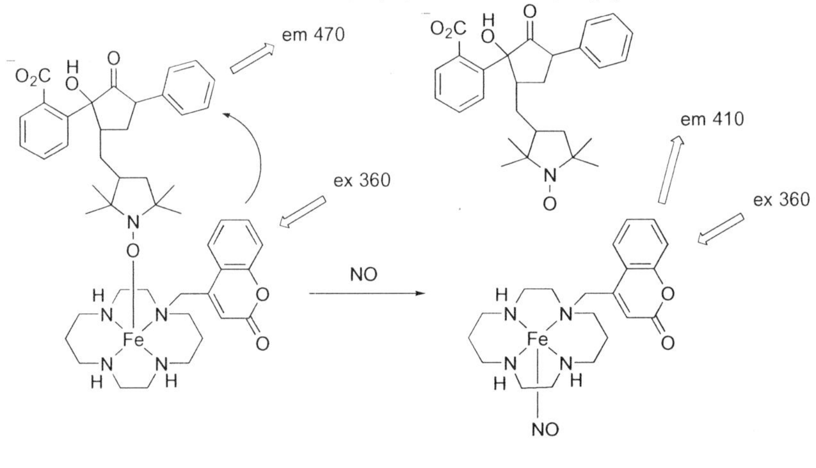

Scheme 4.9. An Fe(II)-cyclam-based ratiometric NO sensor.

4.10. COBALT COMPLEXES

4.10.1. Cobalt DATI Sensors

Results of prior research in our laboratory on the reactivity of NO with Fe,[87, 89] Mn,[88, 89] and Co[90] tropocoronand complexes suggested that fluorescence-based NO sensors using the related aminotroponimine scaffold may be feasible. Accordingly, the Co(II) complexes of the dansyl containing aminotroponimine ligands, H^{R}DATI, were prepared.[91] The Co(II) complexes are not oxygen sensitive either in solution or as solids. Although air-stable, the complexes are susceptible to hydrolysis in the presence of trace moisture in solution. The Co(II) complex of the H$_2$DATI-4 ligand (Figure 4.6) is prepared in a similar fashion.[91, 92]

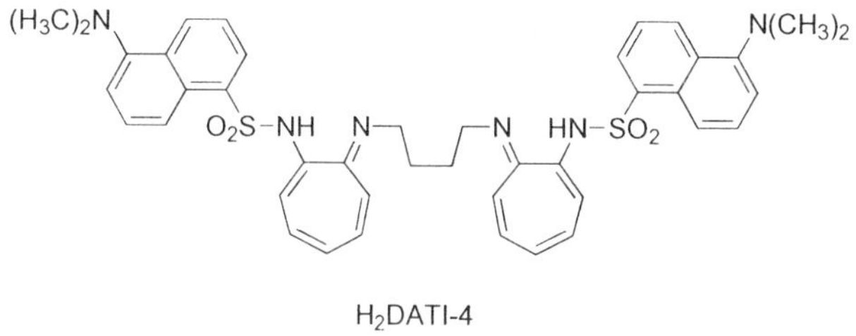

Figure 4.6. The butylene-linked tropocoronand ligand, H$_2$DATI-4.

Structural studies of [Co(^{Bz}DATI)$_2$] (**5**), [Co(^{iPr}DATI)$_2$] (**7**), and [Co(^{tBu}DATI)$_2$] (**8**) reveal them to be pseudo-tetrahedral complexes (Figure 4.7). Dihedral angles of 73.8°, 76.1°, and 81.4° between the 5-member chelate rings in **5**, **7**, and **8**, respectively, reflect the differing steric requirements imposed by the alkyl groups on the ligands. The dihedral angle of 62.2° in [Co(DATI-4)] (**6**) (Figure 4.7) is significantly smaller than the corresponding values in **5**, **7**, and **8**, because of the added strain caused by the *n*-butyl linker between the two aminotroponiminate moieties.[91] The two dansyl groups in **6** are nearly parallel and have an average separation of 3.5(1) Å, which suggests a π-π stacking interaction.[93]

The H^{R}DATI and H$_2$DATI-4 ligands display typical dansyl fluorescence. When excited at 350 nm, a broad emission centered at 500 nm is observed for each ligand. The fluorescence of the Co(II) complexes **5-8** is significantly quenched relative to that of the free ligands, as evidenced by the approximately 20-fold diminution in the emission intensity of **3** when compared to that of free H^{iPr}DATI in CH$_2$Cl$_2$ (Figure 4.8).[91] This observed transition metal quenching of fluorescence is a well-known phenomenon and can occur by energy or electron transfer between the fluorophore excited state and the empty or partially filled d-orbitals of the metal ion.[94, 95]

The reactions of **5**, **7**, and **8** with excess NO in CH$_2$Cl$_2$ are slow. Several hours after exposure of **7** to NO, IR bands at 1838 and 1760 cm^{-1} can be observed, indicating the formation of a dinitrosyl complex.[96-103] Although it was not possible to crystallize the **5**, **7**, and **8** nitrosyl complexes, the structure of the related non-dansylated dinitrosyl, [Co(*i*-Pr$_2$ATI)(NO)$_2$], which has IR bands at 1809 and 1730 cm^{-1}, suggests that **5**, **7**, and **8** are pseudotetrahedral dinitrosyl species. When the reaction of **7** with NO is monitored by ^{1}H NMR spectroscopy, signals for the free ligand, H^{iPr}DATI, grow in over time. The reaction proceeds a reductive nitrosylation mechanism in which [Co(^{R}DATI)$_2$], NO, and an electron combine to form a Co(I) dinitrosyl species (Eq. 1).[91] The species is designated {Co(NO)$_2$}^{10}, where the superscript denotes the sum of the metal d-electrons and the unpaired π* electrons on the nitrosyl ligands.[104]

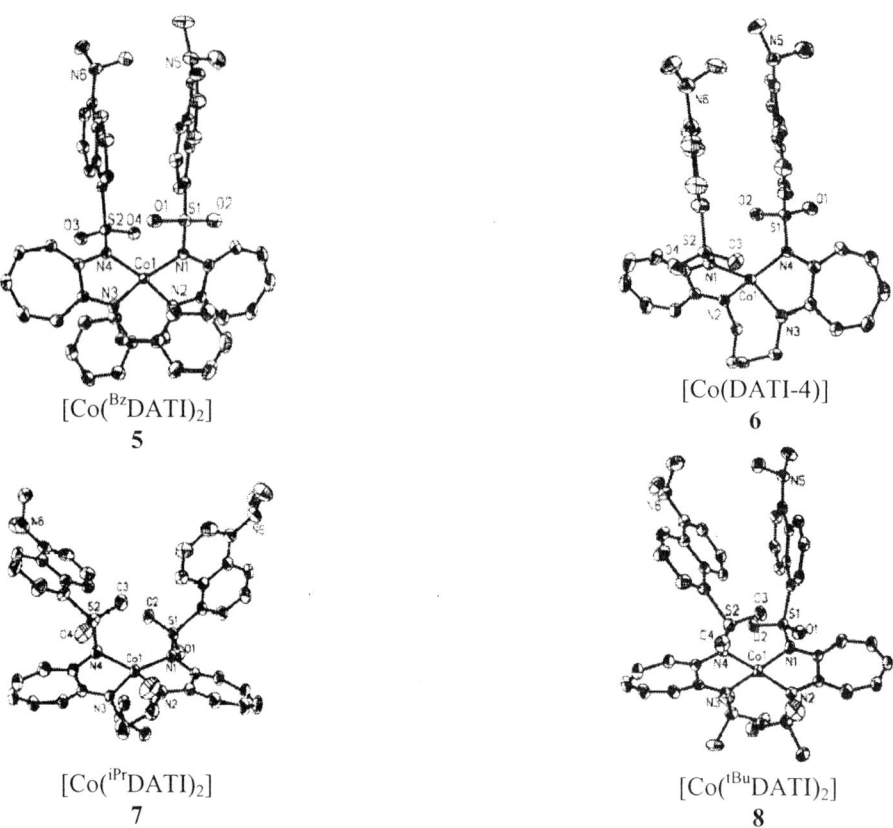

Figure 4.7. ORTEP diagrams showing 50% thermal ellipsoids for the cobalt complexes [Co(^{Bz}DATI)$_2$], [Co(DATI-4)], [Co(^{iPr}DATI)$_2$], and [Co(^{tBu}DATI)$_2$] as reported in reference 91.

$$[Co(^{iPr}DATI)_2] \xrightarrow[e^-, H^+]{NO} [Co(NO)_2(^{iPr}DATI)] + H^{iPr}DATI \qquad (1)$$

Treatment of **7** in CH$_2$Cl$_2$ with excess NO gradually increases the fluorescence emission at 505 nm.[91] After 6 h the intensity increase is 6-fold. Complex **6**, which has a more strained tetrahedral geometry, also reacts with NO and displays a moderate increase in fluorescence. The fluorescence emission of **6** doubles in the first 3 min after exposure to NO and reaches a greater than 4-fold increase after 6 h (Figure 4.9).[91, 92] The initial reactivity of this complex is significantly faster than that observed with **7**. This difference in reaction rate may be the result of the added steric strain imparted by the linker. Unfortunately, the 50-100 μM detection limit of **6** is probably insufficient for detection of biological NO.

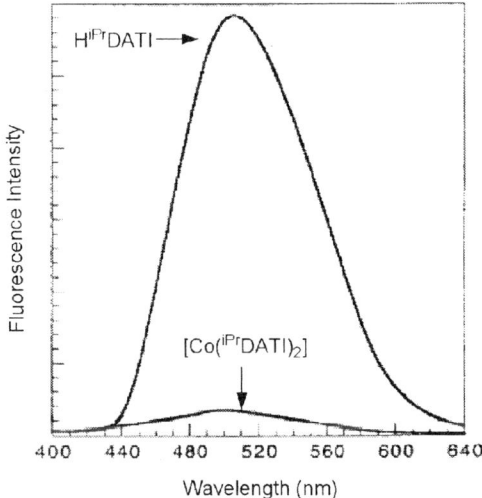

Figure 4.8. Fluorescence emission spectra of 46 μM CH$_2$Cl$_2$ solutions of H^{iPr}DATI and [Co(^{iPr}DATI)$_2$] as reported in reference 91. Fluorescence excitation is at 350 nm.

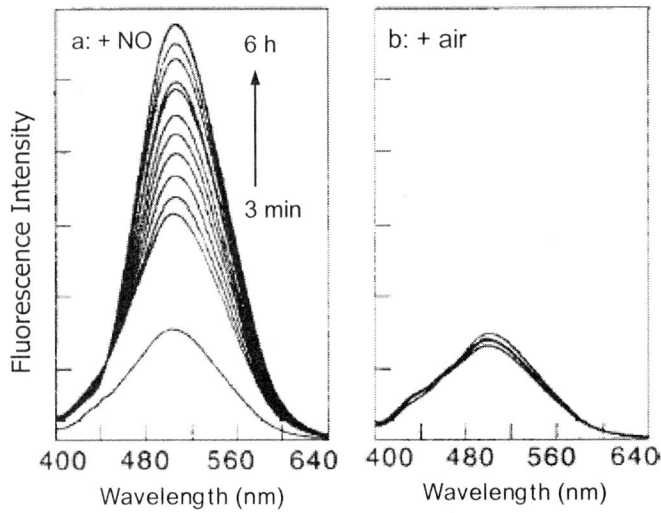

Figure 4.9. (a) Fluorescence emission spectra showing a greater than 4-fold increase in integrated emission (excitation 350 nm) over 6 h upon addition of 1 atm of NO to a 40 μM CH$_2$Cl$_2$ solution of [Co(DATI-4)]. (b) Fluorescence emission spectra of a 40 μM CH$_2$Cl$_2$ solution of [Co(DATI-4)] showing no change in emission over a 6 h period.

4.10.2. Cobalt FATI Sensors

To extend the utility of the dansyl aminotroponiminate ligands, a family of Co(II) complexes was designed in which the ligands were $H_2^{iPr}FATI$-n (n = 3 or 4).[105] Here the construct employs the N,N'-disubstituted aminotroponimine class of ligands coupled to 7'-chloro-4'-fluoresceincarboxaldehyde (Figure 4.10).

Although not characterized by X-ray crystallography, mass spectrometric data for [Co(^{iPr}FATI-3)] (**9**) and [Co(^{iPr}FATI-4)] (**10**) indicate both complexes to be mononuclear. In the case of **10**, a dinuclear structure might have been formed by analogy to the related complex [Co$_2$(^{iPr}SATI-4)$_2$], which contains salicylaldimine instead of fluorescein.[105]

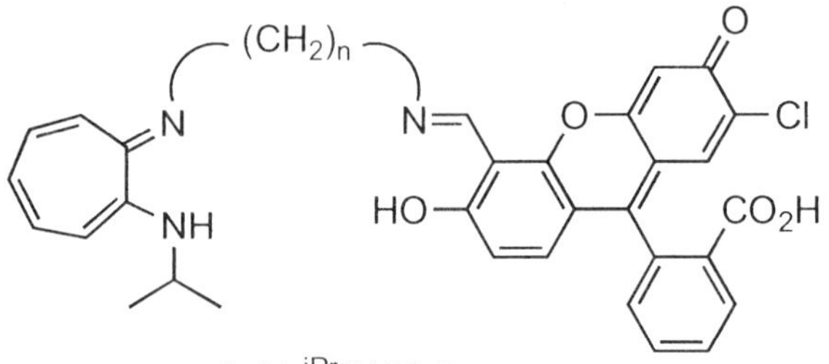

n = 3, $H_2^{iPr}FATI$-3
n = 4, $H_2^{iPr}FATI$-4

Figure 4.10. The ligands containing an aminotroponimine moiety linked by an alkyl chain to a fluorescein, $H_2^{iPr}FATI$-n.

When the reaction of **9** in DMSO with excess NO was followed by IR spectroscopy, within five minutes a new band that is consistent with a nitrosyl stretch appeared at 1630 cm^{-1}. This value is in good agreement with the nitrosyl band at 1621 cm^{-1} that appears during the reaction of [Co(^{iPr}SATI-3)] with NO.[105] After 1 h, another band began to appear at 2114 cm^{-1}, which may be attributed to a cobalt-dinitrogen species.[90, 105, 106] No evidence for formation of a dinitrosyl species is observed, but the intense feature from the fluorescein carboxylic acid at 1759 cm^{-1} may obscure any bands from a dinitrosyl species. Complex **10** also reacts with excess NO in CH$_3$CN and, like **9**, displays a new band at 2117 cm^{-1} corresponding to a dinitrogen adduct.[105]

Exposure of **9** to excess NO in MeOH increases the fluorescence emission to 120% of the original value at 530 nm four hours after addition of NO, upon excitation at 505 nm.[105] The absence of a large fluorescence increase can be explained by the IR data, which indicate that there is no observable formation of a dinitrosyl species. A dinitrosyl adduct could only form if the fluorescein moiety were to dissociate. Treatment of a MeOH solution of **10** with excess NO produces a 3-fold increase in fluorescence intensity after 22 h.[105] This fluorescence response is significantly slower than that of **6**.[91] Because multiple products appear, it is unclear which species gives rise to the increase in fluorescence. Possibilities include the dinitrogen species, a dinitrosyl, a mononitrosyl, or some other unidentified product. However, based on the small increase in fluorescence emission when **9** is treated with NO, it is unlikely that the fluorescence response of **10** results from the formation of the dinitrogen or mononitrosyl adducts. Although the Co(II)

FATI systems confer greater water compatibility than the Co(II) DATI sensors, and represent progress towards a sensor that direactly detects NO, their response time is too slow to be of potential use in a biological setting.

4.11. DIRHODIUM TETRACARBOXYLATES AS REVERSIBLE NO SENSORS

Dirhodium tetracarboxylates are air-stable compounds that coordinate a variety of ligands at the axial positions of the tetra-bridged dimetallic core.[107] In 1963, Johnson showed that treatment of solid [Rh$_2$(μ–O$_2$CMe)$_4$] with NO affords a nitrosyl adduct. This reaction can be reversed by heating to 120 °C, but the nature of the products was uncertain.[108] We designed dirhodium tetracarboxylate complexes containing bound fluorophores as potential NO sensors.[109] The reactions of [Rh$_2$(μ–O$_2$CR)$_4$] (R = Me, **11**, Et, **12**, and Pr, **13**) with NO are extremely fast. Stopped-flow experiments for the reaction of **13** with excess NO show complete formation of the dinitrosyl species within the mixing time of the instrument at -80 °C, corresponding to an approximate on rate greater than 4×10^6 s^{-1} at 40 °C. This result demonstrates the possibility of real-time imaging of NO using a dirhodium tetracarboxylate platform.[109]

The fluorophores dansyl-imidazole (Ds-im) and dansyl-piperazine (Ds-pip) bind to the dirhodium core. X-ray studies of dirhodium tetraacetate complexes of Ds-im or Ds-pip reveal coordination to the axial sites of the dirhodium unit through the imidazole or piperazine nitrogen atom, respectively (Figure 4.11).[109]

Solutions of **11** with Ds-im or Ds-pip in 1,2-dichloroethane (DCE) emit only weakly when excited at 365 nm or 345 nm, respectively. Exposure of 10 μM Ds-im and 40 μM **11** in DCE to 100 equiv of NO immediately increases the integrated fluorescence emission by 16-fold. A 26-fold increase is observed after introducing 100 equiv of NO to 10 μM Ds-pip and 20 μM **11** in DCE. In both systems, the fluorescence response is reversible with a detection limit of 4-8 μM NO (Figure 4.12).[109] Although dirhodium tetraacetate is water-soluble and air-stable, the fluorophore complexes of **11** are incompatible with aqueous environments. One potential strategy for preparing NO biosensors is to sequester the non-aqueous dirhodium fluorophore solution from the surrounding aqueous environment with a NO-permeable polymer membrane. Initial experiments demonstrate the potential utility of this approach for the detection of aqueous NO.[109]

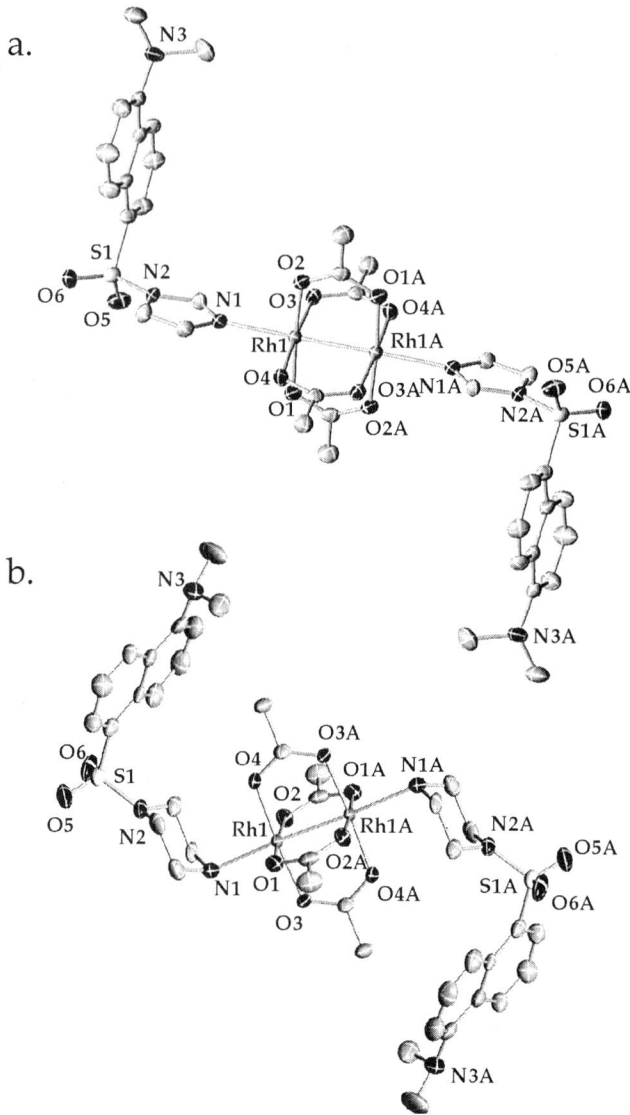

Figure 4.11. ORTEP diagrams showing 50% probability ellipsoids for (a) [Rh$_2$(μ–O$_2$CMe)$_4$(Ds-im)$_2$] and (b) [Rh$_2$(μ–O$_2$CMe)$_4$(Ds-pip)$_2$].

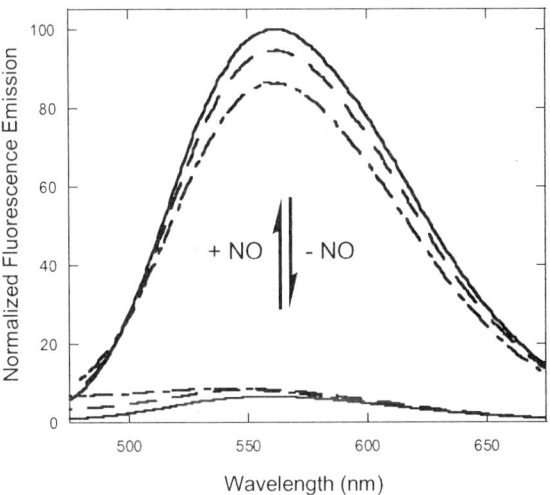

Figure 4.12. Emission spectra showing the reversible fluorescence response of 10 μM Ds-im and 40 μM [Rh$_2$(μ-O$_2$CMe)$_4$] in 1,2-dichloroethane. The lower set of traces: before the addition of the first 100 equiv of NO (solid line) and after the first and second 30 min Ar purges (dashed and dashed-dot lines). The upper set of traces: after the first, second, and third additions of 100 equiv of NO (solid, dashed, and dashed-dot lines). Fluorescence excitation is at 365 nm.

4.12. SUMMARY

Over the past 15 years, multiple strategies for the development of NO biosensors have been investigated, ranging from electrochemical and EPR based techniques to those that make use of fluorescence spectroscopy. The advancement of our understanding of NO in biology and the ability to design ever improving fluorescence-based sensors are remarkable considering the many challenges associated with handling and observing the elusive diatomic gas, nitric oxide. Early research adapting the fluorescence-based nitrite sensor DAN for NO detection was only a beginning. This work ultimately led to development of the diaminofluorescein sensors and many other *o*-diamine-based sensors that are capable of detecting sub-μM NO levels in living cells. Unfortunately, these systems require O$_2$ to react with NO to form a more reactive nitrosylating species such as N$_2$O$_3$. Other organic-based sensors such as dichlorofluorescin, the FNOCTs, and RBH have been developed and investigated for their NO-sensing abilities. All of these organic sensors, whether or not they react directly with NO, are incapable of reversibly binding and detecting nitric oxide.

Some of the more recent successes in NO detection utilize transition metal ions. In these systems, the fluorescence emission properties of an attached fluorophore are altered upon treatment of the complex with NO. One successful strategy utilizes sGC or cytochrome c' labeled with a fluorescent reporter embedded in fiber optic probes, enabling reversible NO detection. Another route is to use small molecule transition metal complexes such as those of Katayama, the Co(II) complexes, and dirhodium tetracarboxylate complexes prepared in our laboratory These complexes are capable of detecting NO over a wide spatial field like the o-diamine-based sensors, but react with NO directly. The potential is great for further modification and development of small molecule transition metal complexes for studies in living cells capable of reversible, fast, and direct detection of NO by fluorescence methodologies.

4.13. REFERENCES

1. R. F. Furchgott, Endothelium-Derived Relaxing Factor: Discovery, Early Studies and Identification as Nitric Oxide, *Angew. Chem. Int. Ed.* **38**, 1870-1880 (1999).
2. L. J. Ignarro, G. M. Buga, K. S. Wood, R. E. Byrns, and G. Chaudhuri, Endothelium-Derived Relaxing Factor Produced and Released from Artery and Vein is Nitric Oxide, *Proc. Natl. Acad. Sci. USA* **84**, 9265-9269 (1987).
3. R. M. J. Palmer, A. G. Ferrige, and S. Moncada, Nitric Oxide Release Accounts for the Biological Activity of Endothelium-Derived Relaxing Factor, *Nature* **327**, 524-526 (1987).
4. D. A. Wink, Y. Vodovotz, J. Laval, F. Laval, M. W. Dewhirst, and J. B. Mitchell, The Multifaceted Roles of Nitric Oxide in Cancer, *Carcinogenesis* **19**, 711-721 (1998).
5. S. Moncada, R. M. J. Palmer, and E. A. Higgs, Nitric Oxide: Physiology, Pathophysiology, and Pharmacology, *Pharmacol. Rev.* **43**, 109-142 (1991).
6. J. F. Kerwin, Jr., J. R. Lancaster, Jr., and P. L. Feldman, Nitric Oxide: A New Paradigm for Second Messengers, *J. Med. Chem.* **38**, 4343-4362 (1995).
7. P. L. Feldman, O. W. Griffith, and D. J. Stuehr, The Surprising Life of Nitric Oxide, *J. Chem. Eng. News* **71**, 26-38 (1993).
8. D. S. Bredt and S. H. Snyder, Nitric Oxide: A Physiologic Messenger Molecule, *Annu. Rev. Biochem.* **63**, 175-195 (1994).
9. F. Murad, Discovery of Some of the Biological Effects of Nitric Oxide and Its Role in Cell Signaling, *Angew. Chem. Int. Ed.* **38**, 1856-1868 (1999).
10. L. J. Ignarro, Nitric Oxide: A Unique Endogenous Signaling Molecule in Vascular Biology, *Angew. Chem. Int. Ed.* **38**, 1882-1892 (1999).
11. A. R. Butler and D. L. H. Williams, The Physiological Role of Nitric Oxide, *Chem. Soc. Rev.*, 233-241 (1993).
12. H. Rubbo, V. Darley-Usmar, and B. A. Freeman, Nitric Oxide Regulation of Tissue Free Radical Injury, *Chem. Res. Toxicol.* **9**, 809-820 (1996).
13. J. R. Lancaster, Jr., in: *Nitric Oxide Biology and Pathobiology*, edited by L. J. Ignarro (Academic Press, San Diego, 2000), pp. 209-224.
14. R. F. Furchgott and P. M. Vanhoutte, Endothelium-Derived Relaxing and Contracting Factors, *FASEB J.* **3**, 2007-2018 (1989).
15. S. Pfeiffer, B. Mayer, and B. Hemmens, Nitric Oxide: Chemical Puzzles Posed by a Biological Messenger, *Angew. Chem. Int. Ed.* **38**, 1714-1731 (1999).
16. D. S. Bredt, P. M. Hwang, and S. H. Snyder, Localization of Nitric Oxide Synthase Indicating a Neural Role for Nitric Oxide, *Nature* **347**, 768-770 (1990).
17. A. M. Leone, R. M. J. Palmer, R. G. Knowles, P. L. Francis, D. S. Ashton, and S. Moncada, Constitutive and Inducible Nitric Oxide Synthases Incorporate Molecular Oxygen into Both Nitric Oxide and Citrulline, *J. Biol. Chem.* **266**, 23790-23795 (1991).
18. F. Laval and D. A. Wink, Inhibition by Nitric Oxide of the Repair Protein, O^6-Methylguanine-DNA-Methyltransferase, *Carcinogenesis* **15**, 443-447 (1994).
19. V. Calabrese, T. E. Bates, and A. M. G. Stella, NO Synthase and NO-Dependent Signal Pathways in Brain Aging and Neurodegenerative Disorders: The Role of Oxidant/Antioxidant Balance, *Neurochem. Res.* **25**, 1315-1341 (2000).
20. E. M. Schuman and D. V. Madison, A Requirement for the Intracellular Messenger Nitric Oxide in Long-Term Potentiation, *Science* **254**, 1503-1506 (1991).
21. F. T. Bonner and G. Stedman, in: *Methods in Nitric Oxide Research*, edited by M. Feelisch and J. S. Stamler (John Wiley & Sons, New York, 1996), pp. 3-18.
22. J. R. Lancaster, Jr., A Tutorial on the Diffusibility and Reactivity of Free Nitric Oxide, *Nitric Oxide: Biol. and Chem.* **1**, 18-30 (1997).

23. J. Wood and J. Garthwaite, Models of the Diffusional Spread of Nitric Oxide: Implications for Neural Nitric Oxide Signalling and Its Pharmacological Properties, *Neuropharm.* **33**, 1235-1244 (1994).
24. E. M. Schuman and D. V. Madison, Nitric Oxide and Synaptic Function, *Annu. Rev. Neurosci.* **17**, 153-183 (1994).
25. G. A. Böhme, C. Bon, M. Lemaire, M. Reibaud, O. Piot, J.-M. Stutzmann, A. Doble, and J.-C. Blanchard, Altered Synaptic Plasticity and Memory Formation in Nitric Oxide Synthase Inhibitor-Treated Rats, *Proc. Natl. Acad. Sci. USA* **90**, 9191-9194 (1993).
26. H. H. H. W. Schmidt and M. Kelm, in: *Methods in Nitric Oxide Research*, edited by M. Feelisch and J. S. Stamler (John Wiley & Sons, New York, 1996), pp. 491-497.
27. V. Hampl, C. L. Walters, and S. L. Archer, in: *Methods in Nitric Oxide Research*, edited by M. Feelisch and J. S. Stamler (John Wiley & Sons, New York, 1996), pp. 309-318.
28. L. Mao, Y. Tian, G. Shi, H. Liu, and L. Jin, A New Ultramicrosensor for Nitric Oxide Based on Electropolymerized Film of Nickel Salen, *Anal. Lett.* **31**, 1991-2007 (1998).
29. F. Bedioui and N. Villeneuve, Electrochemical Nitric Oxide Sensors for Biological Samples - Principle, Selected Examples and Applications, *Electroanalysis* **15**, 5-18 (2003).
30. A. Ciszewski and G. Milczarek, Electrochemical Detection of Nitric Oxide Using Polymer Modified Electrodes, *Talanta* **61**, 11-26 (2003).
31. Y. Kotake, T. Tanigawa, M. Tanigawa, I. Ueno, D. R. Allen, and C.-S. Lai, Continuous Monitoring of Cellular Nitric Oxide Generation by Spin Trapping with an Iron-Dithiocarbamate Complex, *Biochim. Biophys. Acta* **1289**, 362-368 (1996).
32. A. M. Komarov and C.-S. Lai, Detection of Nitric Oxide Production in Mice by Spin-Tapping Electron Paramagnetic Resonance Spectroscopy, *Biochim. Biophys. Acta* **1272**, 29-36 (1995).
33. T. Yoshimura, S. Fujii, H. Yokoyama, and H. Kamada, In Vivo Electron Paramagnetic Resonance Imaging of NO-Bound Iron Complex in a Rat Head, *Chem. Lett.* 309-310 (1995).
34. S. Fujii and T. Yoshimura, A New Trend in Iron-Dithiocarbamate Complexes: As an Endogenous NO Trapping Agent, *Coord. Chem. Rev.* **198**, 89-99 (2000).
35. T. Nagano and T. Yoshimura, Bioimaging of Nitric Oxide, *Chem. Rev.* **102**, 1235-1269 (2002).
36. A. P. de Silva, H. Q. N. Gunaratne, T. Gunnlaugsson, A. J. M. Huxley, C. P. McCoy, J. T. Rademacher, and T. E. Rice, Signaling Recognition Events with Fluorescent Sensors and Switches, *Chem. Rev.* **97**, 1515-1566 (1997).
37. J. R. Lakowicz, *Principles of Fluorescence Spectroscopy*, (Kluwer Academic / Plenum Publishers, Boston, 1999).
38. G. K. Walkup, S. C. Burdette, S. J. Lippard, and R. Y. Tsien, A New Cell-Permeable Fluorescent Probe for Zn^{2+}, *J. Am. Chem. Soc.* **122**, 5644-5645 (2000).
39. S. C. Burdette, G. K. Walkup, B. Spingler, R. Y. Tsien, and S. J. Lippard, Fluorescent Sensors for Zn^{2+} Based on a Fluorescein Platform: Synthesis, Properties, and Intracellular Distribution, *J. Am. Chem. Soc.* **123**, 7831-7841 (2001).
40. S. C. Burdette, C. J. Frederickson, W. Bu, and S. J. Lippard, ZP4, an Improved Neuronal Zn^{2+} Sensor of the Zinpyr Family, *J. Am. Chem. Soc.* **125**, 1778-1787 (2003).
41. C. J. Chang, J. Jaworski, E. M. Nolan, M. Sheng, and S. J. Lippard, A Tautomeric Zinc Sensor for Ratiometric Fluorescence Imaging: Application to Nitric Oxide-Induced Release of Intracellular Zinc., *Proc. Natl. Acad. Sci. USA* **101**, 1129-1134 (2004).
42. J. H. Wiersma, 2,3-Diaminonaphthalene as a Spectrophotometric and Fluorometric Reagent for the Determination of Nitrite Ion, *Anal. Lett.* **3**, 123-132 (1970).
43. P. Damiani and G. Burini, Fluorometric Determination of Nitrite, *Talanta* **33**, 649-652 (1986).
44. C. R. Sawicki, Fluorimetric Determination of Nitrate, *Anal. Lett.* **4**, 761-775 (1971).
45. D. Ralt, J. S. Wishnok, R. Fitts, and S. R. Tannenbaum, Bacterial Catalysis of Nitrosation: Involvement of the *nar* Operon of *Escherichia coli*, *J. Bacteriology* **170**, 359-364 (1988).
46. X.-B. Ji and T. C. Hollocher, Mechanism for Nitrosation of 2,3-Diaminonaphthalene by *Escherichia coli*: Enzymatic Production of NO Followed by O_2-Dependent Chemical Nitrosation, *Appl. Environ. Microbiol.* **54**, 1791-1794 (1988).
47. H. Kosaka, J. S. Wishnok, M. Miwa, C. D. Leaf, and S. R. Tannenbaum, Nitrosation by Stimulated Macrophages. Inhibitiors, Enhancers, and Substrates., *Carcinogenesis* **10**, 563-566 (1989).
48. T. P. Misko, W. M. Moore, T. P. Kasten, G. A. Nickols, J. A. Corbett, R. G. Tilton, M. L. McDaniel, J. R. Williamson, and M. G. Currie, Selective Inhibition of the Inducible Nitric Oxide Synthase by Aminoguanidine, *Eur. J. Pharm.* **233**, 119-125 (1993).
49. T. P. Misko, R. J. Schilling, D. Salvemini, W. M. Moore, and M. G. Currie, A Fluorometric Assay for the Measurement of Nitrite in Biological Samples, *Anal. Biochem.* **214**, 11-16 (1993).
50. A. M. Miles, Y. Chen, M. W. Owens, and M. B. Grisham, Fluorometric Determination of Nitric Oxide, *Methods* **7**, 40-47 (1995).

51. D. J. Kleinhenz, X. Fan, J. Rubin, and C. M. Hart, Detection of Endothelial Nitric Oxide Release with the 2,3-Diaminonaphthalene Assay, *Free Rad. Biol. Med.* **34**, 856-861 (2003).
52. T. Nagano, H. Takizawa, and M. Hirobe, Reactions of Nitric Oxide with Amines in the Presence of Dioxygen, *Tet. Lett.* **36**, 8239-8242 (1995).
53. H. Kojima, K. Sakurai, K. Kikuchi, S. Kawahara, Y. Kirino, H. Nagoshi, Y. Hirata, T. Akaike, H. Maeda, and T. Nagano, Development of a Fluorescent Indicator for the Bioimaging of Nitric Oxide, *Biol. Pharm. Bull.* **20**, 1229-1232 (1997).
54. R. Y. Tsien, A Non-Disruptive Technique for Loading Calcium Buffers and Indicators into Cells, *Nature* **290**, 527-528 (1981).
55. C. Munkholm, D.-R. Parkinson, and D. R. Walt, Intramolecular Fluorescence Self-Quenching of Fluoresceinamine, *J. Am. Chem. Soc.* **112**, 2608-2612 (1990).
56. H. Kojima, N. Nakatsubo, K. Kikuchi, S. Kawahara, Y. Kirino, H. Nagoshi, Y. Hirata, and T. Nagano, Detection and Imaging of Nitric Oxide with Novel Fluorescent Indicators: Diaminofluoresceins, *Anal. Chem.* **70**, 2446-2453 (1998).
57. H. Kojima, Y. Urano, K. Kikuchi, T. Higuchi, Y. Hirata, and T. Nagano, Fluorescent Indicators for Imaging Nitric Oxide Production, *Angew. Chem. Int. Ed.* **38**, 3209-3212 (1999).
58. W.-C. Sun, K. R. Gee, D. H. Klaubert, and R. P. Haugland, Synthesis of Fluorinated Fluoresceins, *J. Org. Chem* **62**, 6469-6475 (1997).
59. H. Kojima, M. Hirotani, N. Nakatsubo, K. Kikuchi, Y. Urano, T. Higuchi, Y. Hirata, and T. Nagano, Bioimaging of Nitric Oxide with Fluorescent Indicators Based on the Rhodamine Chromophore, *Anal. Chem.* **73**, 1967-1973 (2001).
60. H. Kojima, M. Hirotani, Y. Urano, K. Kikuchi, T. Higuchi, and T. Nagano, Fluorescent Indicators for Nitric Oxide Based on Rhodamine Chromophore, *Tet. Lett.* **41**, 69-72 (2000).
61. M. J. Plater, I. Greig, M. H. Helfrich, and S. H. Ralston, The Synthesis and Evaluation of o-Phenylenediamine Derivatives as Fluorescent Probes for Nitric Oxide Detection, *J. Chem. Soc., Perkin Trans. 1*, 2553-2559 (2001).
62. X. Zhang, H. Wang, J.-S. Li, and H.-S. Zhang, Development of a Fluorescent Probe for Nitric Oxide Detection Based on Difluoroboradiaza-s-indacene Fluorophore, *Anal. Chim. Acta* **481**, 101-108 (2003).
63. X. Zhang, H. Wang, S.-C. Liang, and H.-S. Zhang, Spectrofluorimetric Determination of Nitric Oxide at Trace Levels with 5,6-Diamino-1,3-Naphthalene Disulfonic Acid, *Talanta* **56**, 499-504 (2002).
64. S. Schuchmann, D. Albrecht, U. Heinemann, and O. v. B. u. Halbach, Nitric Oxide Modulates Low-Mg^{2+}-Induced Epileptiform Activity in Rat Hippocampal-Entorhinal Cortex Slices, *Nerobiol. of Disease* **11**, 96-105 (2002).
65. N. Nakatsubo, H. Kojima, K. Kikuchi, H. Nagoshi, Y. Hirata, D. Maeda, Y. Imai, T. Irimura, and T. Nagano, Direct Evidence of Nitric Oxide Production from Bovine Aortic Endothelial Cells Using New Fluorescence Indicators: Diaminofluoresceins, *FEBS Lett.* **427**, 263-266 (1998).
66. J. F. Leikert, T. R. Räthel, C. Müller, A. M. Vollmar, and V. M. Dirsch, Reliable In Vitro Measurement of Nitric Oxide Released From Endothelial Cells Using Low Concentrations of the Fluorescent Probe 4,5-Diaminofluorescein, *FEBS Lett.* **506**, 131-134 (2001).
67. D. Jourd'Heuil, Increased Nitric Oxide-Dependent Nitrosylation of 4,5-Diaminofluorescein by Oxidants: Implications for the Measurement of Intracellular Nitric Oxide, *Free Rad. Biol. Med.* **33**, 676-684 (2002).
68. A. Imrich and L. Kobzik, Fluorescence-Based Measurement of Nitric Oxide Synthase Activity in Activated Rat Macrophages Using Dichlorofluorescin, *Nitric Oxide: Biol. and Chem.* **1**, 359-369 (1997).
69. P. G. Gunasekar, A. G. Kanthasamy, J. L. Borowitz, and G. E. Isom, Monitoring Intracellular Nitric Oxide Formation by Dichlorofluorescin in Neuronal Cells, *J. Neurosci. Methods* **61**, 15-21 (1995).
70. S. J. Hempel, G. R. Buettner, Y. Q. O'Malley, D. A. Wessels, and D. M. Flaherty, Dihydrofluorescein Diacetate Is Superior for Detecting Intracellular Oxidants: Comparison with 2',7'-Dichlorodihydrofluorescein Diacetate, and 5(and 6)-Carboxy-2',7'-Dichlorodihydrofluorescein Diacetate and Dihydrorhodamine 123, *Free Rad. Biol. Med.* **27**, 146-159 (1999).
71. K.-i. Setsukinai, Y. Urano, K. Kakinuma, H. J. Majima, and T. Nagano, Development of Novel Fluorescence Probes That Can Reliably Detect Reactive Oxygen Species and Distinguish Specific Species, *J. Biol. Chem.* **278**, 3170-3175 (2003).
72. O. Myhre, J. M. Andersen, H. Aarnes, and F. Fonnum, Evaluation of the Probes 2',7'-Dichlorofluorescin Diacetate, Luminol, and Lucigenin as Indicators of Reactive Species Formation, *Biochem. Pharma.* **65**, 1575-1582 (2003).
73. T. Rieth and K. Sasamoto, Detection of Nitric Oxide and Nitrite by Using a Rhodamine-Type Fluorescent Indicatior, *Anal. Commun.* **35**, 195-197 (1998).
74. P. Meineke, U. Rauen, H. de Groot, H.-G. Korth, and R. Sustmann, Cheletropic Traps for the Fluorescence Spectroscopic Detection of Nitric Oxide (Nitrogen Monoxide) in Biological Systems, *Chem. Eur. J.* **5**, 1738-1747 (1999).

75. M. Bätz, H.-G. Korth, and R. Sustmann, A Novel Method for Detecting Nitric Oxide (NO) by Formation of Fluorescent Prodcuts Based on Cheletropic Spin Traps, *Angew. Chem. Int. Ed.* **36**, 1501-1503 (1997).
76. K. Toda, K. Ochi, and I. Sanemasa, NO-Sensing Properties of Au Thin Film, *Sens. Actuators, B* **32**, 15-18 (1996).
77. S. L. R. Barker and R. Kopelman, Development and Cellular Applications of Fiber Optic Nitric Oxide Sensors Based on a Gold-Adsorbed Fluorophore, *Anal. Chem.* **70**, 4902-4906 (1998).
78. S. L. R. Barker, H. A. Clark, S. F. Swallen, R. Kopelman, A. W. Tsang, and J. A. Swanson, Ratiometric and Fluorescence-Lifetime-Based Biosensors Incorporating Cytochrome c' and the Detection of Extra- and Intracellular Macrophage Nitric Oxide, *Anal. Chem.* **71**, 1767-1772 (1999).
79. S. L. R. Barker, Y. Zhao, M. A. Marletta, and R. Kopelman, Cellular Applications of a Sensitive and Selective Fiber-Optic Nitric Oxide Biosensor Based on a Dye-Labeled Heme Domain of Soluble Guanylate Cyclase, *Anal. Chem.* **71**, 2071-2075 (1999).
80. S. L. R. Barker, R. Kopelman, T. E. Meyer, and M. A. Cusanovich, Fiber-Optic Nitric Oxide-Selective Biosensors and Nanosensors, *Anal. Chem.* **70**, 971-976 (1998).
81. Y. Katayama, S. Takahashi, and M. Maeda, Design, Synthesis and Characterization of a Novel Fluorescent Probe for Nitric Oxide (Nitrogen Monoxide), *Anal. Chim. Acta* **365**, 159-167 (1998).
82. A. W. Varnes, R. B. Dodson, and E. L. Wehry, Interactions of Transition-Metal Ions with Photoexcited States of Flavins. Fluorescence Quenching Studies, *J. Am. Chem. Soc.* **94**, 946-950 (1972).
83. Y. Katayama, N. Soh, K. Koide, and M. Maeda, A Novel Molecular System for Nitric Oxide Detection with High Sensitivity, *Chem. Lett.*, 1152-1153 (2000).
84. N. Soh, Y. Katayama, and M. Maeda, A Fluorescent Probe for Monitoring Nitric Oxide Production Using a Novel Detection Concept, *Analyst* **126**, 564-566 (2001).
85. N. Soh, T. Imato, K. Kawamura, M. Maeda, and Y. Katayama, Ratiometric Direct Detection of Nitric Oxide Based on a Novel Signal-Switching Mechanism, *Chem. Commun.* 2650-2651 (2002).
86. Y. Katayama, N. Soh, and M. Maeda, Strategies and Development of Molecular Probes for Nitrogen Monoxide Monitoring, *Bull. Chem. Soc. Jpn.* **75**, 1681-1691 (2002).
87. K. J. Franz and S. J. Lippard, NO Disproportionation Reactivity of Fe Tropocoronand Complexes, *J. Am. Chem. Soc.* **121**, 10504-10512 (1999).
88. K. J. Franz and S. J. Lippard, Disproportionation of Nitric Oxide Promoted by a Mn Tropocoronand, *J. Am. Chem. Soc.* **120**, 9034-9040 (1998).
89. K. J. Franz and S. J. Lippard, Nitrosyl Transfer from Manganese to Iron in Tropocoronand Complexes, *Inorg. Chem.* **39**, 3722-3723 (2000).
90. K. J. Franz, L. H. Doerrer, B. Spingler, and S. J. Lippard, Pentacoordinate Cobalt(III) Thiolate and Nitrosyl Tropocoronand Compounds, *Inorg. Chem.* **40**, 3774-3780 (2001).
91. K. J. Franz, N. Singh, B. Spingler, and S. J. Lippard, Aminotroponiminates as Ligands for Potential Metal-Based Nitric Oxide Sensors, *Inorg. Chem.* **39**, 4081-4092 (2000).
92. K. J. Franz, N. Singh, and S. J. Lippard, Metal-Based NO Sensing by Selective Ligand Dissociation, *Angew. Chem. Int. Ed.* **39**, 2120-2122 (2000).
93. Z.-H. Liu, C.-Y. Duan, J. Hu, and X.-Z. You, Design, Synthesis, and Crystal Structure of a *cis*-Configuration N_2S_2-Coordinated Palladium(II) Complex: Role of the Intra- and Intermolecular Aromatic-Ring Stacking Interaction, *Inorg. Chem.* **38**, 1719-1724 (1999).
94. L. Fabbrizzi, M. Licchelli, and P. Pallavicini, Transition Metals as Switches, *Acc. Chem. Res.* **32**, 846-853 (1999).
95. R. Bergonzi, L. Fabbrizzi, M. Licchelli, and C. Mangano, Molecular Switches of Fluorescence Operating through Metal Centred Redox Couples, *Coord. Chem. Rev.* **170**, 31-46 (1998).
96. R. L. Martin and D. Taylor, Bending of Linear Nitric Oxide Ligands in Four-Coordinate Transition Metal Complexes. Crystal and Molecular Structure of Dinitrosyldithioacetylacetonatocobalt(-I), $Co(NO)_2(SacSac)$, *Inorg. Chem.* **15**, 2970-2977 (1976).
97. J. S. Field, P. J. Wheatley, and S. Bhaduri, Crystal Structure of {[2-(Diphenylphosphino)ethyl]diphenyl phosphine oxide}iodonitrosylcobalt(0), *J. Chem. Soc. Dalton Trans.* 74-79 (1974).
98. B. L. Haymore, J. C. Huffman, and N. E. Butler, Linear vs. Bent Nitrosyl Ligands in Pseudotetrahedral Nitrosyl Complexes. Low-Temperature Structure of $CoI(NO)_2(P(C_6H_5)_3)$, *Inorg. Chem.* **22**, 168-170 (1983).
99. J. A. Kaduk and J. A. Ibers, Structure of Dinitrosyl(1,2-bis(diphenylphosphino)ethane)cobalt Hexafluorophosphate, $[Co(NO)_2((C_6H_5)_2PC_2H_4P(C_6H_5)_2)][PF_6]$, *Inorg. Chem.* **16**, 3283-3287 (1977).
100. B. E. Reichert, Dinitrosylbis(triphenylphosphine)cobalt Hexafluorophosphate, *Acta Crystallogr., Sect. B* **B32**, 1934-1936 (1976).
101. J.-L. Roustan, N. Ansari, Y. Le Page, and J.-P. Charland, Molecular Geometry of $M(NO)_2$ Complexes: Single Crystal X-ray Structure of Cobalt Nitrosyl Complexes $Co(NO)_2(C_5H_5N)_2^+BF_4^-$, Lability of the

Pyridine Ligands of $Co(NO)_2(C_5H_5N)_2^+$, and Its Relevance to the Formation of the $Co_2(NO)_3^+$ Bimetallic Core, *Can. J. Chem.* **70**, 1650-1657 (1992).
102. M. Aresta, D. Ballivet-Tkatchenko, M. C. Bonnet, R. Faure, and H. Loiseleur, Synthesis and Structural Characterization of $Co(NO)_2[PhP(OCH_2CH_2)_2NH]Cl$: A Novel Carbon Dioxide Carrier, *J. Am. Chem. Soc.* **107**, 2994-2995 (1985).
103. A. R. Hendrickson, R. K. Y. Ho, and R. L. Martin, Four- and Five-Coordinated Nitrosyls of Cobalt Dithioacetylacetonate, *Inorg. Chem.* **13**, 1279-1281 (1974).
104. J. H. Enemark and R. D. Feltham, Principles of Structure Bonding, and Reactivity for Metal Nitrosyl Complexes, *Coord. Chem. Rev.* **13**, 339-406 (1974).
105. S. A. Hilderbrand and S. J. Lippard, Cobalt Chemistry with Mixed Aminotroponiminate Salicylaldiminate Ligands: Synthesis, Characterization, and Nitric Oxide Reactivity, *Inorg. Chem.* **43**, 4674-4682 (2004).
106. J. Chatt, J. R. Dilworth, and R. L. Richards, Recent Advances in the Chemistry of Nitrogen Fixation, *Chem. Rev.* **78**, 589-625 (1978).
107. E. B. Boyar and S. D. Robinson, Rhodium(II) Carboxylates, *Coord. Chem. Rev.* **50**, 109-208 (1983).
108. S. A. Johnson, H. R. Hunt, and H. M. Neumann, Preparation and Properties of Anhydrous Rhodium(II) Acetate and Some Adducts Thereof, *Inorg. Chem.* **2**, 960-962 (1963).
109. S. A. Hilderbrand, M. H. Lim, and S. J. Lippard, Dirhodium Tetracarboxylate Scaffolds as Reversible Fluorescence-Based Nitric Oxide Sensors, *J. Am. Chem. Soc.* **126**, 4972-4978 (2004).

FLUORESCENT REDOX-SWITCHABLE DEVICES

Ute Resch-Genger and Gunther Hennrich[#]

5.1. INTRODUCTION

The miniaturized dimensions and high degree of control of molecular design offered by chemical synthesis render organic molecules attractive candidates for molecular electronics and photonics, especially for digital processing and communication.[1-5] This includes molecular and supramolecular devices whose optical properties can be switched or modulated by external stimuli such as chemical,[6-12] electrochemical,[13-15] thermal,[16,17] or optical[18-27] inputs thereby producing a detectable signal, in some cases even on a single molecule level.[28-31] Such systems can be obtained by combining specific components suitably arranged and connected either via covalent bonds or intermolecular interactions with the interplay of the elementary acts performed by the respective modules determining the device´s function.[32,33]

The key issue for the construction of favorable devices is the efficient transduction of an event into a measurable signal. This can be achieved upon combining a properly selected, externally addressable control module which can exist in two defined states of comparable and sufficient stability, that – dependent on the desired application – should or should not be reversibly convertible into each other, and a reporting module, efficiently and selectively communicating the switching event, i.e., external stimulus. The latter implies that, aside from the desired strong changes of the optical properties, the generated output should strongly differ from any signals from an unspecific background, the sensor in a different (for instance oxidation or analyte binding) state or the switch in the „*ZERO*" or „*OFF*" position. A particularly attractive signaling feature for sensing and switching applications is luminescence. Luminescence, that requires light absorption by the chromophore, is characterized by its intensity, spectral, and temporal distribution as well as (de)polarization, thus enabling multiparameter communication, and offers an

---

[#] Dr. Ute Resch-Genger, Federal Institute for Materials Research and Testing, Richard-Willstätter-Str. 11, D-12489 Berlin, Germany. Dr. Gunther Hennrich, Dpto. de Química Orgánica, L 101, Facultad de Ciencias, Universidad Autónoma de Madrid, Cantoblanco 28049-Madrid, Spain.

intrinsic selectivity typically via the two experimental parameters excitation and emission wavelength in combination with a high sensitivity down to the single molecule level.[34] Furthermore, signal accessing via remote control is possible.

In the growing field of sensory devices and switches, fluorescent redox systems, whose emission properties can be controlled by electrochemically or chemically induced redox processes,[35-39] have attracted increasing interest for applications in biology, (bio)chemistry, environmental analysis, and material sciences. This includes for instance fluorescent probes to map the electric potential across neurons, determine synaptic activity, and measure membrane potentials,[40,41] the field of redox fluorometry, i.e., redox imaging of cells, tissues, and organs with biotic and abiotic fluorescent molecules reporting on cellular metabolisms,[42-44] fluorosensors for the detection of redox-active analytes and radicals,[45] fluorescent redox switches,[46] devices for electrochemical and optical sensing,[47,48] and systems with oxidation state dependent analyte affinity and emission.[49] Aiming at a broad coverage of the fascinating field of composite fluorescent redox-active devices, this review illustrates the fundamental working principles of systems with oxidation state dependent luminescence and redox-switchable and analyte responsive fluorosensors, thereby focusing on multicomponent systems connected via covalent bonds. In the case of fluorescent redox systems that incorporate for instance redox systems such as ferrocene or redox-active transition metal ions, this definition focuses on the covalent bond between the chromophore and the ligand stabilizing the metal ion. The concepts of redox-activated and analyte triggered fluorescence signal generation are highlighted and chemical strategies employed so far for their realization are presented. The intension here is to provide a selective overview rather than an exhaustive review. Special emphasis is dedicated to our own research on redox-switchable fluorosensors for metal ions and anions that offer redox control of analyte recognition and emission thereby combining both methodologies. We do not attempt to cover neither the field of redox fluorometry nor devices employing redox-active receptors for electrochemically recognizing and sensing analytes via a change in redox potential[50,51] or redox switches in which electrochemistry causes conformational changes inducing gross mechanical motions.[52-54] Also, polymeric systems are beyond the scope of this article.[55,56]

5.2. REDOX-ACTIVATED FLUORESCENCE SIGNAL GENERATION

Fluorescent redox devices efficiently transduce a redox process into a measurable fluorescence signal. They are typically designed by a composite approach combining a bistable redox-active control unit (C) which can exist in different oxidation states (C_{ox}/C_{red}) of comparable and defined stability, and a luminescent signaling module (F) whose absorption and/or emission properties are selectively affected by the control unit's oxidation state.[46] The oxidation states of C are connected by a one or two-electron redox process with the access to each state triggered by an external input. In the majority of systems, control and signaling unit are electronically decoupled, often via a short alkyl spacer (S). The desired redox control of the emission properties requires interaction between both modules to a different extent in both oxidation states through an inter-component process such as electron (*ET*) or energy transfer (*EnT*). This difference in interaction generates the function that can be turned *ON/OFF* or modulated through an external electrochemical or chemical stimulus that adjusts the oxidation state of C. Aiming at devices of optimized selectivity and sensitivity, the advantage of such a composite construction lays within the independent control and tuning of the electro-active unit, i.e., its redox potential and quenching ability of excited states, and reporting moiety, i.e., dye architecture and photophysics. Such systems principally present an extension of the *ET* concept succesfully employed for the construction of pH and cation responsive fluorescent devices.[57,58]

The design of efficient fluorescent redox systems commonly requires one of the oxidation states of the control unit to act as a quencher of the emission of the neighboring chromophore, whereas the other oxidation state should not, only slightly or to a strongly different extent affect its luminescence depending on the desired application. Typically, for use as simple fluorescent redox switches or sensors, *OFF/ON* or *ON/OFF* systems are desired with *OFF/ON* and *ON/OFF* corresponding to the (thermodynamically determined) occurrence/nonoccurence of the fluorescence quenching pathway. However, the latter operating condition can be for instance advantageous for the construction of fluorosensors with oxidation and binding state controlled emission. An example for a redox-activated, *ET* or *EnT* operated two state fluorescent *OFF/ON* switch, that combines an electro-active with a light emitting component via a spacer, is illustrated in scheme 1. In this three component device, the oxidized form of the control unit, C_{ox}, quenches the emission of the adjacent excited fluorophore, whereas the reduced form C_{red} does not influence its luminescence. Redox switching of the chromophore emission can be achieved in two ways, either electrochemically via the adjustment of the potential of a working electrode or chemically upon addition of reducing and oxidizing agents. The device's change in oxidation state that is reported as a change in luminescence can be (more or less) reversible as is desired for an efficient fluorescent redox switch or irreversible yielding for instance chemodosimeter-type[59] sensors for the detection of redox active analytes.[60]

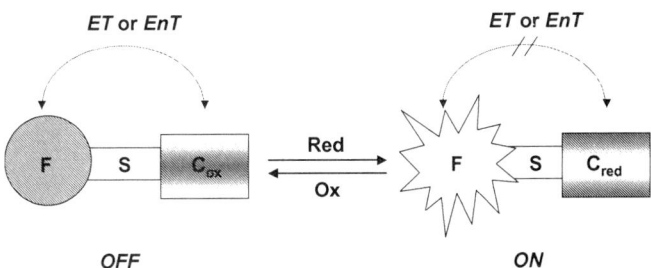

Scheme 5.1
Composite fluorescent redox switch with oxidation state controlled emission; oxidized (C_{ox}) and reduced (C_{red}) state of the redox-active control unit, fluorescent signaling module (F), and spacer (S).

5.2.1. Redox-Active Control Units

The key point for the construction of efficient fluorescent redox switches and sensors are properly chosen redox-active modules. Well characterized and bistable redox systems (C_{ox}/C_{red}) successfully employed in such devices are depicted in Figure 5.1. For the realization of favorable redox switches, the oxidation states of the control unit, i.e., C_{ox} and C_{red}, should be connected through a fast and reversible redox change thereby permitting multi-cycle switching, whereas for fluorescence sensing of redox-active analytes, an ideally selective reversible or irreversible fast chemical redox reaction with the species to be detected is desired. The virtue of the quinone/hydroquinone redox couple **1/2**, used for the construction of the first fluorescent redox switch, lays within its reversible electrochemical interconversion in protic media by the exchange of two protons and two electrons[*] and the potential of the electron deficient quinone to diminish the emission of numerous chromophores, commonly via photoinduced *ET*.[61] Furthermore, the reduced hydroquinone form typically does not promote fluorescence quenching. The electro-active tetrathiafulvalene (TTF) unit that can exist in three stable electrochemical states, i.e., **3** (TTF^{2+}), **4** ($TTF^{+\bullet}$) and **5** (TTF^0), connected via one-electron redox changes, has been frequently used as a building block in molecular and supramolecular switches[62,63] and electrochemical sensing devices.[64-66] Here, the reduced (TTF^0) form, a strong electron donor, often leads to diminished fluorescence. Similarly, the (one-electron) ferrocenium/ferrocene redox couple **6/7** has been incorporated into a variety of luminescent systems with the reduced electron donating ferrocene form being a classical quencher of excited states, either via *ET* or *EnT*.[67-69] Aside from the metallocenes, another class of redox-active control units are ligand stabilized redox-active transition metal ions, i.e., (one-electron) metal centered redox couples ($M^{(n+1)+}/M^{n+}$) with the metal ion (M) for instance equaling copper or nickel.[46] Proper functioning of these redox systems requires receptors capable of stabilizing two adjacent oxidation states of the transition metal ion to a comparable extent.[70] A representative example is the Cu^{II}/Cu^{I} redox couple **8/9** stabilized by a tetrathia macrocycle. For most metal centered redox systems, at least one of the oxidation states of the transition metal ion can quench the emission of a neighboring chromophore via either *ET* or *EnT* (double electron exchange mechanism, Dexter type *EnT*).[71]

Principally, the addressability of the control unit by chemically induced redox processes can be also exploited for sensing of redox-active species. Analyte recognition requires an ideally selective chemical redox reaction between the redox-active building block and the compound to be detected. The chemical selectivity is basically redox potential controlled, which can be limited as compared to other recognition mechanisms such as for instance a complexation reaction between the sensor molecule and the guest.[6,7,11] An example for a redox-active control unit suited for the reaction with redox-active analytes as well as radicals is the nitroxide group, here a part of **10**, which is a strong quencher of fluorescence predominantly via either (electron exchange induced)

---

[*] For applications as redox-active control unit in the fluorescent switches presented here, the intermediate state, the semihydroquinone form, is typically not considered.

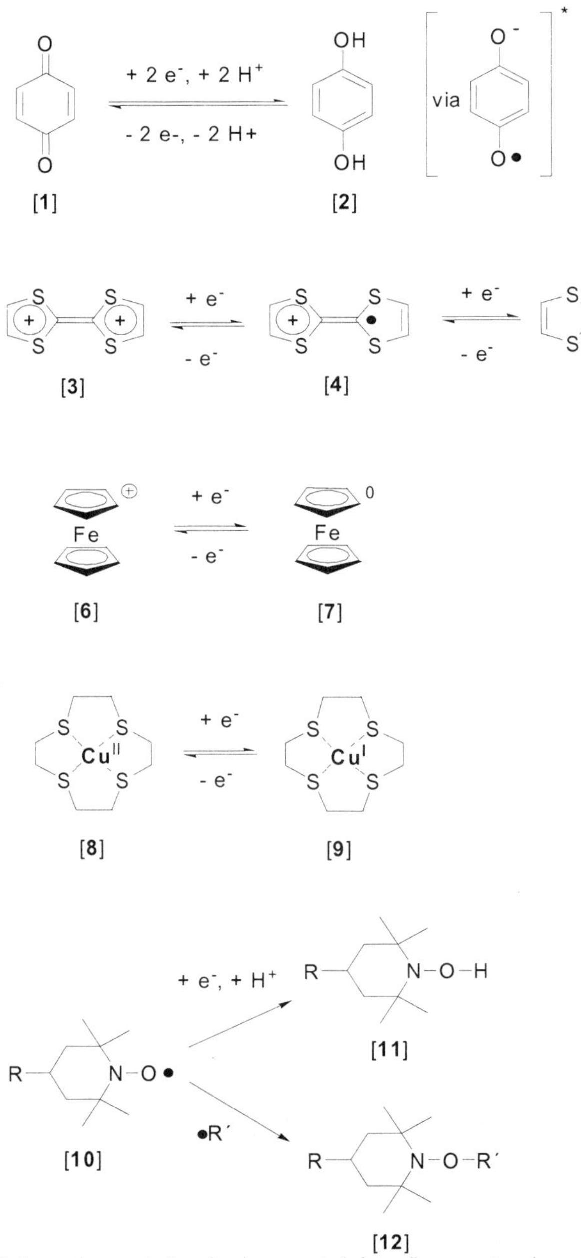

Figure 5.1. Redox-active control units incorporated into fluorescent redox switches and sensors. The semihydroquinone form* shown in brackets is typically not considered for applications of the quinone/hydroquinone redox couple as redox-active control unit in the fluorescent switches presented here.

intersystem crossing (ISC), enhanced internal conversion (IC) or ET. - For sensing applications, typically its reduction to diamagnetic hydroxylamine 11 and the addition of (carbon centered) radicals yielding O-alkylhydroxylamine 12 are considered.

5.2.2. Composite Fluorescent Redox Switches

For the construction of simple fluorescent redox switches, see scheme 1, the redox systems shown in Figure 5.1 have been linked to a variety of chromophores such as for instance porphyrins or ruthenium complexes, either via a spacer or directly. The function of the redox-active building block in such a device is restricted to the control of the luminescence output, i.e., redox-activated optical signal generation. In the majority of these systems, both modules are (more or less) electronically decoupled and interact only via *ET* and/or *EnT*. Such systems are in many aspects analogous to model systems of the photosynthetic reaction center, in which *ET* occurs depending on the redox state of the electron acceptor.[61]

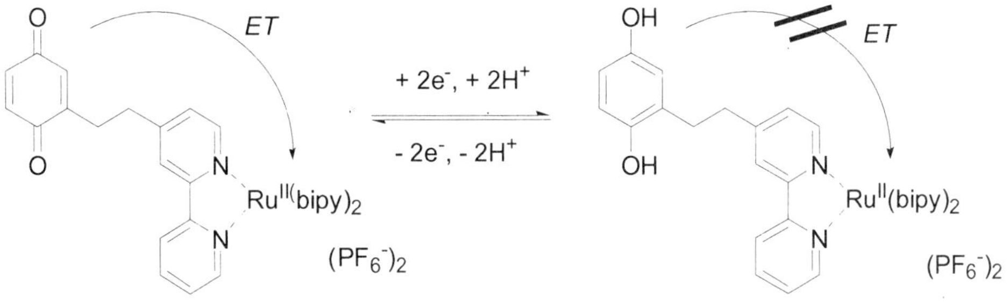

Figure 5.2. First fluorescent redox switch based on the quinone/hydroquinone redox couple.

The first electrochemically addressable, bistable, and reversible *OFF/ON* fluorescent redox switch **13**, a three component system, was obtained upon tethering a quinone/hydroquinone moiety to a [Ru(bpy)$_3$]$^{2+}$ chromophore via an ethylene spacer, see Figure 5.2.[72] In **13**, luminescence quenching through *ET* from the reductive triplet state of the ruthenium(II) complex to the quinone electron acceptor, C$_{ox}$, can be reversed upon its electrochemical reduction to the corresponding hydroquinone **14** thereby leading to an increase in emission by a factor of *ca*. 5. Other examples for quinone-type redox-activated and reversible luminescence *OFF/ON* switching devices include a ruthenium complex incorporating a quinone-fused dipyridophenazine ligand[73] and an anthracene-benzodifurane quinone system.[74]

Prominent tetrathiafulvalene-based fluorescent redox switches, that often only exploit the TTF$^{+\bullet}$/TTF^0 redox couple, are a TTF-ruthenium complex[75] and mixtures of TTF- and anthracene-substituted polystyrenes,[76] TTF-phthalocyanine dyads,[77] TTF-porphyrin dyads,[78] and TTF-porphyrin-fullerene triads.[79] In the case of the porphyrin-

FLUORESCENT REDOX-SWITCHABLE DEVICES 195

based devices, the redox-active module is either annulated to the porphyrin core, electronically decoupled via an orthogonal *meso*-linkage or attached via a saturated spacer. In all these systems, the diminution of the chromophore emission in the reduced state of the control unit is attributed to *ET* from electron donating TTF^0 to the excited chromophore. However, contrary to the quinone/hydroquinone redox couple, oxidation of TTF^0 can open up a new fluorescence quenching pathway such as *ET* to the electron deficient radical cation $TTF^{+\bullet}$ that also results in emission quenching thereby yielding an unefficient *OFF/OFF* switch. To the best of our knowledge, the only TTF-based fluorescent redox switch where the strong diminution of the porphyrin emission by TTF^0, here by 98 %, can be oxidatively reversed is mono-TTF-annulated porphyrin **15**, a two component device with no significant electronic interaction between both modules shown in Figure 5.3.[80]

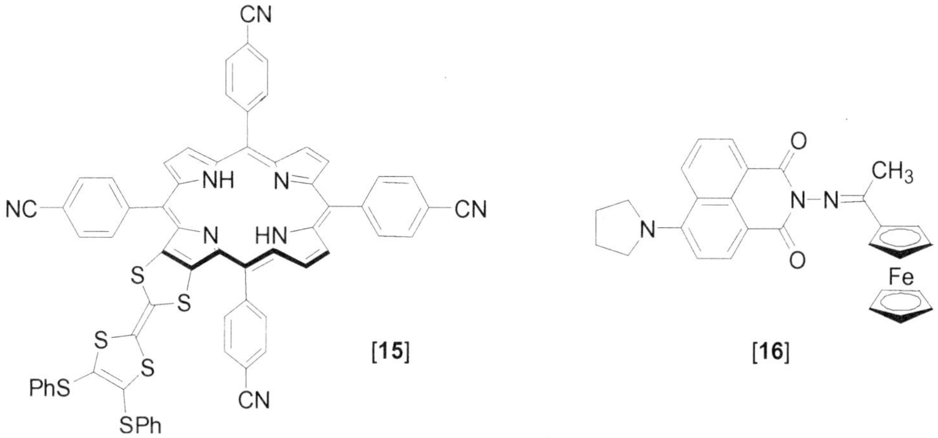

Figure 5.3. Molecular structures of selected TTF- and ferrocene-type fluorescent redox switches.

A representative example for an *OFF/ON* fluorescent redox switch employing the ferrocenium/ferrocene redox couple is the ferrocene-naphthalimide dyad **16** that undergoes a large fluorescence enhancement upon conversion of electron donating ferrocene into electron deficient ferrocenium.[81] Plausible fluorescence quenching pathways are both *ET* and *EnT* with the authors favoring *ET* due to the small spectral overlap between the naphthalimide emission and the ferrocene absorption. However, not only ferrocene but ferrocenium as well can act as fluorescence quencher, often via *EnT*.[82]

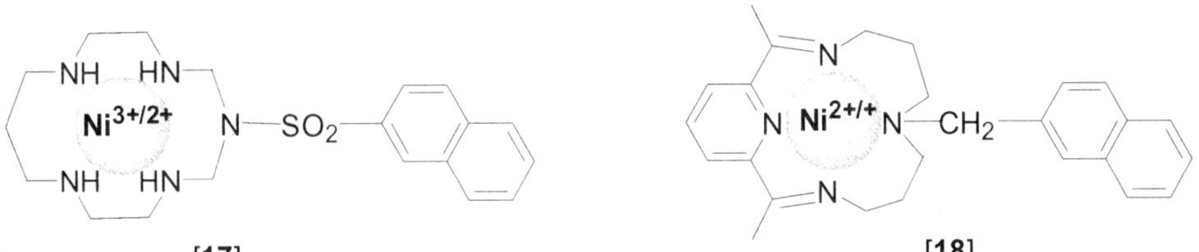

Figure 5.4. Examples for *OFF/ON* and *ON/OFF* fluorescent redox switches operated through ligand stabilized redox-active transition metal ions; Ni^+ equals $[Ni^{II}(L^{\bullet})]^+$.

The vast majority of fluorescent redox switches is operated via ligand stabilized redox-active transition metal ions acting as (oxidation state dependent) luminescence quenchers via either *ET* or *EnT*. These devices are typically three component systems in which the fluorescent signaling unit and the multidentate receptor coordinatively binding the electro-active transition metal ion are electronically decoupled by a spacer, commonly a short alkyl linkage.[83] Common switches combine an organic chromophore such as anthracene, naphthalene or dansyl and a tetrathia or tetraaza macrocycle bound to either copper or nickel.[84] Here, typically only the oxidized metal ion, i.e., Cu^{II} or Ni^{III}, diminishes the emission of the adjacent fluorophore. Representative *ET*-type *OFF/ON* and rare *ON/OFF* devices are molecules **17** and **18** shown in Figure 5.4.[87,85] However, also for metal centered redox couples, there exist examples of unfavorable *OFF/OFF* switches.[87]

Advantages of this class of modular fluorescent redox-switchable systems are the potential of the (uncomplexed) molecules to act additionally as fluorosensors for heavy and transition metal ions,[71] the possibility to modulate the redox potential of the metal centered redox couples by varying the nature of the hosting coordinative environment, and redox-switching of the properties of the metal center such as for instance stereochemical preferences, electron transfer tendencies, binding affinity towards a donor atom or magnetism. The latter can be for instance exploited for the design of sophisti-cated molecular and supramolecular multicomponent devices such as for instance systems undergoing electrochemically switched anion translocation,[86,87] intramolecular pH-driven translocation of metal ions,[88,89] or intramolecular motion.[90] An interesting extension of this type of fluorescent electro-active device are for example redox-switchable *ET*-systems based on trimetallic complexes.[91]

5.2.3. Fluorescence Signalling of Redox-Active Analytes

Modular fluorescent redox-active systems have also considerable analytical potential such as the sensitive and ideally selective fluorescence communication of redox-active analytes and radicals. Fluorosensors with oxidation state dependent emission play for instance an important role in the investigation of metabolisms and for the detection of reactive oxygen and nitrogen species, often in combination with imaging techniques.[92] The majority of abiotic reporter molecules are not built via a composite strategy but are of fluorescent indicator type design with reactive groups attached to the actual chromophore, for instance a fluorescein or rhodamine core, and are typically only highly fluorescent after reaction with the species to be signalled.[93-95] For composite architectures emerging for sensing applications, following the design strategy introduced in scheme 1, an efficient control unit should undergo a fast and ideally selective redox reaction with the analyte of interest thereby generating a detectable and specific change in the emission of the signaling unit. Analytically favored is here a strong fluorescence enhancement as compared to an analyte induced decrease in fluorescence since the former yields an improved signal-to-noise ratio. Contrary to fluorescent redox switches requiring reversibility of the switching event, reversibility is of no importance here, i.e., reversible and irreversible redox reactions can be equally considered.

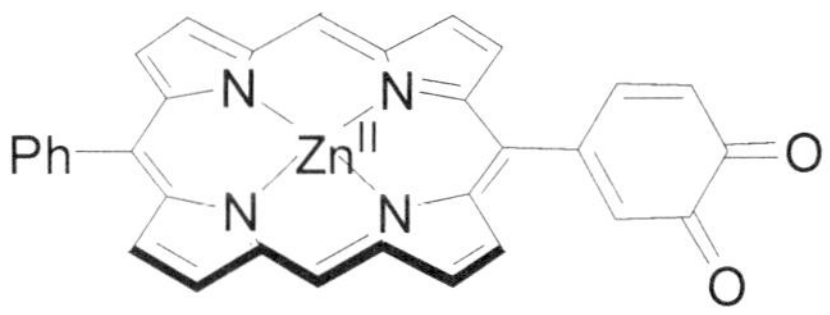

[19]

Figure 5.5. Fluorescent redox switch acting as fluorosensor for redox-active analytes, here 1,2-dihydroxybenzenes.

A straigthforward approach to modular fluorescent redox sensors relies on the exploitation of the chemical addressability of the control unit of a fluorescent redox switch such as for instance the *o*-benzoquinone appended zinc porphyrin **19** shown in Figure 5.5.[96] The emission of **19**, characteristic of a *meso*-aryl substituted zinc porphyrin, which is di-minished via *ET* in the oxidized state of C, can be revived upon reduction of C with 1,2-dihydroxybenzene derivatives. Herewith, **19** acts as (reversible) fluorosensor for these molecules. The magnitude of the fluorescence enhancement, i.e., a factor of maximum 3, is affected by the oxidation potential of the reducing agents reflecting the expected redox potential controlled (limited) selectivity of such a fluorescent redox sensor.

Another class of fluorescent redox sensors with emerging analytical potential are fluorophore-nitroxide (F-S-N-O•) or so-called fluorescence-spin molecules that undergo an irreversible redox reaction with the species to be signaled and, as a unique feature, enable simultaneous communication via optical spectroscopy and electron spin resonance. As is illustrated in scheme 2 and Figure 5.6, these molecules are constructed from an organic fluorophore such as for example a naphthalene, stilbene, pyrene, dansyl or coumarin chromophore, a spacer, and a stable paramagnetic nitroxide-type control unit (N-O•), a strong quencher of fluorescence predominantly via either *ISC*, *IC* or *ET*.[72-74] These generally weakly emissive fluorescent redox sensors respond to any changes of the nitroxide fragment with a change in emission. For instance, reduction of the nitroxide fragment or addition of a (carbon centred) radical yield a stable diamagnetic hydroxylamine or *O*-alkylhydroxylamine, respectively, thereby eliminating the intramolecular fluorescence quenching pathway, with the resulting analytically favorable fluorescence enhancement, typically a factor of several ten, reflecting radical/redox scavenging.

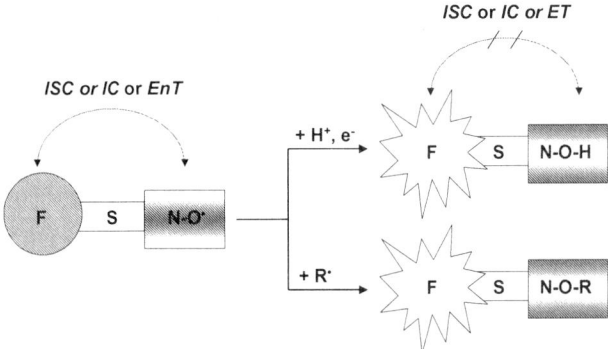

Scheme 5.2
Fluorophore-nitroxide-type fluorosensor (F-S-N-O•) for (carbon centered) radicals (upper part) and reducing agents (lower part); F: fluorophore, S: spacer, N-O•: paramagnetic nitroxide acting as redox-active control unit.

Advantageous of this class of fluorosensors is the tremendous flexibility offered by the composite architecture in combination with the *ET*-type design, with the chemical selectivity and sensitivity predominantly connected to the nitroxide module and the signaling features to dye architecture and photophysics. Fluorophore-nitroxide molecules enable for instance the quantification of (carbon centered) radicals[97,98] and the antioxidant ascorbate[99,100] in solution and in biological liquids as well as monitoring of nitric oxide production.[101] Trapping of singlet oxygen can be performed with a fluorophore-heterocyclic amine thereby converted into the corresponding only weakly emissive fluorophore-nitroxide molecule.[102] Other areas of application are emerging in materials sciences with fluorosensors such as **20**, see Figure 5.6, serving as mechanistic tools for the study of radical polymerization reactions and photoinitiated radical processes in polymer films or the preparation of functional fluorescent images.[103,104]

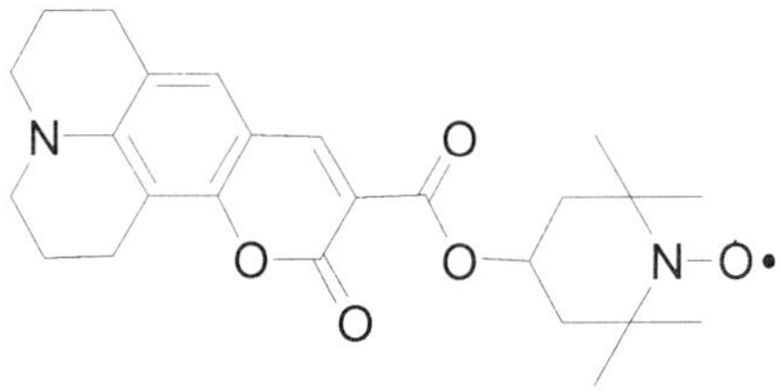

[20]

Figure 5.6. Fluorophore-nitroxide-type fluorosensor.

5.3. FLUORESCENT REDOX-ACTIVE AND ANALYTE-RESPONSIVE DEVICES

For applications as advanced sensory devices, modular systems are desired that contain more than one receptor and/or reporter unit.[105] Such constitution which for instance permits simultaneously or cooperatively sensing of different species, signaling of analyte coordination by two or more macroscopically observable events or a combination of both, allows tuning of sensitivity and selectivity by the choice of the functional components. In the field of fluorescent redox-switchable molecules, of special interest are analyte responsive systems for simultaneous optical and electrochemical guest communication and, as an extension of the presented fluorescent redox switches and sensors, fluorescent redox-switchable devices for the sensitive and selective detection of charged and neutral species with oxidation state dependent guest affinity, optical properties, and fluorescence signaling.[47-49] Analyte recognition implies here reversible binding of a guest by a receptor, followed by the transduction of the binding event into a measurable signal, not a chemical redox reaction between both compounds yielding a fluorometrically detectable product as illustrated in the last section. Composite systems for fluorometric and electrochemical analyte sensing are only briefly discussed here whereas the redox-switchable and analyte responsive fluorosensors designed and studied by us are presented in detail. In any case, i.e., combining either electrochemical and optical sensing methodologies or redox-activated and analyte triggered fluorescence signal generation, the design of analytically favorable systems requires consideration of the well established design strategies of analyte responsive fluorescent sensors and switches.[106] Common fluorosensors are either of *ET*, charge transfer (*CT*) or exciplex/excimer forming type, at least for the detection of protons, metal ions, and many anions. They typically consist of an analyte coordinating receptor acting as control unit (free/bound) and a chromophore generating the fluorescence signal, with the nature of the spacer, i.e., saturated or unsaturated, covalently linking both modules depending on the underlying photophysical mechanism. *ET* probes with their electronically decoupled building blocks report the binding event commonly only by a change in fluorescence quantum yield and lifetime, similarly to redox-activated optical signal generation in *ET* or *EnT* based fluorescent redox switches. In *CT*-active, so-called intrinsic sensor molecules with their electronically connected subunits, typically chosen to act as donor and acceptor moieties, analyte coordination to the donor or acceptor part of the molecule yields an analyte mediated change of the *CT* process. This change is reflected by spectral shifts in absorption and, depending on the actual binding site involved, also in emission, as well as moderate (compared to *ET* probes) alterations in fluorescence intensity and lifetime. Exciplex or excimer forming fluorosensors that are based on a flexible architecture with the receptor unit(s) integrated into the spacer allow for control of the exciplex/excimer-to-monomer-emission by the receptor's binding state. The sensing properties of such systems generally depend on the receptor controlled (chemical) selectivity and the analyte mediated signaling (spectroscopic) selectivity as well as sensitivity. Furthermore, signal generation is affected by the receptor's binding mode, the electronic nature of the analyte, i.e., its potential to act as fluorescence quencher for instance via the heavy atom effect such as closed shell diamagnetic Hg^{II} or via enhanced spin-orbit coupling as well as *ET* or *EnT* as is typical for open shell and redox active transition metal ions such as copper and nickel,[71] and complexation-induced structural effects including counterproductive cation-fluorophore interactions.[107] The design of efficient devices also demands consideration of the latter effects.

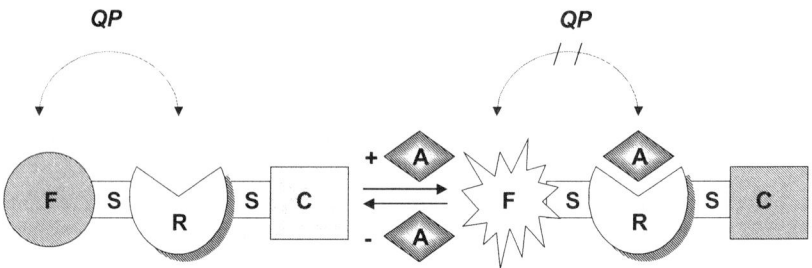

Scheme 5.3

Fluorescent redox-active device for electrochemical and optical analyte (A) sensing. Here, the binding state of the receptor (R) controls the output of the electro-active (C) and fluorescent (F) module. The underlying fluorescence quenching process (QP) can be for instance *ET*. The chelation induced change in redox potential of C is depicted by a change in color.

Fluorescent redox-active devices for (simultaneous) optical and electrochemical analyte sensing and redox-switchable fluorosensors with oxidation state dependent guest affinity, optical properties, and signaling are both constructed from an analyte binding, a redox-active, and a fluorescent component properly arranged with/without spacers. In the former type of system, depicted in scheme 3, the analyte coordinating module, via its binding state, acts as control unit for the two signaling units, i.e., the electro-active and fluorescent moiety.[108] Selected examples for sensor molecules designed for electrochemical and fluorometric analyte communication of metal ions are shown in Figure 5.7. **21**[109] and triad **22**, combining two redox-active bisthiophene-hydroquinone chromophores and the acceptor-type terpyridine receptor,[110,111] act as heavy metal ion sensors whereas **23** was constructed for signaling of alkali metal ions.[112] This design strategy can be extended to anion sensors, for instance by incorporating a redox-active and emissive ruthenium or osmium bipyridyl complex into an acyclic or macrocyclic receptor.[113,114] For this type of device, the design of efficient systems undergoing both strong shifts in redox potential and modulations in emission upon the binding event, i.e., favorably a fluorescence increase, requires also minimization of fluorescence quenching interactions between the fluorophore and the electrochemically sensing moiety.

FLUORESCENT REDOX-SWITCHABLE DEVICES

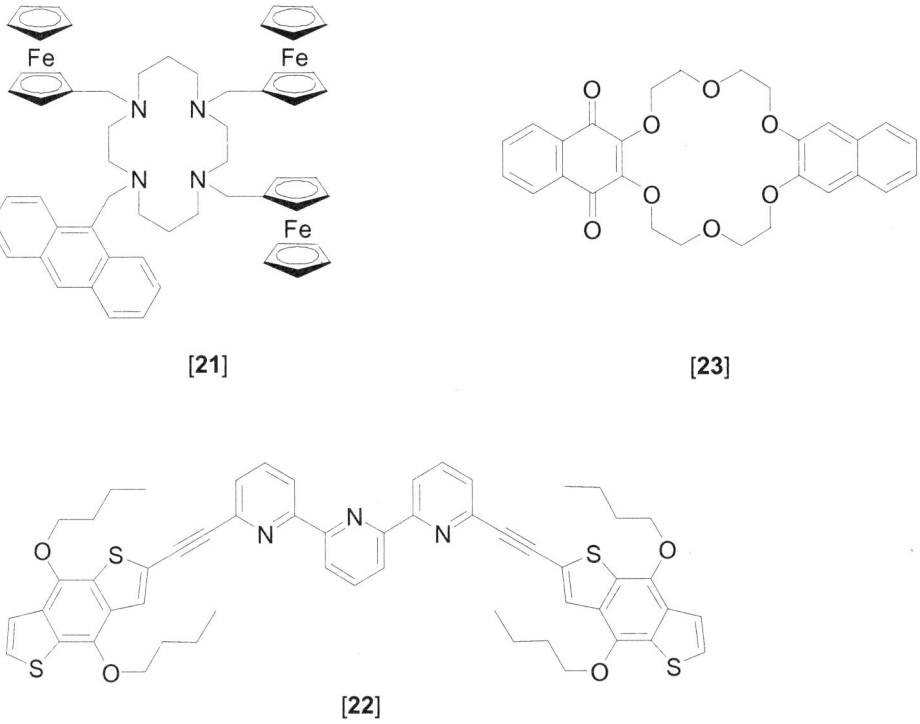

Figure 5.7. Fluorescent redox systems designed for electrochemical and optical analyte sensing.

The working principle of redox-switchable fluorosensors with oxidation state triggered analyte selectivity and oxidation and binding state dependent emission is illustrated in scheme 4. Here, redox chemistry is not only used to control optical signal generation but also to enhance or diminish the binding ability of ligands for guests. The key point for the construction of such devices is a properly selected bistable redox-active control unit that simultaneously acts as analyte binding site. A favorable redox-switchable receptor module should strongly differ in the electronic nature, i.e., donor capacity and hydrogen bonding ability, and geometric arrangement of the analyte coordinating heteroatoms in its two accessible oxidation states and should ideally affect the emission of the neighboring fluorophore to a clearly distinguishable extent in the diffe-rent oxidation and binding states. The latter is the prerequisite for the desired spectroscopic discrimination of the two oxidation states of the unbound probe and fluorometric analyte detection in each oxidation state.

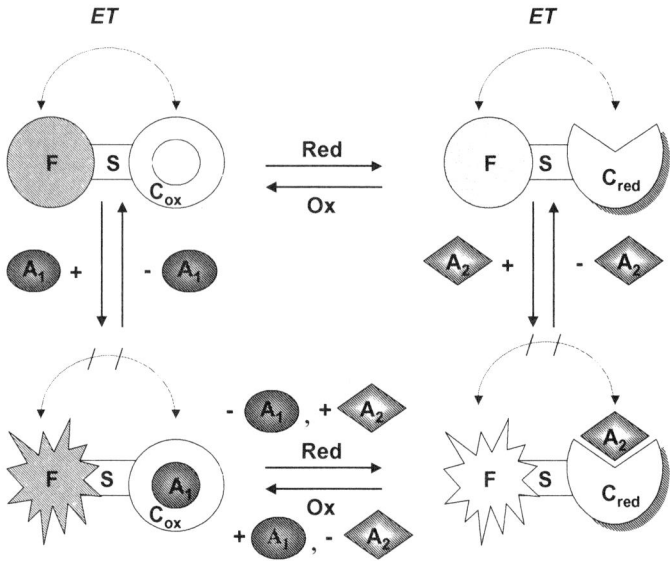

Scheme 5.4
Operating principle of a fluorosensor, here of *ET*-type, with redox control of guest selectivity and emission constructed from a redox-active control module (C), simulta-neously acting as analyte binding site, and a fluorophore (F), linked via a spacer (S). A_1 and A_2 present different analytes and the different shapes and colors of the signaling unit different fluorescence properties.

5.3.1. Fluorosensors with Redox Control of Analyte Recognition and Fluorescence

As an extension of our research on *ET* and *CT*-operative fluorescent probes for metal ions, with the ultimate goal of multiparameter analyte sensing and fast sensor regeneration, we followed the idea of constructing redox-switchable devices for the oxidation state selective fluorometric detection of heavy and transition metal ions and anions via incorporation of properly designed redox-addressable analyte binding sites into the molecular architecture of analyte responsive fluorosensors. These systems should basically present four-state fluorescent redox switches responding to two different types of external stimuli, i.e., chemically induced redox reactions and chemical inputs, see scheme 4. As follows from the fluorophores selected, i.e., anthracene and naphthalene chromophores, the systems illustrated here are built for "proof of principle" of this design concept and not for "real life" analytical applications typically requiring water soluble molecules emitting in the visible or NIR region.

The key building block for the rationalization of redox-switchable and analyte responsive fluorosensors is a bifunctional receptor, here suited for binding either heavy and transition metal ions or anions, that enables both redox control of its guest affinity and transduction of its oxidation and binding state via the emission of the adjacent fluorophore. The coordination of soft heavy and transition metal ions requires ligands

FLUORESCENT REDOX-SWITCHABLE DEVICES 203

containing soft donor atoms such as nitrogen or sulfur spatially arranged to meet the coordinative needs of the cation to be bound[115,116] whereas binding of anions with their broad variety of molecular geometries demands receptors with ligating sites whose geometric arrangement enables electrostatic, hydrogen bonding, coordinative, and/or hydrophobic host guest interactions.[9,117] Contrary to the receptors often used in electrochemical sensing systems such as coordination sites functionalized with redox-active groups, where the redox-induced production of charged groups modifies the binding ability of the adjacent ligating site,[118] or other typical ligands with redox-switchable binding affinity for guests,[119] we chose a different approach that relies on the manipulation of the receptor's molecular architecture by reversible formation and cleavage of covalent bonds upon oxidation/reduction. As initially demonstrated in switchable cation receptors, the requirements for such a switching process can most easily be achieved by incorporating heteroatom-heteroatom bonds with comparably low bond energies into the molecular scaffold. Upon oxidation or reduction, the respective bond is closed or opened, delivering two distinct, chemically different, and stable forms of the binding site. Prominent examples for such redox-switchable building blocks are the thiol-dithiol moiety[120] or heterocyclic units that contain a sulfur-nitrogen bond as is for instance found in the thiapentalenes.[121,122] Due to the presence of analyte coordinating nitrogen and sulfur heteroatoms and the redox-activated change in their donor capacity, geometric arrangement, and hydrogen bonding ability, such systems should provide the desired redox-switchable affinity of the receptor for heavy metal ions and anions as well as, not exploited here, for neutral molecules. This redox-generated change in chemical nature should principally also assure a different influence of both oxidation states of the control unit on the fluorophore emission.

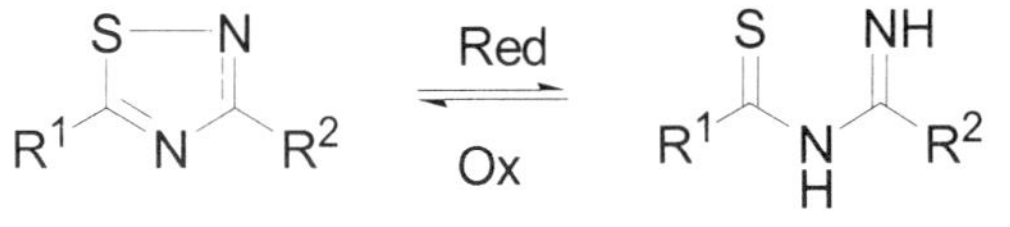

[24] ⇌ (Red/Ox) [25]

Figure 5.8. Reductive conversion of 1,2,4-thiadiazole into thiocarbonyl amidine and oxidative back reaction.

For the rationalization of our goal, we focused our attention on 1,2,4-thiadiazoles and 1,6,6aλ^4-trithia-3,4-diapentalenes. In 1,2,4-thiadiazoles, the heterocylic N-S bond can be chemically reduced to give the open N-thiocarbonyl amidine, see Figure 5.8. Upon reversion of this process, the thiadiazole heterocycle is oxidatively closed again. The straight forward synthesis of both, the 1,2,4-thiadiazole[123] as well as the corresponding ring opened form in combination with their interesting redox[124] and coordination chemistry[125] made this system a particularly attractive candidate for use as a redox-active and analyte coordinating building block. Additionally, both forms contain notably different binding sites, regarding steric features as well as electron donating capacity and hydrogen bonding ability of the S- and N-heteroatoms and were expected to distinguishably affect the emission of the fluorophore, in ET-operative systems due to their different redox potentials and in *CT*-operative systems due to their different donor/acceptor strength. In cation responsive fluorosensors, the electron rich thiadiazole can serve as coordination unit for soft metal cations, and *N*-thioacyl amides and *N*-iminoyl thioureas are well known ligands for heavy and transition metals for over twenty years.[118,119] The most intriguing features connected to the thiadiazole/*N*-iminoyl thiourea system for the equally desired construction of anion sensing systems are the reduction-induced increase in the number of free NH groups and the conversion of the thiadiazole sulfur into a sulfur atom of a thiocarbonyl group thereby changing the donor capacity of this heteroatom.[9] Also, the guanidyl, amide, (thio)urea or thioamide motives are often used for anion coordination.[9,126,127] In parallel, we constructed redox-switchable fluorosensors from 1,6,6aλ^4-trithia-3,4-diapentalenes and their reduced trithiourea derivatives, which have been for instance successfully employed as redox-active building blocks in macrocyclic receptors showing an oxidation state dependent cation complexation behavior.[128] However, the resulting fluorosensor systems were strongly hampered by the limited stability of either one of the partners of the respective redox couple, usually the ring opened form. The thiapentalene-based systems are thus not further discussed here.

5.3.1.1. Fluorosensor Systems for the Detection of Heavy and Transition Metal Ions

For the construction of the first fluorosensor system for heavy and transition metal ions with redox-controlled analyte recognition and fluorescence output, we chose the 1,2,4-thiadiazole/*N*-iminoyl thiourea analyte binding and redox-active control system with its two oxidation states being interconvertible into each other via chemically induced (2-electron) redox processes and an *ET*-type design. To achieve an analytically favorable fluorescence enhancement upon metal ion coordination, a strong, but distinguishable diminution of the fluorophore emission was desired for both analyte-free oxidation states of the receptor. Furthermore, despite of the rather simple design, we hoped for at maximum minimal contraproductive fluorophore-metal ion interactions. This is a prerequisite for the observation of a chelation induced increase in emission especially with classical quenchers of fluorescence such as most heavy and transition metal ions.

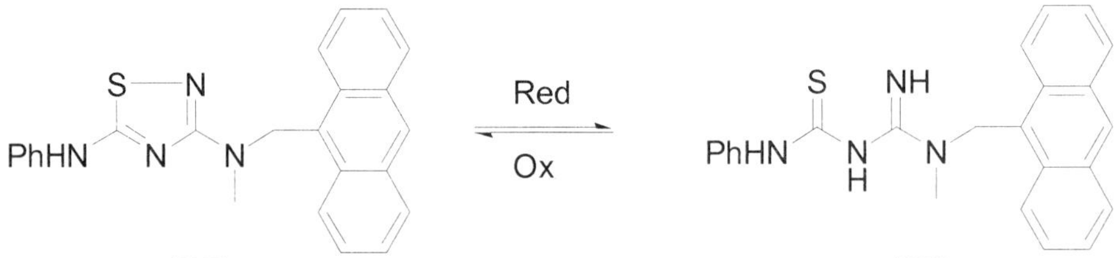

Figure 5.9. *ET*-operative redox-switchable fluorosensor system **26/27** for heavy and transition metal ions.

Tethering the 1,2,4-thiadiazole (C_{ox}) or *N*-iminoyl thiourea (C_{red}) receptor module via an alkyl spacer to anthracene yields fluorosensors **26** and **27** shown in Figure 5.9

presenting the two components or oxidation states of the first fluorescent redox-switchable system for the detection of heavy and transition metal ions.[129] As a consequence of the notably different binding sites and different redox potentials of the receptor modules, **26** and **27** display clearly distinguishable fluorescence and cation complexation properties as well as signaling features, see Figure 5.10. As typical for an *ET*-active device, the emission spectra are barely affected by the oxidation and binding state. However, the fluorescence quan-tum yields (ϕ_f; **26**: $\phi_f = 0.0027$, **27**: $\phi_f = 0.039$) display both the desired diminution and notable differences, here a factor of 14. Amongst a series of bivalent heavy and transition metal cations tested, the heterocyclic oxidized form **26** of the fluorosensor system signals selectively the presence of Hg^{II} with a 44-fold fluorescence enhancement. Herewith, this molecule belongs to the comparatively few cation responsive fluorescent probes showing an increase in emission even with the classical fluorescence quencher mercury.[130-134] The size of the fluorescence enhancement almost approaching the maximum achievable effect, i.e., an increase in fluorescence by a factor of 47 obtained upon protonation, and the very similar emission spectra of the mercury bound and free fluorosensor suggest the desired minimum interaction between the receptor coordinated metal ion and the anthracene π-system.[110]

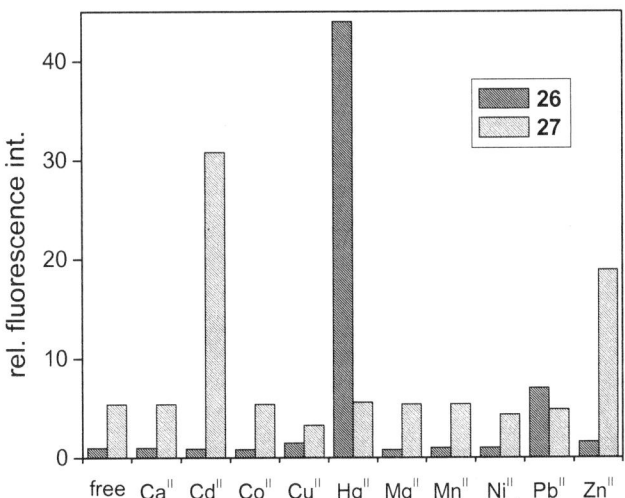

Figure 5.10. Selectivity pattern of the response of **26** and **27** to various heavy and transition metal ions in acetonitrile based on the cation induced changes in fluorescence intensity of the free fluorosensor; complete complexation, metal-to-ligand concentration ratio of 100:1 and only in the case of cadmium of 164:1.

Contrary to **26**, the ring opened reduced form **27** responds predominantly to the presence of Cd^{II} and Zn^{II} with distinct changes, i.e., a 6- and a 3.6 fold increase in emission, see Figure 5.10. As initially favored, for this redox-switchable and metal ion responsive fluorosensor system, not only the fluorescence properties of the unbound molecules **26** and **27** and the fluorometrically transduced metal ions preferences clearly differ, but also the fluorescence quantum yields of the Hg^{II} complex of **26** and the Cd^{II} complex of **27** with values of 0.12 and 0.22, respectively.

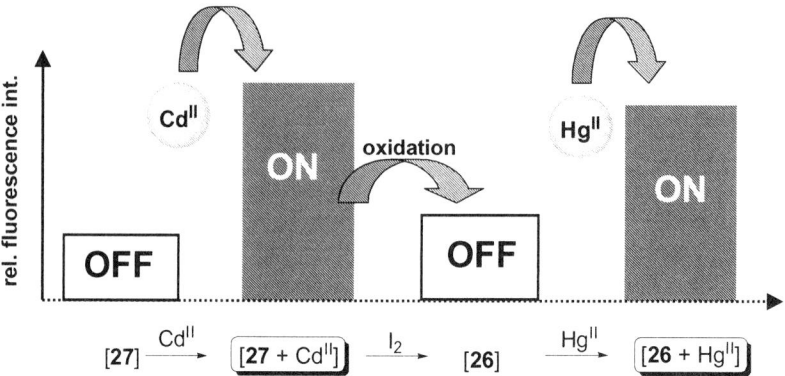

Figure 5.11. Redox-switchability of the analyte affinity and signaling behavior of the fluorosensor system **26/27**.

The redox-switchability of the analyte recognition and emission of fluorosensor system **26/27** could be demonstrated with a simple cuvette experiment. As follows from Figure 5.11, the weak fluorescence of **27** is switched *ON* upon addition of Cd^{II}, i.e. forma-tion of the Cd^{II}-**27**-complex, and *OFF* again by adding iodine thereby oxidizing cadmium coordinated **27** to the corresponding weakly emissive thiadiazole **26**. Spectroscopically there is no hint for the complexation of **26** to Cd^{II}. Finally, the fluorescence is switched *ON* again upon addition of Hg^{II} yielding the moderately fluorescent Hg^{II}-**26**-complex. This presents also an elegant example for the desired fast sensor regeneration.

To the best of our knowledge, fluorosensor system **26/27** that basically consists of two analyte responsive *ET*-active *OFF/ON* switches connected via a chemical redox reaction, presents the first example for a device permitting redox-switchable analyte recognition and fluorescence. In this four state system, see scheme 4, the redox-active receptor module controls not only the analyte affinity, optical properties, and fluorescence output, but also the maximum achievable signal, i.e., chelation induced fluorescence enhancement in each of its two accessible oxidation states. Due to the *ET*-type design, the clearly distinguishable oxidation and binding state dependent fluorescence properties are restricted to differences in fluorescence quantum yield. Also, though not measured, notable differences in fluorescence lifetime are to be expected assuming similar radiative rate constants.

FLUORESCENT REDOX-SWITCHABLE DEVICES

[Structures of compounds 28, 29, and 30 with Red/Ox equilibrium arrow between 28 and 29]

[28] ⇌ (Red/Ox) ⇌ **[29]**   **[30]**

Figure 5.12. Metal ion responsive redox-switchable fluorosensor system **28/29** and urea compound **30**.

A structurally more simple 1,2,4-thiadiazole/N-iminoyl thiourea-based redox-switchable fluorosensor system for the detection of heavy and transition metal ions present molecules **28** and **29** shown in Figure 5.12.[135] Here, the redox-active control unit is directly linked to the nitrogen atom of an 1-aminonaphthalene chromophore. Both oxida-tion states of the fluorosensor system are only weakly emissive (**28**: ϕ_f = 0.002; **29**: ϕ_f = 0.005) and notably differ in the spectral position of the emission bands suggesting electronic interactions between the redox-active and signaling module.

As depicted in Figure 5.13 (left part), addition of Cu^{II} to weakly emissive **28** selectively leads to both spectral shifts and a remarkable increase in fluorescence by a factor of 46 (0.5 equivalents Cu^{II}). Chelation enhanced emission is comparably rarely observed for the classical fluorescence quencher Cu^{II}.[136-139] The strongly red shifted emission spectrum occurring at higher Cu^{II} concentrations (Figure 5.13, left part, 2 equivalents Cu^{II}) is attributed to excimer formation, here interaction of two ligands in the 1:2 Cu^{II}-**28**-complex. The spectroscopic observation of different coordination modes, i.e., different Cu^{II} complexes, points to various possible binding sites and conformers with Cu^{II} coordination by **28** occuring at the NH group of the 1-aminonaphthalene chromophore as well as the thiadiazole ring.

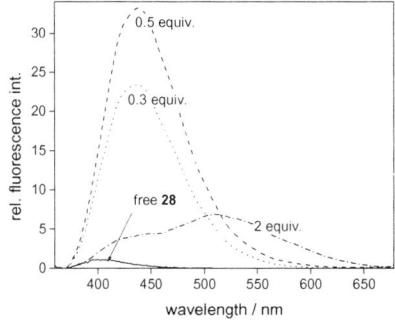
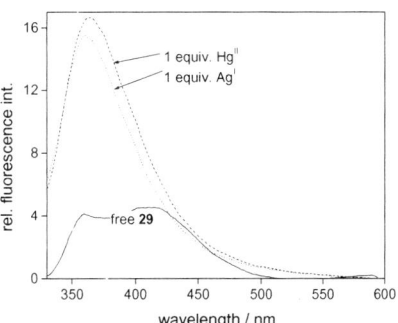

Figure 5.13. Left: Cu^{II} induced modulations in emission of **28** in acetonitrile, addition of various Cu^{II} concentrations; right: Change of fluorescence of **29** after reaction with Hg^{II} or Ag^{I} yielding the respective cation-**30**-complexes. The emission spectra are not weigthed by fluorescence quantum yield.

The reduced form **29** signals selectively the presence of thiophilic Hg^{II} and Ag^{I}, see Figure 5.13 (right part), with a strong increase in emission by a factor of about 30 observed for both cations. Here, desulfurization of iminoyl thiourea **29** by silver and mercury cations yielding the urea derivative **30**, see Figure 5.12, followed by its complexation to the respective metal ions, accounts for the observed effects. Thus, **29** acts as selective chemo-dosimeter for Hg^{II} and Ag^{I}.[140]

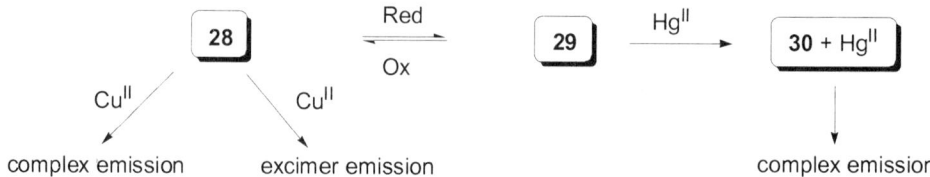

Figure 5.14 Schematic presentation of the cation triggered fluorescence response of the two oxidation states of the redox-switchable fluorosensor system **28/29**.

Summarizing the overall oxidation state dependent and analyte triggered fluorescence response of the redox-switchable fluorosensor system **28/29**, as is illustrated in Figure 5.14, also these structurally simple molecules permit redox controlled analyte recog-nition, emission, and signaling. The four oxidation and binding states involved differ not only in fluorescence quantum yield as is observed for the fluorosensor system **26/27** but also spectrally.

5.3.1.2. Fluorosensor Systems for the Detection of Anions

After having successfully combined the concepts of redox-activated and analyte triggered fluorescence signaling for the detection of heavy and transition metal ions, for further "proof of principle" of this design concept, we aimed at extending this design strategy to anion responsive fluorosensors with oxidation and binding state dependent emission. This led us incorporate the redox-switchable receptor system 1,2,4-thiadiazole/*N*-iminoyl thiourea into a bichromophoric molecular architecture thereby taking advantage of the oxidation state controlled donor capacity, hydrogen bonding ability, and geometric arrangement of the heteroatoms of this module for redox-controlled anion recognition. The straightforward synthetically accessible and structurally simple fluorosensors **31** and **32** containing two naphthalene chromophores and the corresponding fluorosensor system **33/34** equipped with two receptor side arms are shown in Figure 5.15.[141]

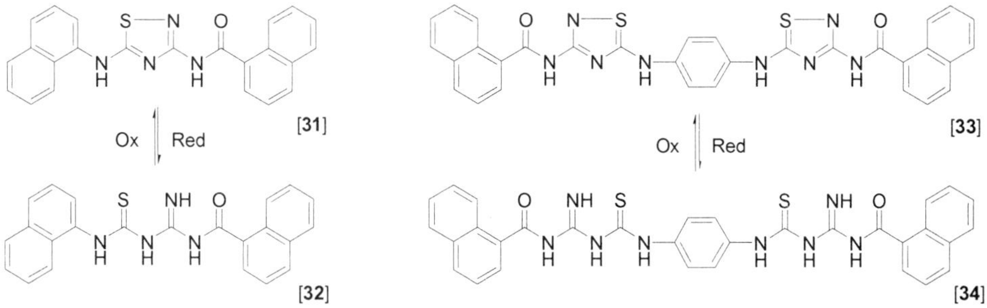

FLUORESCENT REDOX-SWITCHABLE DEVICES

Figure 5.15. Anion-responsive redox-switchable fluorosensors **31** and **32** and the corresponding fluorosensor system **32/33** equipped with two receptor side arms.

Due to the presence of potential hydrogen bond donors such as the guanidyl, amide, (thio)urea or thioamide motives,[9,120] the ring opened thiourea form was expected to be particularly useful for the coordination of simple oxo-anions such as hydrogen phosphate, hydrogen carbonate, and carbonate. For the desired oxidation state controlled fluorometric anion recognition, the most intriguing features connected to the reduction of the control unit are the increase in the number of free NH groups and the conversion of the thiadiazole sulfur into a sulfur atom of a thiocarbonyl group. These changes strongly alter the hydrogen bonding ability as well as the sulfur's donor capacity upon conversion of the thiadiazole-based fluorosensors **31** and **33** to the corresponding *N*-iminoyl thiourea systems **32** and **34**.

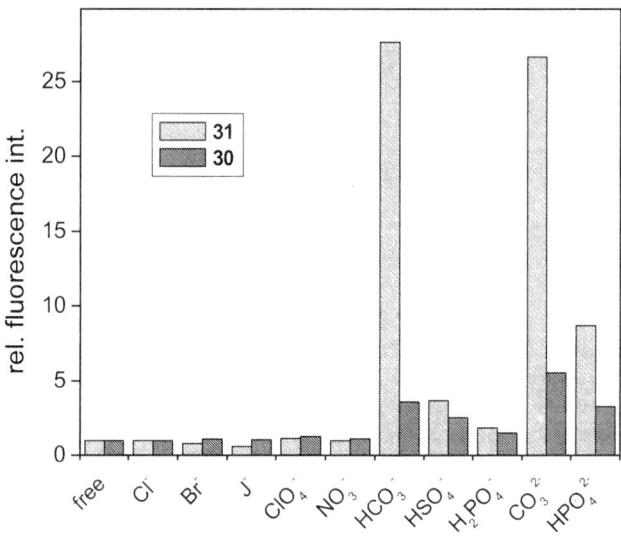

Figure 5.16. Fluorescence intensity of **31** and **32** in methanol in the presence of a 1000-excess of various anions.

The spectroscopic features of the fluorosensors **31**, **31**, **33**, and **34** in the protic solvent methanol chosen for the anion complexation studies are characterized by a weak and broad emission with the position of the fluorescence maximum being slightly affected by the oxidation state, i.e., red shifted by 9 or 5 nm for the oxidized state, respectively. Amongst the anions studied, all the sensor molecules respond selectively to the oxo-anions dihydrogen phosphate, hydrogen carbonate, and carbonate with a change in emission, predominantly an increase in intensity. The observed selectivity pattern clearly depend on the oxidation state of the fluorosensor systems investigated. In Figure 5.16, the anion-induced fluorescence effects are exemplarily shown for **31** and **32**. Hydrogen bond mediated complexation of dihydrogen phosphate, hydrogen carbonate, and carbonate leads to an exceptionally high fluorescence enhancement with factors of 9, 27, and 27 in the case of compound **32** or even 12, 39, and 53 for its bidentate analogue **34**. Herewith, both molecules are amongst the very few anion responsive fluorescent probes undergoing a considerable chelation induced increase in emission,[142-146] with **34** showing a slightly improved selectivity.

The spectroscopic and complexation behavior of **31/32** and **33/34** clearly demonstrates the feasibility of transferring our design strategy successfully employed for the rationalization of redox-switchable and cation responsive fluorosensors to similarly operating systems for anions. Even with the structurally simple fluorosensors systems **31/32** and **33/34**, redox control of the anion affinity and fluorescence can be achieved with the reduced N-iminoyl thiourea forms showing improved anion recognition behavior, i.e., an analytically favorable strong chelation induced fluorescence increase and enhanced selectivity over their thiadiazole counterparts. In addition to the notable differences in fluorescence quantum yield, the two analyte-free oxidation states of the fluorosensor systems and the anion complexed molecules slightly differ spectrally.

5.4. CONCLUSION AND OUTLOOK

The combination of redox activity and fluorescence signaling seems to be particularly intriguing in the context of molecular electronics and information transfer and storage at the molecular level as demonstrated by the introduced composite fluorescent redox systems. Here, molecular devices appear at the moment the most realistic, and at the same time the most potent link between chemistry on the molecular scale with "real-life" technological applications. Although considerable efforts need to be dedicated to push the limits on the molecular level, the state of the art on the pure chemistry side appears to be perfectly ready to take it on further towards application in an interdisciplinary fashion.

As follows from our own research on redox-switchable fluorosensors for heavy and transition metal ions as well as anions - employing properly designed redox-switchable analyte binding sites - even with compartimental fluorescent devices of fascinating structural simplicity, stunning operations can be conducted such as the demonstrated redox-control of analyte recognition and emission. These novel systems offer multiparameter sensing, i.e., signaling of the molecules's oxidation and binding state, and enable fast sensor regeneration. Based on the design strategies presented by us, for "real life" applications, further efforts need to be devoted to a second generation of fluorescent redox-switchable and analyte responsive devices, for a broader variety of analytes and ideally capable of multiparameter sensing in water in the Vis/NIR spectral region as well as multi-cycle reversible redox-switching. This requires the incorporation of suited fluorophores and sufficiently stable, easily switchable, and properly functionalized building blocks offering oxidation state dependent analyte affinity incorporated into the fluorosensor architecture. Furthermore, relying on the presented construction principles, the rationalization of bichromophoric *EnT* and *ET* operative redox-switchable devices seems to be particularly attractive for redox-controlled analytically favorable ratiometric analyte sensing.

5.5. ACKNOWLEDGEMENT

Financial support by the German Research Council (DFG) is gratefully acknowledged.

5.6. REFERENCES

1. B. L. Feringa, *Molecular Switches* (Wiley-VCH GmbH, Weinheim, 2001).

2. F. M. Raymo, Digital processing and communication with molecular switches, *Adv. Mater.* **14**, 401-414 (2002).
3. G. J. Brown, A. P. De Silva, and S. Pagliari, Molecules that add up, *Chem. Commun.*, 2461-2463 (2002).
4. A. R. Pease, J. O. Jeppesen, J. F. Stoddart, Y. Luo, C. P. Collier, and . R. Heath, Switching devices based on interlocked molecules, *Acc. Chem. Res.* **34**, 433-444 (2001).
5. R. Ziessel, M. Hissler, A. El-ghayoury, and A. Harriman, Multifunctional transition metal complexes: Information transfer at the molecular level, *Coord. Chem. Rev.* **178-180**, 1251-1298 (1998).
6. A. P. de Silva, H. Q. N. Gunaratne, T. Gunnlaugson, A. J. M. Huxley, C. P. McCoy, J. T. Rademacher, and T. E. Rice, Signaling recognition events with fluorescent sensors, *Chem. Rev.* **97**, 1515-1566 (1997).
7. K. Rurack, Flipping the light switch 'ON' - the design of sensor molecules that show cation-induced fluorescence enhancement with heavy and transition metal ions, S*pectrochim. Acta, Part A* **57A**, 2161-2195 (2001).
8. C. Suksai and T. Tuntulan, Chromogenic anion sensors, *Chem. Rev.* **32**(4), 192-202 (2003).
9. P. D. Beer and P. A. Gale, Anion Recognition and Sensing: The State of the Art and Future Perspectives, *Angew. Chem. Int Ed.* **40**, 486-516 (2001).
10. John J. Lavigne and Eric V. Anslyn, Sensing a paradigm in the field of molecular recognition: From selective to differential receptors, *Angew. Chem. Int Ed.* **40**, 486-516 (2001).
11. Special issue on "Luminescent sensors", *Coord. Chem. Rev.* **205** (2000).
12. S. Houbrechts, Y. Kubo, T. Tozawa, S. Tokita, T. Wada, and H. Sasabe, Second-order nonlinear optical properties of functionalized ionophores: Cation-steered modulation of the first hyperpolarizability, *Angew. Chem. Int. Ed.* **39**, 3859-3862 (2000).
13. R. K. Lammi, R. W. Wagner, A. Ambroise, J. R. Diers, D. F. Bocian, D. Holten, and J. S. Lindsey, Mechanisms of excited-state energy-transfer gating in linear versus branched multiporphyrin arrays, *J. Phys. Chem. B.* **105**, 5341-5352 (2001).
14. S. H. Kawai, S. L. Gilat, R. Posinet, and J.-M. Lehn, A dual mode molecular switching device: Bisphenolic diarylethenes with integrated photochromic and electrochromic properties, *Chem. Eur. J.* **1**, 285-293 (1995).
15. R. Deans, A. Niemz, E. C. Breinlinger, and V. M. Rotello, Electrochemical control of recognition processes. A three-component molecular switch, *J. Am. Chem. Soc.* **119**, 10863 – 10864 (1997).
16. M. Engeser, L. Fabbrizzi, M. Licchelli, and D. Sacchi, A fluorescent molecular thermometer based on the nickel(II) high-spin/low-spin interconversion, *Chem. Commun.*, 1191-1192 (1999).
17. M. A. Herranz, N. Martín, J. Ramey, and D. M. Guldi, Thermally reversible C60-based donor–acceptor ensembles, *Chem. Commun.*, 2968-2969 (2002).
18. I. Willner and B. Willner, Photoswitchable biometrials as grounds for optobioelectronic devices, *Bio-electrochem. Bioenerg.* **42**, 43-57 (1997).
19. M. Irie, Diarylethenes for memories and switches, *Chem. Rev.* **100**, 1685-1716 (2000).
20. G. Berkovic, V. Krongauz, and V. Weiss, Spiropyrans and spirooxazines for memories and switches, *Chem. Rev.* **100**, 1741-1753 (2000).
21. D. S. Tyson, C. A. Bignozzi, and F. N. Castellano, Metal-organic approach to binary optical memory, *J. Am. Chem. Soc.* **124**, 4562-4563 (2002).
22. Y. Liang, A. S. Dvornikov, and P. M. Rentzepis, New near infrared-sensitive photochromic fluorescing molecules, *J. Mater.* **13**, 286-290 (2002).
23. J. Daub, T. Mrozek, and A. Ajayagosh, Signaling and expression of electronic effects by dihydroazulene-vinylheptafulvene photochromism, *Mol. Cryst. Liq. Cryst.* **344**, 41-50 (2000).
24. S. Murase, M. Teramoto, H. Furukawa, Y. Miyashita, and K. Horie, Photochemically induced fluorescence control with intermolecular energy transfer from a fluorescent dye to a photochromic diarylethene in a polymer film, *Macromolecules* **36**, 964-966 (2003).
25. A. Fernandez-Acebes and J.-M. Lehn, Optical switching and fluorescence modulation properties of photochromic metal complexes derived from dithienylethene ligands, *Eur. Chem. J.* **5**, 3285-3292 (1999).
26. L. Giordano, T. M. Jovin, M. Irie, and E. A. Jares-Erijman, Diheteroarylethenes as thermally stable photoswitchable acceptors in photochromic fluorescence resonance energy transfer (pcFRET), *J. Am. Chem. Soc.* **124**, 7481-7489 (2002).
27. M. P. Debreczeny, W. A. Svec, and M. R. Wasielewski, Optical control of photogenerated ion pair lifetimes: An approach to a molecular switch, *Science* **274**, 584-587 (1996).

28. M. Irie, T. Fukaminato, T. Saski, N. Tamai, and T. Kawai, A digital fluorescent molecular photoswitch, *Nature* **420**, 759-760 (2002).
29. R. A. G. Cinelli, V. Pellegrini, A. Ferrari, P. Faraci, R. Nifosi, M. Tyagi, M. Giacca, and F. Beltram, Green fluorescent proteins as optically controllable elements in bioelectronics, *Appl. Phys. Lett.* **79**, 3353-3355 (2001).
30. J. J. La Clair, An atmospherically driven optical switch, *Angew. Chem. Int. Ed.* **38**(20), 3045-3047 (1999).
31. L. Zang, R. Liu, M. W. Holman, K. T. Nguyen, and D. M. Adams, A single-molecule probe based on intramolecular electron transfer, *J. Am. Chem. Soc.* **124**, 10640-10641 (2002).
32. J.-M. Lehn, Supramolecular chemistry - scope and perspectives molecules, supermolecules, and molecular devices (Nobel Lecture), *Angew. Chem. Int. Ed.* **27**, 89-112 (1988).
33. J.-M. Lehn, Perspectives in supramolecular chemistry - from molecular recognition towards molecular information processing and self-organization, *Angew. Chem. Int. Ed.* **29**, 1304-1319 (1990).
34. J. R. Lakowicz, *Principles of Fluorescence Spectroscopy, Second Edition* (Kluwer Academic/Plenum Publishers, New York, 1999)
35. E. W. Tsai, L. Phan, and K. Rajeshwar, Electrochemical modulation of luminescence from an interfacial probe during redox switching of polypyrrole, *J. Chem. Soc. Chem. Commun.*, 771-773 (1988).
36. J. Otsuki, M. Tsujino, T. Iizaki, K. Araki, M. Seno, K. Takatera, and T. Watanabe, Redox-responsive molecular switch for intramolecular energy transfer, *J. Am. Chem. Soc.* **119**, *119*, 7895-7896 (1997).
37. T. Komura, G. Y. Niu, T. Yamaguchi, and M. Asano, Redox and ionic binding switched fluorescence of phenosafranine and thionine included in Nafion films, *Electrochim. Acta* **48**, 631-639 (2003).
38. R. W. Wagner, J. S. Lindsey, J. Seth, V. Palaniappan, and D. F. Bocian, Molecular optoelectronic gates, *J. Am. Chem. Soc.* **118**, 3996-3997 (1996).
39. H. Ostergaard, A. Henriksen, F. G. Hansen, and J. R. Winther, Shedding light on disulfide bond formation: engineering a redox switch in green fluorescent protein, *Eur. Mol. Biol. Org. J.* **20**, 5853-5862 (2001).
40. S. Rohr and B. M. Salzberg, Multiple site optical recording of transmembrane voltage (MSORTV) in patterned growth heart cell cultures: Assessing electrical behavior with microsecond resolution on a cellular and subcellular scale, *Biophys. J.* **67**, 1301-1315 (1994).
41. P. T. Toth, V. P. Bindokas, D. Bleakman, W. F. Colmers, and R. J. Miller, Mechanism of presynaptic inhibition by neuropeptide Y at sympathetic nerve terminals, *Nature* **364**, 635-639 (1993).
42. B. R. Masters and B. Chance in: *Biological Techniques: Fluorescent and Luminescent Probes for Biological Activity*, edited by W. T. Mason, (Academic Press, San Diego 1999), pp. 44-57.
43. C.-S. Chen and K. R. Gee, Redox dependent trafficking of 2,3,4,5,6-pentafluorodihydrotetramethylrosamine, a novel fluorogenic indicator of cellular oxidative activity, *Free Rad. Biol. Med.* **28**, 1266-1278 (2000).
44. Y. Gu, B. Chance, C.-H. Tung, and R. Weissleder, Study the characteristic of protease-activated NIR fluorescent probes in tumor by using 3D high resolution imaging system, *Proc. SPIE* **4250**, 196-203 (2001).
45. B. Li, P. L. Gutierrez, and N. V. Blough, Trace determination of hydroxyl radical using fluorescence detection, *Methods. Enzymol.* **300**, 202-216 (1999).
46. R. Bergonzi, L. Fabbrizzi, M. Licchelli, and C. Mangano, Molecular switches of fluorescence operating through metal centered redox couples, *Coord. Chem. Rev.* **170**, 31-46 (1998).
47. P. D. Beer, Transition-metal receptor systems for the selective recognition and sensing of anionic guest species, *Acc. Chem. Res.* **31**, 71-80 (1998).
48. P. D. Beer, P. A. Gale, and Z. Chen, Electrochemical recognition of charged and neutral guest species by redox-active receptor molecules, *Adv. Phys. Org. Chem.* **31**, 1-89 (1998).
49. Ph.D thesis G. Hennrich, Berlin 2000.
50. P. D. Beer, Charged guest recognition by redox responsive ligand systems, *Adv. Mater.* **6**, 607-609 (1994).
51. P. D. Beer, P. A. Gale, and G. Z. Chen, Mechanism of electrochemical recognition of cations, anions, and neutral guest species by redox-active receptor molecules, *Coord. Chem. Rev.* **185-186**, 3-36 (1999).
52. P. R. Ashton, R. Ballardini, V. Balzani, S. E. Boyd, A. Credi, M. R. Ggandolfi, M. Gomez-Lopez, S. Iqbal, D. Philp, J. A. Preece, L. Prodi, H. G. Ricketts, J. F. Stoddart, M. S. Tolley, M. Venturi, A. J. P. White, and D. J. Williams, Simple mechanical molecular and supramolecular machines: photochemical and electrochemical control of switching processes, *Chem. Eur. J.* **3**, 157-170 (1997).

53. C. Canevet, J. Libman, and A. Shanzer, Molecular redox-switches by ligand exchange, *Angew. Chem. Int. Ed. Engl.* **35**, 2657-2660 (1996).
54. A. C. Benniston, Photo- and redox-active [2]rotaxanes and [2]catenanes, *Chem. Soc. Rev.* **1996**, 427-435 (1996).
55. T. M. Swager, The molecular wire approach to sensory signal amplification, *Acc. Chem. Res.* **31**, 201-207 (1998).
56. D. T. McQuade, A. E. Pullen, and T. M. Swager, Conjugated polymer-based chemical sensors, *Chem. Rev.* **100**, 2537-2574 (2000).
57. A. P. de Silva, H. Q. N. Gunaratne, T. Gunnlaugsson, A. J. M. Huxley, J. T. Rademacher, and T. E. Rice in: *Chemosensors of Ion and Molecule Recognition*, edited by J. P. Desvergne and A. W. Czarnik (Kluwer Academic Publishers, Netherlands, 1997), pp. 143–157.
58. A. W. Czarnik in: *Topics in Fluorescence Spectroscopy; Probe Design and Chemical Sensing*, edited by J. R. Lakowicz, (Plenum Press, New York, 1994), pp.49 -70.
59. J. L. Bricks, K. Rurack, R. Radeglia, G. Reck, B. Schulz, H. Sonnenschein, and U. Resch-Genger, Reaction of a *N*-anthrylcarbonylthiourea derivative with Cu^{2+} or H^+: Unusual rearrangement to a highly fluorescent S-(9-anthryl)isothiouronium salt, *J. Chem. Soc. Perkin Trans.* 2, 1209-1214 (2000).
60. N. V. Blough, D. J. Simpson, Chemically mediated fluorescence yield switching in nitroxide-fluorophore adducts: Optical sensors of radical/redox reactions, *J. Am. Chem. Soc.* **110**, 1915-1917 (1988).
61. M. R. Wasielewski, Photoinduced electron transfer in supramolecular systems for artificial photosynthesis, *Chem. Rev.* **92**, 435-461 (1992).
62. T. Jorgensen, T. K. Hansen, and J. Becher, Tetrathiafulvalenes as building blocks in supramolecular chemistry, *Chem. Soc. Rev.*, 41-51 (1994).
63. V. Balzani, A. Credi, G. Mattersteig, O. A. Matthews, F. M. Raymo, J. F. Stoddart, M. Venturi, A. J. P. White, and D. J. Williams, Switching of pseudorotaxanes and catenanes incorporating a tetrathiafulvalene unit by redox and chemical inputs, *J. Org. Chem.* **65**, 1924-1936 (2000).
64. A. J. Moore, L. M. Goldenberg, M. R. Bryce, M. C. Petty, J. Moloney, J. A. K. Howard, M. J. Joyce, and S. N. Port, New crown annelated tetrathiafulvalenes: Synthesis, electrochemistry, self assembly of thio derivatives, and metal cation recognition, *J. Org. Chem.* **65**, 8269-8276 (2000).
65. S. G. Liu, H. Y. Liu, K. Bandyopadhyay, Z. Q. Gao, and L. Echegoyen, Dithia-crown-annelated tetrathiafulvalene disulfides: synthesis, electrochemistry, self-assembled films, and metal ion recognition, *J. Org. Chem.* **65**, 3292-3298(2000).
66. K. S. Bang, M. B. Nielsen, R. Zubarev, and J. Becher, Tetrathiafulvalene-phenanthroline macrocycles as redox responsive sensors for metal ions, *Chem. Commun.*, 215-216 (2000).
67. H. Miyaji, S. R. Collinson, I. Proke, and J. H. R. Tucker, A ditopic ferrocene receptor for anions and cations that functions as a chromogenic molecular switch, *Chem. Commun.*, 64-65 (2003).
68. R. Giasson, E. J. Lee, X. Zhao, and M. S. Wrighton, Inter- and intramolecular quenching of the singlet excited state of porphyrins by ferrocene, *J. Phys. Chem.* **97**, *97*, 2596-2601 (1993).
69. S. Fery-Forgues and B. Delavaux-Nicot, Ferrocene and ferrocenyl derivatives in luminescent systems, *J. Photochem. Photobiol. A: Chem.* **132**, 137-159 (2000).
70. G. de Santis, L. Fabbrizzi, M. Licchelli, N. Sardone, and A. H. Velders, Fluorescence redox switching systems operating through metal centers: the Ni(III)/Ni(II) couple, *Chem. Eur. J.* **2**, 1243-1250 (1996).
71. L. Fabbrizzi, and A. Poggi, Sensors and switches from supramolecular chemistry, *Chem. Soc. Rev.* **24**, 197–202 (1995).
72. V. Goulle, A. Hariman, and J.-M. Lehn, An electro-photoswitch: Redox switching of the luminescence of a bipyridine metal complex, *J. Chem. Soc., Chem. Commun.*, 1034-1036 (1993).
73. S. Arounaguiri, and B. G. Maiya, "Electro-photo switch" and "molecular light switch" devices based on ruthenium(II) complexes of modified dipyridophenazine ligands: Modulation of the photochemical function through ligand design, *Inorg. Chem.* **38**, 842-843 (1999).
74. J. Daub, M. Beck, A. Knorr, and H. Spreitzer, New molecular systems for functional dye-based molecular switching of luminescence, *Pure Appl. Chem.* **68**, 199-1404 (1996).
75. V. Goulle, *Ph.D. thesis*, Strasbourg 1992.
76. H. A. De Cremiers, G. Clavier, F. Ilhan, G. Cooke, and V. M. Rotello, Tunable electrochemical interaction between polystyrenes with anthracenyl and tetrathiafulvalenyl.sidechains, *Chem. Commun.*, 2232-2233 (2001).

77. C. Farren, C. A. Christensen, S. FitzGerald, M. R. Bryce, and A. Beeby, Synthesis of novel phthalocyanine-tetrathiafulvalene hybrids; intramolecular fluorescence quenching related to molecular geometry, *J. Org. Chem.* **67**, 9130-9139 (2002).
78. S. Sadaike, K. Takimiya, Y. Aso, and T. Otsubo, TTF–porphyrin dyads as novel photoinduced electron transfer systems, *Tetrahedron Lett.* **44**, 161-165 (2003).
79. P. A. Liddell, G. Kodis, L. de la Garza, J. L. Bahr, A. Moore, T. A. Moore, and D. Gust, Photoinduced electron transfer in tetrathiafulvalene-porphyrin-fullerene molecular triads, *Helv. Chim. Acta* **84**, 2765-2783. (2001)
80. H. Li, J. O. Jeppesen, E. Levillain, and J. Becher, A mono-TTF-annulated porphyrin as a fluorescence switch, *Chem. Commun.*, 846-847 (2003).
81. Z. H. Wang, K. C. Chen, and H. Tian, Intramolecular fluorescence quenching in ferrocene-naphthalimide dyads, *Chem. Lett.*, 423-424 (1999).
82. A. Ambroise, R. W. Wagner, P. D. Polisetti Dharma, J. A. Riggs, P. Hascoat, J. R. Diers, J. Seth, R. K. Lammi, D. F. Bocian, D. Holten, and J. S. Lindsey, Design and synthesis of porphyrin-based optoelectronic gates, *Chem. Mat.* **13**, 1023-1034 (2001).
83. L. Fabbrizzi, M. Licchelli, P. Pallavicini, and D. Sacchi, Supramolecular functions related to the redox activity of transition metal ions, *Supramol. Chem.* **13**, 569-582 (2001).
84. L. Fabbrizzi, M. Licchelli, and P. Pallavicini, Transition metals as switches, *Acc. Chem. Res.* **32**, 846-853 (1999).
85. M. Di Casa, L. Fabbrizzi, M. Licchelli, A. Poggi, D. Sacchi, and M. Zema, A novel fluorescence redox switch based on the formal Ni^{II}/Ni^{I} couple, *J. Chem. Soc. Dalton Trans.*, 1671-1675 (2001).
86. G. De Santis, L. Fabbrizzi, D. Iacopino, P. Pallavicini, A. Perotti, and A. Poggi, Electrochemically switched anion translocation in a multicomponent coordination compound, *Inorg. Chem.* **36**, 827-832 (1997).
87. V. Amendola, M. Di Casa, L. Fabbrizzi, M. Licchelli, C. Mangano, P. Pallavicini, and A. Poggi, Mechanical switches of fluorescence, *J. Incl. Phenom. Macrocyclic Chem.* **41**, 13-18 (2001).
88. V. Amendola, L. Fabbrizzi, C. Mangano, H. Miller, P. Pallavicini, A. Perotti, and A. Taglietti, Signal amplification by a fluorescent indicator of a pH-driven intramolecular translocation of a copper(II) ion, *Angew. Chem. Int. Ed.* **41**, 2553-2556 (2002).
89. V. Amendola, L. Fabbrizzi, P. Pallavicini, E. Sartirana, and A. Taglietti, Monitoring the redox-driven assembly/disassembly of a dicopper(I) helicate with an auxiliary fluorescent probe, *Inorg. Chem.* **42**, 1632-1636 (2003).
90. L. Fabbrizzi, M. Licchelli, P. Pallavicini, and L. Parodi, Controllable intramolecular motion that generate fluorescent signals for a metal scorpionate complex, *Angew. Chem. Int. Ed.* **37**, 800-802 (1998).
91. E. Zahavy and M. A. Fox, An Os^{II}-Ni^{II}-Pd^{II} trimetallic complex as an electro-switchable-photoinduced electron- transfer device, *Chem Eur. J.* **4**, 1647-1652 (1998).
92. L. Zuo and T. L. Clanton, Detection of reactive oxygen and nitrogen species in tissues using redox-sensitive fluorescent probes, *Methods Enzymol.* **352**, 307-325 (2000).
93. K. Setsukinai, Y. Urano, K. Kikuchi, T. Higuchi, and T. Nagano, Fluorescence switching by O-dearylation of 7-aryloxycoumarins. Development of novel fluorescence probes to detect reactive oxygen species, *J. Chem. Soc. Perkin Trans 2*, 2453-2457 (2000).
94. S. L. Hempel, G. R. Buettner, Y. Q. O'Malley, D. A. Wessels, and D. M. Flaherty, Dihydrofluorescein diacetate is superior for detecting intracellular oxidants: Comparison with 2',7'-dichlorodihydrofluorescein diacetate, 5 (and 6)-carboxy-2',7'-dichlorodihydrofluorescein diacetate, and dihydrorhodamine 123, *Free Radical Biol. Med.* **27**, 146-159 (1999).
95. K. Tanaka, T. Miura, N. Umezawa, Y. Urano, K. Kikuchi, T. Higuchi, and T. Nagano, Rational design of fluorescein-based fluorescence probes. Mechanism-based design of a maximum fluorescence probe for singlet oxygen, *J. Am. Chem. Soc.* **123**, 2530-2536 (2001).
96. G. R. Deviprasad, B. Keshavan, and F. D'Souza, O-benzoquinone appended zinc(II) porphyrin: a new fluorescent sensor for catechols, *J. Chem. Soc., Perkin Trans. 1*, 3133-3135 (1998).
97. B. Li, P. Gutierrez, and N. V. Blough, Trace determination of hydroxyl radical in biological systems, *Anal. Chem.* **69**, 4295-4302 (1997).
98. G. Moad, D. A. Shipp, T. A. Smith, and D. H. Solomon, Measurement of primary radical concentrations generated by pulse laser photolysis using fluorescence detection, *J. Phys. Chem. A* **103**, 6580-6586 (1999)

99. E. Lozinsky, V. V. Martin, T. A. Berezina, A. I. Shames, A. L. Weis, and G. I. Likhtenshtein, Dual fluorophore-nitroxide probes for analysis of vitamin C in biological liquids, *J. Biophys. Biochem. Methods* **38**, 29-42 (1999).
100. P. P.-B. Arye, N. Strashnikovy, G. I. Likhtenstein, Stilbene photochrome-fluorescence-spin molecules: Covalent immobilization on silica plate and applications as redox and viscosity probes, *J. Biochem. Biophys. Methods* **51**, 1-15 (2002).
101. N. Soh, Y. Katayama, and M. Maeda, A novel fluorescent probe for monitoring nitric oxide production using a novel detection concept, *Analyst* **126**, 564-566 (2001).
102. T. Kalai, E. Hideg, E. Vass, and K. Hideg, Double fluorescent and spin sensors for detection of reactive oxygen species in the thylakoid membrane, *Free Radical Biol. Med.* **24**, 649-652 (1998).
103. O. G. Ballesteros, L. Maretti, R. Sastre, and J. C. Scaiano, Kinetics of cap separation in nitroxide-regulated "living" free radical polymerization: Application of a novel methodology involving a prefluorescent nitroxide switch, *Macromolecules* **34**, 6184-6187 (2001).
104. C. Coenjarts, O. García, L. Llauger, J. Palfreyman, A. L. Vinette, and J. C. Scaiano, Mapping photogenerated radicals in thin polymer films: Fluorescence imaging using a prefluorescent radical probe, *J. Am. Chem. Soc.* **125**, *125*, 620-621 (2003).
105. G. J. Brown, A. P. De Silva, S. Pagliari, Molecules that add up, *Chem. Commun.*, 2461-2464 (2002)
106. K. Rurack and U. Resch-Genger, Rigidization, preorientation and electronic decoupling - the magic triangle for the design of highly efficient fluorescent sensors and switches, *Chem. Soc. Rev.* **31**, 116-127 (2002).
107. J. L. Bricks, R. Radeglia, G. Reck, U. Resch-Genger, K. Rurack, B. Schulz, J. L. Slominskii, and M. Spieles, Comparative solution and solid-state study of fluorophore-functionalized recogbnition modules and their Ag^I and Hg^{II} complexes: Effect of complex conformation, nature of the metal ion, and cation-π-interactions, to be submitted.
108. P. D. Beer, Anion selective recognition and optical/electochemical sensing by novel transition-metal receptor systems, *Chem. Commun.*, 689-696 (1996).
109. F. Sancenon, A. Benito, F. J. Hernandez, J. M. Lloris, R. Martinez-Manez, T. Pardo, and J. Soto, Difunctionalized chemosensors containing electroactive and fluorescent signalling subunits, *Eur. J. Inorg. Chem.* **2002**, 866-875 (2002).
110. K. Rurack, M. Büschel, and J. Daub, Synthesis, spectroscopic and electrochemical features of benzodithiophene hydroquinone (BTQ) dyads, triads, and BTQ terpyridine conjugates: from charge-transfer fluorescence to metal ion-mediated electronic coupling, to be submitted.
111. M. Büschel, M. Helldobler, and J. Daub, Electronic coupling in 6,6-donor-substituted terpyridines: tuning of the mixed valence state by proton and metal ion complexation, *Chem. Commun*, 1338-1339 (2002).
112. A. M. Bond, K. P. Ghiggino, C. F. Hogan, J. A. Hutchinson, S. J. Langford, E. Lygris, and M. N. Paddon-Row, Synthesis and electrochemical studies on a crown ether bearing a naphthoquinone acceptor, *Aust. J. Chem.* **54**, 735-738 (2001).
113. F. Szemes, D. Hesek, Z. Chen, S. W. Dent, M. G. B. Drew, A. J. Goulden, A. R. Graydon, A. Grieve, R. J. Mortimer, T. Wear, J. S. Weightman, and P. D. Beer, Synthesis and characterization of novel acyclic, macrocycle, and calix[4]arene ruthenium(II) bipyridyl receptor molecules that recognize and sense anions, *Inorg. Chem.* **35**, 5868-5879 (1996).
114. P. D. Beer, F. Szemes, V. Balzani, C. M. Salà, M. I. B. Drew, S. W. Dent, and M. Maestri, Anion selective recognition and sensing by novel macrocyclic transition metal receptor systems. ^{1}H NMR, electrochemical, and photophysical investigations, *J. Am. Chem. Soc.* **119**, 11864-11875 (1997).
115. R. D. Hancock and A. E. Martell, Ligand design for selective complexation of metal ions in aqueous solution, *Chem. Rev.* **89**, 1875-1914 (1989).
116. B. P. Hay, R. D. Hancock, The role of donor group orientation as a factor in metal ion recognition by ligands, *Coord. Chem. Rev.* **212**, 61-78 (2001).
117. F. P. Schmidtchen and M. Berger, Artificial organic hosts for anions, *Chem. Rev.* **97**, 1609-1646 (1997).
118. R. Martinez-Manez, J. Soto, J. M. Lloris, and T. Pardo, Receptors for the electrochemical sensing of transition metal ions and anions in aqueous environments, *Trends Inorg. Chem.* **5**, 183-203 (1998).
119. P.L. Boulas, M. Gomez-Kaifer, and L. Echegoyen, Electrochemistry of supramolecular systems, *Angew. Chem. Int. Ed.* **37**, 216-247 (1998).

120. T. Nabeshima, H. Furusawa, and Y. Yano Redox control of the recognition of Ag^+ ions in a macrocycle containing two SH groups or an S-S bridge inside thew cavity, *Angew. Chem. Chem. Int. Ed. Engl..x*33, 1750-1751 (1994).
121. H. Graubaum, F. Tittelbach, G. Lutze, K. Gloe, M. Mackrodt, T. Krüger, N. Krauss, A. Deege, and H. Hinrichs, Novel crown ethers with a trithiadiaza-pentalene-trithioutriuret redox system, *Angew. Chem. Int. Ed.* **1997**, *36*, 1648-1650 (1997).
122. H. Graubaum, F. Tittelbach, G. Lutze, K. Gloe, M. Mackrodt, T. Krüger, N. Krauss, A. Deege, and H. Hinrichs, Macrocyclic and open-chain ligands with the redox switchable trithiadiazapentalene unit: Synthesis, structures and complexation phenomena, *Coord. Chem. Rev.* **222**, 103-126 (2001).
123. H. Sonnenschein, E. Schmitz, E. Gründemann, and E. Schröder, Boulton-Katritzky Rearrangement of 5-(Cyanoimino)-1,2,4-thiadiazolines, *Liebigs Ann. Chem.*, 1177-1180 (1994).
124. Y. Kihara, S. Kabashima, K. Uno, T. Okawara, T. Yamasaki, and M. Furukawa, Oxidative heteocyclization using diethyl azodicarboxylate, *Synthesis*, 1020-1029 (1990).
125. G. Hennrich, H. Sonnenschein, and U. Resch-Genger, Redox switchable ionophores for heavy and transition metal cations, *Eur. J. Org. Chem.*, 539-542 (2000).
126. S. Nishizawa, K. Shigemori, and N. Teramae, A thiourea-functionalized benzo-15-crown-5 for cooperative binding of sodium ions and anions, *Chem. Lett.*, 1185-1186 (1999).
127. B. H. M. Snellink-Ruel, M. M. G. Antonisse, J. F. J. Engbersen, P. Timmerman, and D. N. Reinhoudt, Neutral anion receptors with multiple urea-binding sites, *Eur. J. Org. Chem.*, 165-170 (2000).
128. N. Matsumura, R. Hirase, and H. Inoue, Synthesis of new azacrown and azathiacrown ethers using a hypervalent sulfur-containing tetraazapentalene as a ring-building block, *Tetrahedron Le*tt. **35**, 899-902 (1994).
129. G. Hennrich, H. Sonnenschein, and U. Resch-Genger, Redox switchable fluorescent probe selective for either Hg(II) or Cd(II) and Zn(II), *J. Am. Chem. Soc.* **1999**, *121*, 5073-5074.
130. K. Rurack, M. Kollmannsberger, U. Resch-Genger, and J. Daub, A selective and sensitive fluoroionophore for Hg^{II}, Ag^{I}, and Cu^{II} with virtually decoupled fluorophore and receptor units, *J. Am. Chem. Soc.* **122**, 968-969 (2000).
131. K. Rurack, U. Resch-Genger, J. L. Bricks, and M. Spieles, Cation-triggered 'switching on' of the red/near infra-red (NIR) fluorescence of rigid fluorophore-spacer-receptor ionophores, *Chem. Commun.*, 1741-1742 (2000).
132. K. Rurack, J. L. Bricks, B. Schulz, M. Maus, G. Reck, and U. Resch-Genger, Substituted 1,5-diphenyl-3-benzothiazol-2-yl-2-pyrazolines: Synthesis, X-ray structure, photophysics, and cation complexation properties, *J. Phys. Chem A* **104**, 6171-6188 (2000).
133. L. Prodi, C. Bargossi, M. Montalti, N. Zaccheroni, N. Su, J. S. Bradshaw, R. M. Izatt, and P. B. Savage, An effective fluorescent chemosensor for mercury ions, *J. Am. Chem. Soc.* **122**, 6769-6770 (2000).
134. M. Sandor, F. Geistmann, and M. Schuster, An anthracene substituted benzoylthiourea for the selective determination of Hg(II) in micellar media, *Anal. Chim Acta* **388**, 19-26 (1999).
135. G. Hennrich, W. Walther, U. Resch-Genger, and H. Sonnenschein, Cu(II)- and Hg(II)-Induced modulation of the fluorescence behavior of a redox-active sensor molecule, *Inorg. Chem.* **40**, 641-644 (2001).
136 B. Ramachandram, G. Saroja, N. B. Sankaran, and A. Samanta, Unusually high fluorescence enhancement of some 1,8-naphthalimide derivatives induced by transition metal salts, *J. Phys. Chem. B* **104**, 11824-11832 (2000).
137. K. Rurack, Y. L. Bricks, Y. L. Slominski, and U. Resch, Cation complexing fluorescence probes containing the benz[c,d]indole fluorophore, *Dyes Pigm.* **1998**, *6*, 121-138.
138. K. Kubo, T. Sakurai, and A. Mori, Complexation and fluorescence behavior of 9,10-bis[bis(β- hydroxyethyl)aminomethyl]anthracene, *Talanta* **50**, 73-77 (1999).
139. G. Das, P. K. Bharadwaj, M. B. Roy, and S. Ghosh, Transition metal cryptate-enhanced fluorescence in a trianthroyl cryptand: effect of spacer on the photoinduced electron transfer process, *J. Photochem. Photobiol. A: Chem.* **135**, 7-11 (2000).
140. M.-Y. Chae and A. W. Czarnik, Fluorimetric chemodosimetry. Mercury(II) and silver(I) indication in water via enhanced fluorescence signaling, *J. Am. Chem. Soc.* **114**, 9704-9705 (1992).
141. G. Hennrich, H. Sonnenschein, and U. Resch-Genger, Fluorescent anion receptors with iminoylthiourea binding sites - selective hydrogen bond mediated recognition of CO_3^{2-}, HCO_3^- and HPO_4^{2-}, *Tetrahedron Lett.* **42**, 2805-2808 (2001).

142. M.-P. Teulade-Fichou, J.-P. Vigneron, and J.-M. Lehn, Detection of organic anions in water through complexation enhanced fluorescence of a macrobicyclic tris-acridine cryptand, J. Chem. Soc., Perkin Trans. 2, 2169-2175 (1996).
143. H. Fenniri, M. W. Hosseini, and J.-M. Lehn, Molecula recognition of NADP(H) and ATP by macrocyclic polyamines bearing acridine groups, Helv. Chim. Acta **80**, 786-803 (1997).
144. Y. Kubo, S. Ishihara, M. Tsukahara, and S. Tokita, Isothiouronium-derived simple fluorescent chemosensors of anions, J. Chem. Soc., Perkin Trans. 2, 1455-1460 (2002).
145. J.-H. Liao, C.-T. Chen, and J.-M. Fang, A novel phosphate chemosensor utilizing anion-induced fluorescence change, Org. Lett. **4**, 561-564 (2002).
146. S. Nishizawa, Y.-Y. Cui, M. Minagawa, K. Morita, Y. Kato, S. Taniguchi, R. Kato, and N. Teramae, Conversion of thioureas to fluorescent isothiouronium-based photoinduced electron transfer sensors for oxoanion sensing, J. Chem. Soc., Perkin Trans. 2, 866-870 (2002).

PET SENSORS

Kanji Kubo[*]

6.1. INTRODUCTION

Molecular recognition is one of the important fields in supramolecular chemistry.[1] Given any substrate (molecule, cation or anion), the supramolecular approach is that an appropriate receptor, possessing structural and chemical features suitable for recognition of that substrate, can be designed. In the supramolecular molecular recognition system, receptor molecules bearing photosensitive groups may display marked modification of their photophysical properties upon binding to their substrate species, leading to changes in their light absorption or emission features and allowing their detection by spectroscopic measurements. They represent molecular devices for substrate-selective optical signal generation and optical reading-out of the recognition process. Such photochemical sensors make it possible to develop sensitive analytical methods for the detection of a specific substrate. The design of fluorescent sensors is of major importance because of the high demand in analytical chemistry, clinical biochemistry, medicine, the environment, etc.[2,3]

Numerous chemical and biochemical analytes can be detected by fluorescence methods: cations (H^+, Li^+, Na^+, K^+, Ca^{2+}, Mg^{2+}, Cu^{2+}, Zn^{2+}, Pb^{2+}, Al^{3+}, Cd^{2+}, Hg^{2+} etc.), anions (halide ions, citrates, carboxylates, phosphates, ATP, etc.), neutral molecules (sugars, e.g. glucose, etc.), gases (O_2, CO_2, NO, etc.), and biochemical analytes (amino acids, coenzymes, carbohydrates, nucleosides, nucleotides, etc.). For example, sodium, potassium, magnesium and calcium are involved in biological processes such as transmission of nerve impulses, muscle contraction, regulation of cell activity, etc. Zinc is an essential component of many enzymes (e.g. carbonic anhydrase and zinc finger proteins), and plays a major role in enzyme regulation, gene expression, neurotransmission, etc. In medicine, monitoring of metal ions (e.g. Na^+, K^+, Mg^{2+}, Ca^{2+}) in blood and urine is of major importance in diagnosis. Regarding the toxicity of some metal ions, it is well known that mercury, lead and cadmium are toxic to organisms, and early detection in the environment is desirable. However, it is very difficult to detect the target guest selectively because many metal ions exist in an organism and in the natural

---

[*] Kanji Kubo, Kyushu University, Fukuoka, Japan 816-8580.

environment. In order to solve this problem, molecular sensors, which consist of a receptor and a signal unit, have been developed. The basic concept of such a sensor is illustrated in Figure 6.1.[4] The assembly of the receptor and the signaling unit constitutes the sensor. Fluorescent molecular sensors will be presented with a classification according to the nature of the photoinduced process (electron transfer (ET), charge transfer (CT), energy transfer, excimer formation, and exciplex formation) that is responsible for the photophysical changes upon cation binding. Photoinduced electron transfer (PET) is a remarkable achievement of photochemistry. On photoexcitation, certain molecules become powerful electron donors or acceptors and an electron migrates between the photoexcited and the ground-state species. PET is often responsible for fluorescence quenching. This process is involved in many organic photochemical reactions. It plays a major role in photosynthesis and in artificial systems for the conversion of solar energy based on photoinduced charge separation. The redox properties of molecules can be enhanced in the excited state. By using these properties, many PET sensors have been synthesized. This chapter describes the recent development of PET sensors.

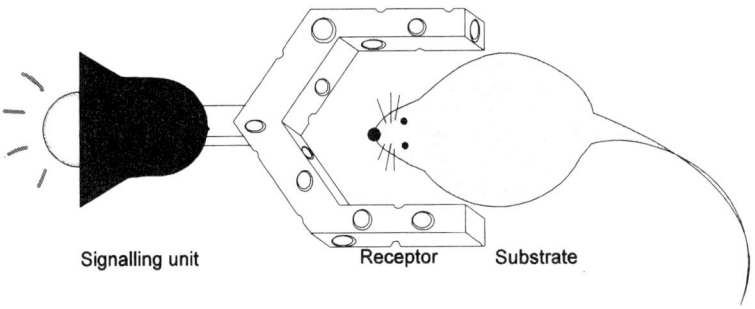

Figure 6.1. Design of fluorescent sensors. (Reproduced from Ref. 4. Copyright 1997 American Chemical Society.)

6.2. PRINCIPLE OF PET SENSORS

6.2.1. "OFF-ON" Switches

Most PET sensors, including pH indicators, consist of a fluorophore linked to an amine moiety via a methylene spacer. PET, which takes place from amino groups to aromatic hydrocarbons, causes fluorescence quenching of the latter. When the amino group is protonated (or strongly interacts with a cation), electron transfer is hindered and a very large enhancement of the fluorescence is observed. Figure 6.2 illustrates the mechanism in terms of molecular orbitals.[5] On excitation of the fluorophore, an electron of the highest occupied molecular orbital (HOMO) is promoted to the lowest unoccupied molecular orbital (LUMO), which enables PET from the HOMO of the donor (proton-free amine or cation-free receptor) to that of the fluorophore, causing fluorescence quenching (OFF). Upon cation binding (or protonation), the redox potential of the donor is raised so that the relevant HOMO becomes lower in energy than that of the fluorophore, and consequently, PET is no longer possible and the fluorescence quenching is suppressed. In other words, the fluorescence intensity is enhanced upon cation binding (ON). This system is an "OFF-ON" switch. Compound (**1**)[6] is the first and representative PET sensor. Compound (**1**) is very weakly fluorescent in methanol solution, and its fluorescence quantum yield increases from 0.003 to 0.14 upon binding of K^+ in methanol and to 0.38 in HCl solution.

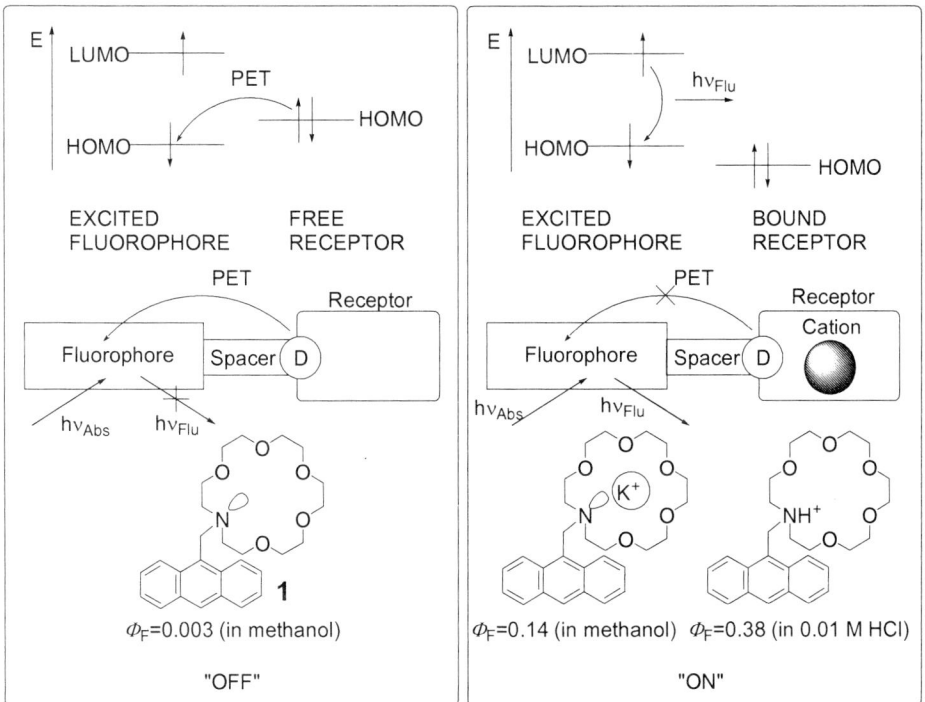

Figure 6.2. Principles of cation sensing by PET sensors (Reproduced from Ref. 5. Copyright 1997 American Chemical Society.) and the representative PET sensor described by de Silva, 1986.

Subsequently, many "OFF-ON" switches containing fluorophore units (anthracene, naphthalene, pyrene etc.) and receptor units (coronands, chelators, podands, cryptands, and cavitands) were synthesized. More than one receptor and/or more than one fluorophore are involved in the structure of PET sensors as shown in Figure 6.3. The fluorescent PET "OFF-ON" system is classified as chelation-enhanced fluorescence (CHEF).[6]

6.2.2. "ON-OFF" Switches

Fluorescent PET "ON-OFF" switches are the opposite of the "OFF-ON" switches. A schematic summary of fluorescent "ON-OFF" switches is shown in Figure 6.4 Compounds **2**[8] and **3**[9] are the first family and are only simple representatives of "ON-OFF" PET pH sensors whose signals cover most of the visible spectrum. The significant absorptiometric pH sensing action relies on the proton-induced conversion of receptor π systems which are hardly or weakly electron-deficient into rather strong PET acceptors. Compound **4**[10] is a PET "ON-OFF" switch with a transition metal ion (Cu^{2+}). The metal-receptor interaction is signaled through fluorescence quenching. Compound **5**[11] is an "ON-OFF" switch involving the electron transfer abilities of the redox-active guests. **5**·Cu^+ is strongly emissive. The chemical and electrooxidation state quenches the fluorescence. The stability of both the copper complexes permits reversible switching of the device. Compound **6**[12] is also an "ON-OFF" fluorescent switch since it shows Zn(II)-induced folding so that the terminals are brought within range for rapid PET. These fluorescent PET "ON-OFF" systems are classified as chelation-enhanced quenching (CHEQ).[13]

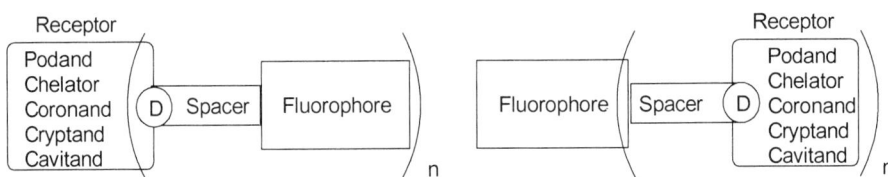

Figure 6.3. Multi-fluorophore-spacer-receptor systems. (Reproduced from Ref. 7. Copyright 2000 The Japan Institute of Heterocyclic Chemistry.)

PET SENSORS

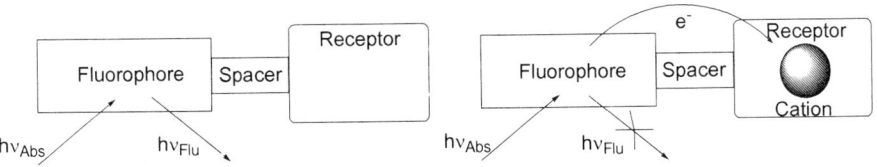

Figure 6.4. Fluorescent PET "ON-OFF" switch. (Reproduced from Ref. 5. Copyright 1997 American Chemical Society.)

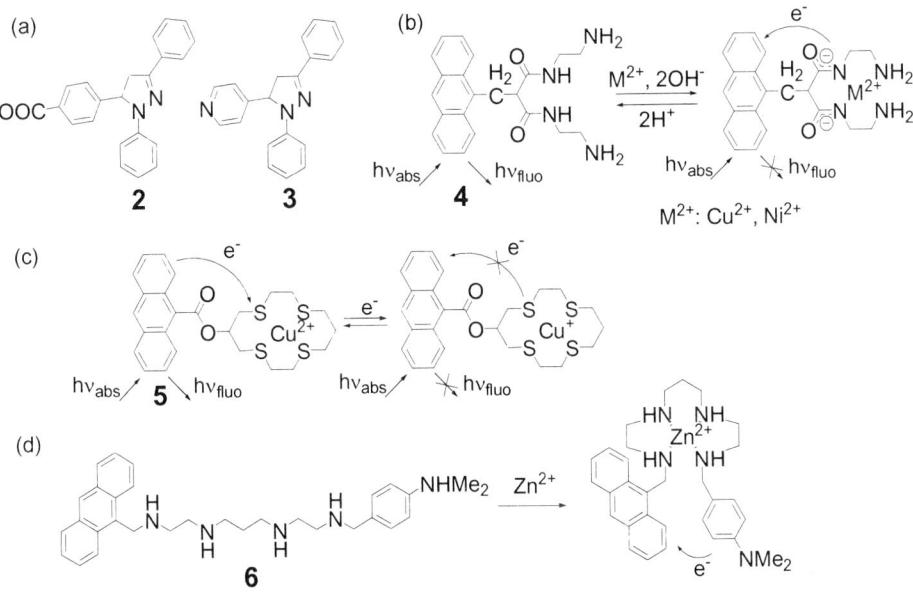

Figure 6.5. Examples of PET "ON-OFF" switches: (a) pH sensors described by de Silva, 1989 & 1995; (b) PET sensor for Cu^{2+} and Ni^{2+}; (c) redox-active PET sensor (Reproduced from Ref. 10. Copyright 1994 American Chemical Society.); and (d) Zn^{2+}-induced folding PET sensor. (Reproduced from Ref. 12. Copyright 1996 American Chemical Society.)

6.2.3. "OFF-ON-OFF" Switches

Compound **7**[14] is an "OFF-ON-OFF" switch which is controlled by two protonations, i.e., **7**, **7-H$^+$**, and **7-2H$^+$** represent fluorescence "OFF", "ON", and "OFF" states with Φ_F of 0.009, 0.38, 0.046, respectively, as shown in Figure 6.6.

Compound **8**[15] is controlled by three cation binding events, i.e., **8-Na$^+$**, **8-Na-H$^+$**, and **8-Na$^+$-2H$^+$** represent fluorescence "OFF", "ON", and "OFF" states with Φ_F of 0.03, 0.54, 0.03, respectively.

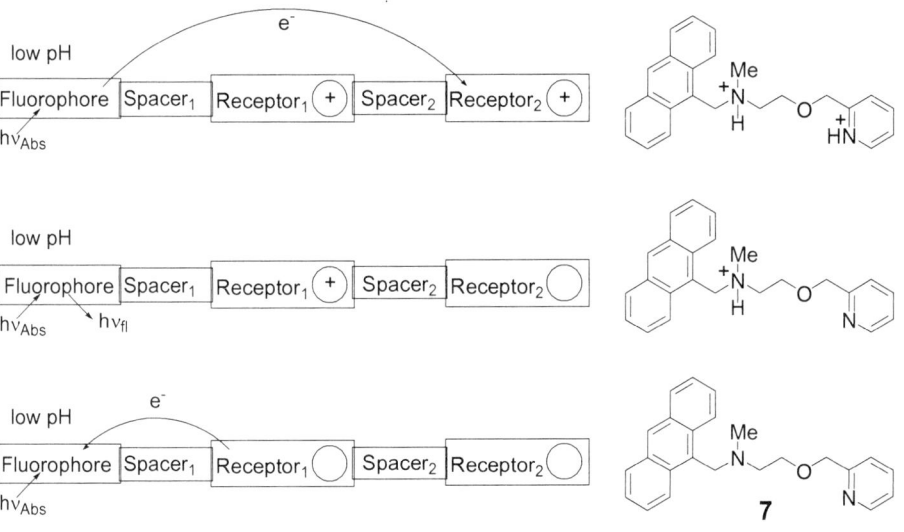

Figure 6.6. Schematic representation of the operation of a fluorescent PET sensor for a pH window. (Reproduced from Ref. 14. Copyright 1996 Royal Chemical Society.)

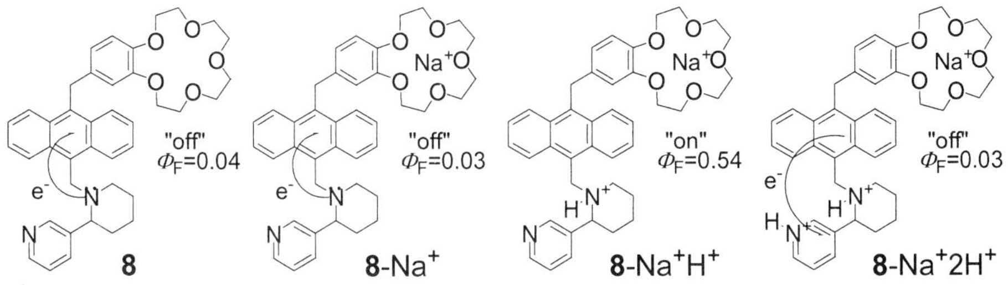

PET SENSORS

Figure 6.7. Fluorescence "OFF" and "ON" states of system **8** with PET processes that correspond to various cation binding events. (Reproduced from Ref. 15. Copyright 2002 Royal Chemical Society.)

Table 6.1. Fluorescence quantum yield of fluorescent molecule

Fluorescent molecules	Φ_F	Fluorescent molecules	Φ_F
Naphthalene	0.21	9-Phenylanthacene	0.49
1-Methylnaphthalene	0.19	9,10-dimethylanthracene	0.89
2-Methylnaphthalene	0.16	9,10-dimethoxyanthracene	0.41
Anthracene	0.11	9,10-diphenylanthracene	0.95
9-Methylanthracene	0.33	Pyrene	0.72
9-Methoxyanthracene	0.17		

(Reproduced from Ref. 16. Copyright 1993 Marcel Dekker, Inc.)

6.3. DESIGN OF PET SENSORS

Typical PET sensors consist of a fluorophore (anthracene, naphthalene, pyrene), a spacer (methylene unit), and a receptor (chelator, podand, coronand, cryptand, cavitand). The substrate (guest, analyte) must be attracted to the receptor portion of the sensor. This is a straightforward molecular recognition event. The binding must be selective for the target substrate in the presence of a range of other potential guest species, depending on the system and its environment. The receptor must also be in communication with a signaling unit that is responsive to the guest binding. The criteria[2] for the construction of useful fluorescent sensors are summarized as follows.

1) Stability
2) Guest (analyte) selectivity
3) Guest affinity
4) Solubility
5) Efficient signal transduction
6) Emission of detectable intensities of UV or visible radiation or other quantifiable signal
7) Kinetically rapid sensitization
8) Ease of delivery to the target system

9) Availability

In addition, to design an effective PET sensor of the "OFF-ON" switch type, it is important that PET sensors have a low fluorescence quantum yield by PET from the donor to the excited fluorophores, and a high fluorescence quantum yield in complexation with the guest cation.[7] The design of sensors is divided into three units, and the function as a PET sensor is explained.

6.3.1. Design Of The Fluorophore Unit

In the development of a sensor, the selection of the fluorescent molecule is important with respect to high sensitivity detection. The fluorescent molecules and their fluorescence quantum yields[16] are shown in Table 6.1. The fluorescence quantum yield of a fluorescent molecule depends on the kind and substituent positions of the substituent group. To design an effective PET sensor, it is desirable that the fluorophore unit in the PET sensor has a large fluorescence quantum yield.

6.3.2. Design of The Spacer Unit

The connection between the receptor and the fluorophore is a very important aspect of sensor design, bearing in mind the search for the strongest perturbation of the photophysical properties of the fluorophore by the cation. Davidson[17] reported fluorescent intermolecular exciplex formation and fluorescence quenching of naphthylalkylamines (as shown in Table 6.2). The fluorescence quantum yield of naphthylalkylamines yields changes with the length of the spacer and the substituted position on naphthalene. The weakest fluorescences are n=1 for α-alkylnaphthalene and n=2 for β-alkylnaphthalene, respectively. The spacer length affects the quenching efficiency in the switch "OFF" state.

6.3.3. Design of the Receptor Unit

In the design of PET sensors, much attention is paid to the characteristics of the receptor moiety and to the expected changes in the fluorescence characteristics of the

Table 6.2. Fluorescence quantum yields of naphthylalkylamines

α-(CH$_2$)$_n$NEt$_2$		β-(CH$_2$)$_n$NEt$_2$	
Φ_M	$\Phi_{Exciplex}$	Φ_M	$\Phi_{Exciplex}$
n=1 0.006	0.001 (495 nm)	n=1 0.018	0.006 (495 nm)
n=2 0.008	0.040 (482 nm)	n=2 0.003	0.028 (470 nm)
n=3 0.012	0.015 (505 nm)	n=3 0.010	0.010 (470 nm)
n=4 0.008	0.002 (520 nm)	n=4 0.012	0.012 (500 nm)

(Reproduced from Ref. 17. Copyright 1976 Royal Chemical Society)

fluorophore moiety upon binding. It should first be remembered that the stability of a complex between a given ligand and a cation depends on many factors: nature of the cation, nature of the solvent, temperature, ionic strength, and pH in some cases.

In ion recognition, complex selectivity (i.e. the preferred complexation of a certain cation when other cations are present) is of major importance. In this regard, the characteristics of the receptor, i.e. the ligand topology and the number and nature of the complexing heteroatoms or groups, should match the characteristics of the substrate such as a cation, i.e. ionic diameter, charge density, coordination number, intrinsic nature (e.g. hardness of metal cations, nature and structure of organic cations, etc.) according to the general principles of supramolecular chemistry. The receptor unit can be a chelator, podand, coronand, cryptand, or cavitands. The stability of a complex between a given receptor and a substrate depends on many factors: nature of the substrate, nature of the solvent, temperature, ionic strength, and pH in some cases. In molecular recognition, complex selectivity and stability are of great importance.

The receptor may be linked to the fluorophore via a spacer, but in many cases some atoms or groups participating in the complexation belong to the fluorophore. Therefore, the selectivity of binding often results from the whole structure involving both the signaling and the recognition moieties.

6.4. CLASSIFICATION OF PET SENSORS

PET sensors are classified according to the chemical structure of the recognition moiety and to the photophysical process.[3]

6.4.1. Coronand (Crown)-Based PET Sensors

Examples of PET sensors containing various kinds of coronands such as crowns are shown in Figure 6.8. Monoazacrown ether **1**[6] is the first and simplest coronand PET sensor. Its fluorescence quantum yield increases from 0.003 to 0.14 upon binding of K^+ in methanol and to 0.38 in HCl solution. Monoazacrown ether **9a**[18] shows a fluorescence increase with bivalent metal thiocyanates and a fluorescence decrease with alkali metal thiocyanates. Monoazacrown ether **9b**[19] also shows a fluorescence decrease with NaSCN. This suggests that the observed quenching is due to PET from the thiocyanate anion to the fluorophore.[20]

The multi-PET systems **10**[21,22] and **11**[23-25] give weak emission and high CHEF. The emission-band intensities of **9a** and **11c** are reduced to approximately one-30th and one-440th of that of 1-methylnaphthalene, respectively. The larger quenching of **11c** means that two nitrogen atoms and two naphthalene rings participate in the intramolecular PET process. Diaza-18-crown-6 ethers **10b** and **11c** display a remarkable fluorescence enhancement in the presence of K^+ for **10b** and Ba^{2+} for **11c**, respectively. Diaza-12-crown-4 ethers **10a** and **11a**, diaza-15-crown-5 ether **11b**, and cyclam **12**[26] exhibit Zn^{2+} fluorescence enhancement selectivity and in the presence of this cation the host fluorescence intensity is increased by a factor of 182, 43, 8.3 and 8.1, respectively. C-Armed crown ethers **13**[27] also show Zn^{2+} fluorescence enhancement selectivity, but the changes are too small. Compound **14**[28] exhibits Cu^{2+} selectivity and in the presence of this cation the host fluorescence intensity is increased by a factor of 7.8. Chiral diaza-9-crown-3 **15**[29] is the first example of a PET sensor designed specifically for Li$^+$ detection. Its fluorescence quantum yield increases from 0.022 to 0.11 upon binding of Li$^+$ (Log K=5.4) in acetonitrile. Compound **16**[30] displays a fluorescence enhancement at 515 nm which was observed on binding of alkali earth metal cations such as Mg^{2+}, Ca^{2+}, and Ba^{2+}. The fluorescence quantum yield of **16a** increased from 0.021 to 0.679 upon binding of Ca^{2+} in acetonitrile. In addition, tetrathiacrown **5**-Cu^{2+} is a redox-active PET sensor with an "ON-OFF" switching property.

Azamacrocycles **17a-e**,[10, 31] containing a water-soluble polyazamacrocycle, were designed for the recognition of soft metal ions like Zn^{2+} and Hg^{2+}. These compounds are sensitive to pH due to the protonability of the aliphatic nitrogen atoms. Azamacrocycles **17b-e** yield large (20- to 190-fold) changes in fluorescence on metal ion complexation in aqueous solution, with large association constants between several transition metals (e.g., Pb^{2+}, Cu^{2+}, Zn^{2+}, Cd^{2+}, Hg^{2+}) and azamacrocycles. Fluorescence titrations with Zn^{2+} and Hg^{2+} (net heavy atom quenching) cause CHEF and CHEQ, respectively. If the macrocyclic polyamine is directly bound to the phenyl group of a fluorophore, as in **18a**

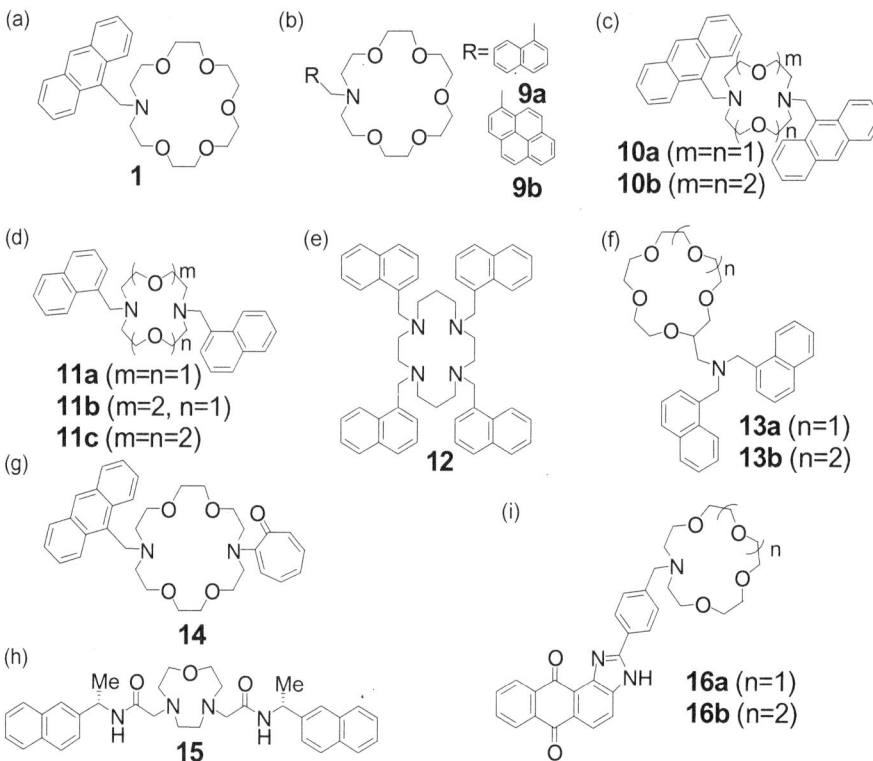

Figure 6.8. Some crown-based PET sensors described by: (a) de Silva, 1986; (b-g) Kubo (ref. 18, 19, 21-28); (h) Gunnlaugsson, 2002; and (i) Yoshida, 1999.

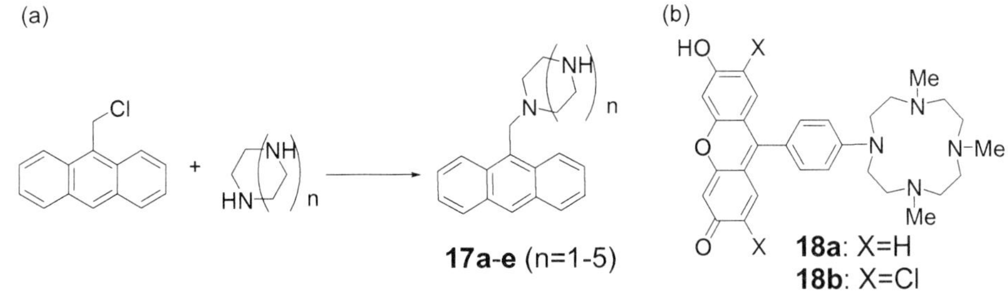

PET SENSORS

Figure 6.9. (a) PET sensors containing a water-soluble polyazamacrocycle (Reproduced from Ref. 10. Copyright 1994 American Chemical Society.) and (b) macrocyclic polyamine directly bound to the phenyl group described by Hirano, 2000.

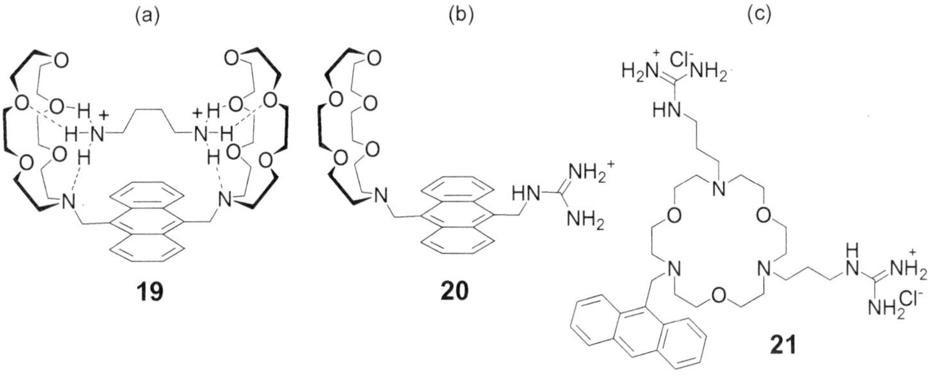

Figure 6.10. PET sensors for (a) alkane diammonium ions described by de Silva, 1990 and amino acids described by (b) de Silva, 1996 and (c) Sasaki, 2002.

and **18b**,[32] the pK is lowered and the intensity changes are comparable at pH 10 and pH 7. PET sensors for Zn^{2+}, **18a** and **18b**, are excited by visible light and can selectively detect Zn^{2+} under physiological conditions, and are thus more appropriate for biological applications.

Suitably positioned pairs of aza-18-crown-6 ether units can signal the presence of butane-1,4-diammonium ions via an intervening anthracene fluorophore due to complex **19**.[33] Hydrogen-bonding blocks both nitrogen lone electron pairs and their possible PET activity.

Compound **20**[34] targets the neurotransmitter γ-aminobutyric acid zwitterion, although improvements will be necessary if this target is to be sensed in its natural habitat. The

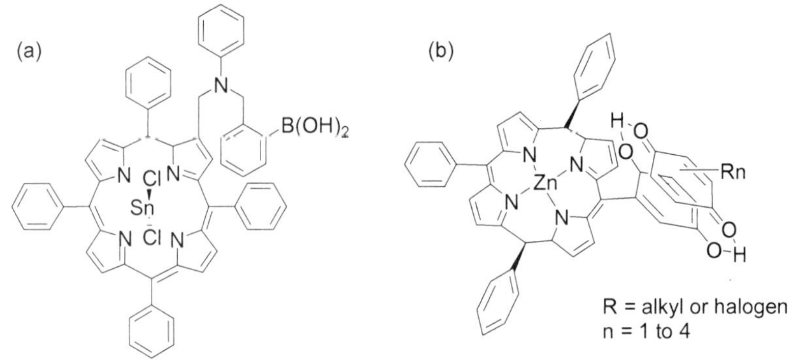

Figure 6.11. Porphyrin-based fluorescent sensors of neutral molecules described by (a) Kijima, 1999 and (b) D'Souza, 1997.

triaza-18-crown-6 ether **21**[35] combined with two guanidium groups could bind several biologically important amino acids (e.g. γ-aminobutyric acid, lysine, glycine) in aqueous methanol solution and showed a fluorescence enhancement response by a PET mechanism. The receptor **21** is expected to be useful as an indicator for amino acids in aqueous solution or as a component of an optode membrane.

Porphyrins are very attractive as a platform for the design of sensors for molecule recognition. The flat and relatively rigid structure of porphyrins offers the possibility of constructing various types of recognition sites. Numerous functionalized porphyrins have been designed for the recognition of hydrophobic molecules, amino acids, nucleobases, etc. Compound **22**[36] associates with a boronic acid moiety for recognition of diols (and in particular saccharides) and a tin-metallated porphyrin as a signaling moiety. This PET sensor changes its fluorescence intensity in an "ON-OFF" manner instead of the usual "OFF-ON" manner. In the absence of an analyte, the boronic acid moiety is positioned within a sterically crowded cleft of the porphyrin. The weak fluorescence emitted by **22** can be explained by the residual boron-nitrogen interaction that prevents complete quenching by electron transfer from the intramolecular amino group to the porphyrin moiety. Upon binding of diols, the boronic moiety is pulled away from the cleft and the diol is suspended over the plane of the porphyrin. This results in a weakening of the boron-nitrogen interaction and thus in an increase in quenching efficiency by the amino group. The most bulky diols induce the greatest fluorescence reduction. Compound **23**[37] with an appended hydroquinone can form complexes with quinones via H-bonds and charge-transfer interactions. In such complexes, PET from the porphyrin to the quinone is possible, and efficient fluorescence quenching was indeed observed upon binding with substituted benzoquinones, naphthaquinones and anthraquinones.

6.4.2. Chelator- And Podand-Based PET Sensors

Nonmacrocyclic receptors (**24-30**)[38-43] play perhaps a larger part in serving as recognition components within fluorescent PET "OFF-ON" signaling systems. The basicity of aliphatic amines is exploited within the pH sensors/switches. Compounds **28a,b**[42] and **29**[42] show fluorescent "OFF-ON" switching in aqueous alcoholic solution with protons. The excitation wavelengths lie in the ultraviolet (λ_{max}=345 and 351 nm) for **28a** and **28b** and in the blue-green (λ_{max}=461, 492, and 528 nm) for **29**, while the emission wavelengths lie in the violet (λ_{max}=408 nm) for **28a**, in the blue (λ_{max}=474 nm) for **28b**, and in the yellow-orange (λ_{max}=543 and 583 nm) for **29**. The pH-sensing behavior of **30**[43] was examined in solution and in polymer membrane matrices, which are crucial for the use of fluorescent PET sensors as optode pH monitors. Compound **30** can be used to locally probe the nanoscale surface structure. When compound **30** was spin-coated onto glass and quartz cover slips, scanning confocal fluorescence images showed bright fluorescent spots of **30** only on glass which contained metal and metal oxide impurities such as TiO_2, ZnO, and Al_2O_3.

Acyclic polyamines **31**[44] and **32**[10] contain polyamine chains (Figure 6.13) and were thus aimed at Zn^{2+} because this soft cation has a strong affinity for the soft nitrogen atoms. However, they operate in a very limited pH range and the affinity for Cu^{2+} is also strong. A clever adaptation of a partially protonated polyamine to serve as a receptor whose PET channel is blocked upon arrival of a HPO_4^{2-} guest can be seen in complex **33**.[45] Two of these receptor units can be placed on the 1- and 8-positions of an anthracene fluorophore as in **34**[46] in order to create a signaling PET system for pyrophosphate in the form of $H_2P_2O_7^{2-}$. Both of these cases require careful pH control for success. As a bonus, **34** will also assay the pyrophosphatase enzyme since the hydrolysis products will be released from the receptor pair with concomitant loss of fluorescence.

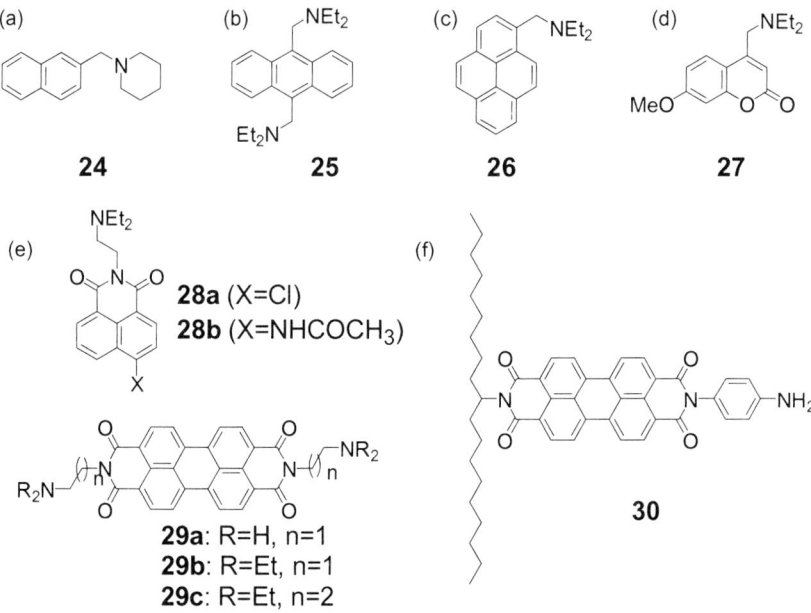

Figure 6.12. Non-macrocyclic pH PET sensors described by (a) Mes, 1984; (b) de Silva, 1985; (c) Draxler, 1995; (d) de Silva, 1993; (e) Daffy, 1998; and (f) Zang, 2002.

Diethanolamine **35**[47, 48] is selective for boronic and boric acids, and shows up to a 19-fold fluorescent intensity enhancement upon binding. The phenyl boronic acid complex of **35** can be used in a competition assay for diols in chloroform. Compound **36**[49, 50] is a PET sensor for boronic acids, boric acids and several transition metals (e.g., Pb^{2+}, Cu^{2+}, Zn^{2+}, Cd^{2+}, Hg^{2+}). Interestingly, the complexation of **36** with quenching metal cations such as Ni^{2+}, Cu^{2+}, and Hg^{2+} enhanced the emission intensity.

Dipicolylamine **37**[51] is an "OFF-ON-OFF" switch which is controlled by two protonations. Compounds **37** and **38**[52] exhibit a fluorescence-intensity enhancement in the presence of Zn^{2+} and their complexes[53] can be used as chemosensors for phosphorylated peptides.

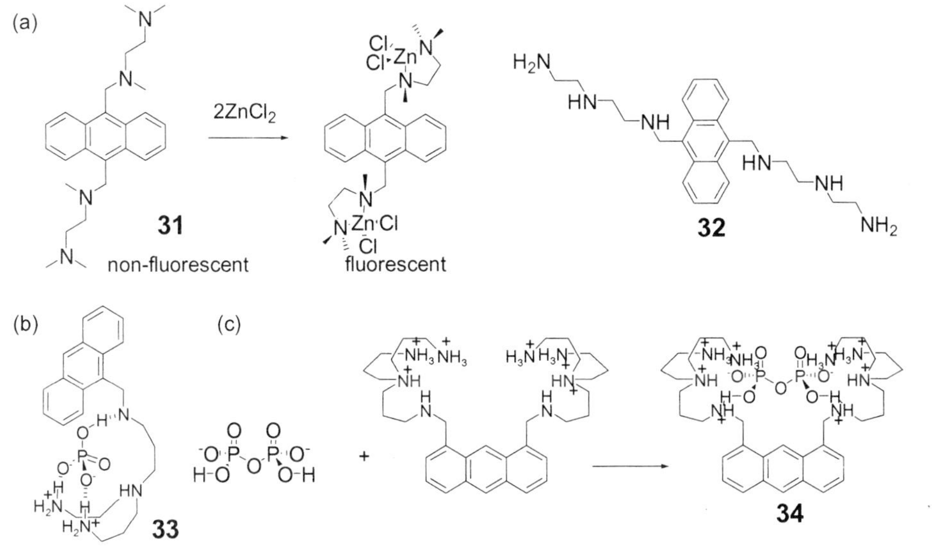

Figure 6.13. PET sensors containing acyclic polyamines described by (a) Czarnik, 1994; (b) Huston, 1989; and Vance, 1994.

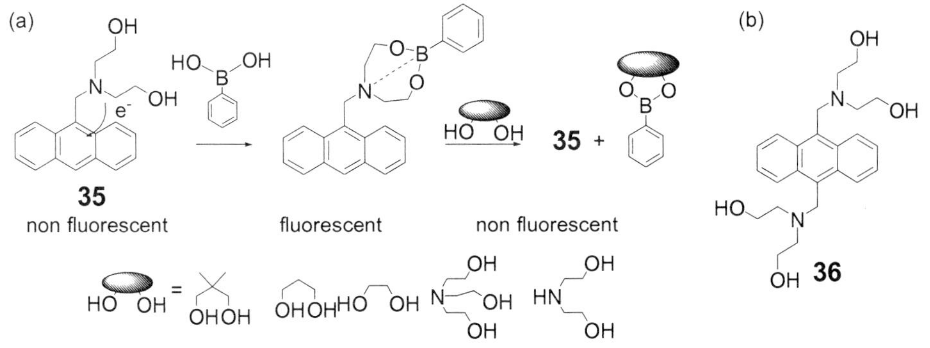

Figure 6.14. PET sensors containing a diethanolamine group. (Reproduced from Ref. 46, 48. Copyright 2002 & 1999 Elsevier Science Ltd.)

Compound **39**[54, 55] containing fluorescein as the fluorophore, and bis(2-pyridylmethyl)amine as a chelating moiety has been designed for probing Zn^{2+} in living cells.

This compound is indeed cell-permeable and has essentially nonmeasurable affinity for Ca^{2+} or Mg^{2+}. Upon addition of Zn^{2+} under physiological conditions, the fluorescence quantum yield increases from 0.39 to 0.87, and concentrations in the nanomolar range can be determined (stability constant: 1.4×10^9).

Chelators with carboxylic groups are known to efficiently bind divalent hard cations like Ca^{2+} and Mg^{2+}. Many selective calcium probes have been designed by Tsien and coworkers for applications in cellular biology, i.e. probing calcium concentrations in the micromolar range. The so-called BAPTA recognition moiety resembles EDTA, but the nitrogen atoms are linked to a phenyl group in order to avoid pH sensitivity in physiological media. Most of them are not based on the PET principle but on photoinduced charge transfer. Figure 6.16 shows examples of chelator-based PET sensors. Chelators **40-43**[56] are selective for calcium. In PET sensors, the changes in fluorescence quantum yield are accompanied by proportional changes in the excited-state lifetime. Therefore, chelators **40-42** were found to be suitable for fluorescence lifetime imaging of calcium. The same design principles apply to the magnesium sensors **43** and **44**,[57] in which the recognition moiety has a smaller 'cavity'. Compound **45**[58] showed ON-OFF switching in 100% aqueous solution with metal ions such as Co^{2+}, Cu^{2+}, Hg^{2+}, Mn^{2+}, Pb^{2+}, and Zn^{2+} via PET and chelatoselective aromatic cadmiation.

A new anthracene derivative **46**[59] bearing two phenylurea groups at the 1,8-position of anthracene shows a selective fluorescence quenching effect with fluoride ions *via* a PET mechanism. Isothiouronium salts have been explored quite recently as a new class of hydrogen-bonding subunit for the purpose of anion recognition. Charge neutral anthracene-based fluorescent sensors **47** and **48**[60, 61], having an aromatic or aliphatic urea or thiourea moiety as an anion, show fluorescence quenching with anions such as AcO^-, $H_2PO_4^-$ and F^- *via* a PET mechanism. The naphthyl isothiouronium **49**[62] with an ethylene spacer shows a significant fluorescence enhancement upon formation of a 1:1 complex with oxoanions in methanol, and the selectivity follows the order of hydrogen phosphate > acetate >> dihydrogen phosphate >> chloride. The isothiouronium derivative (**50**)[63] exhibits a fluorescence enhancement with anions. The selectivity for the anions in CH_3CN seems to run in the following order: $AcO^- > (BuO)_2P(O)O^- > Cl^-$.

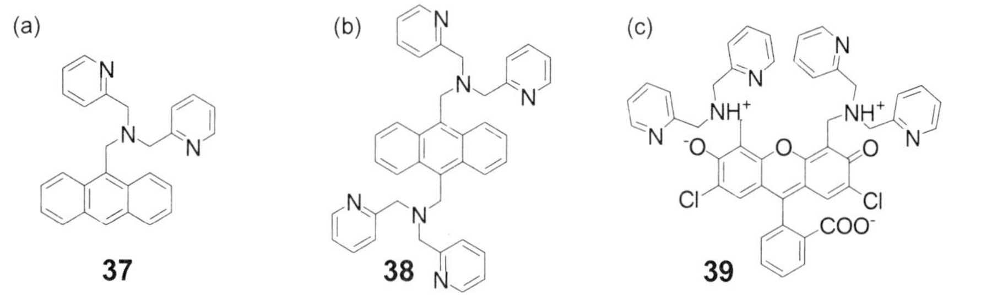

PET SENSORS

Figure 6.15. Some PET sensors containing pyridines described by (a) de Silva, 1997; (b) Kubo, 2003; and (c) Burdette, 2003.

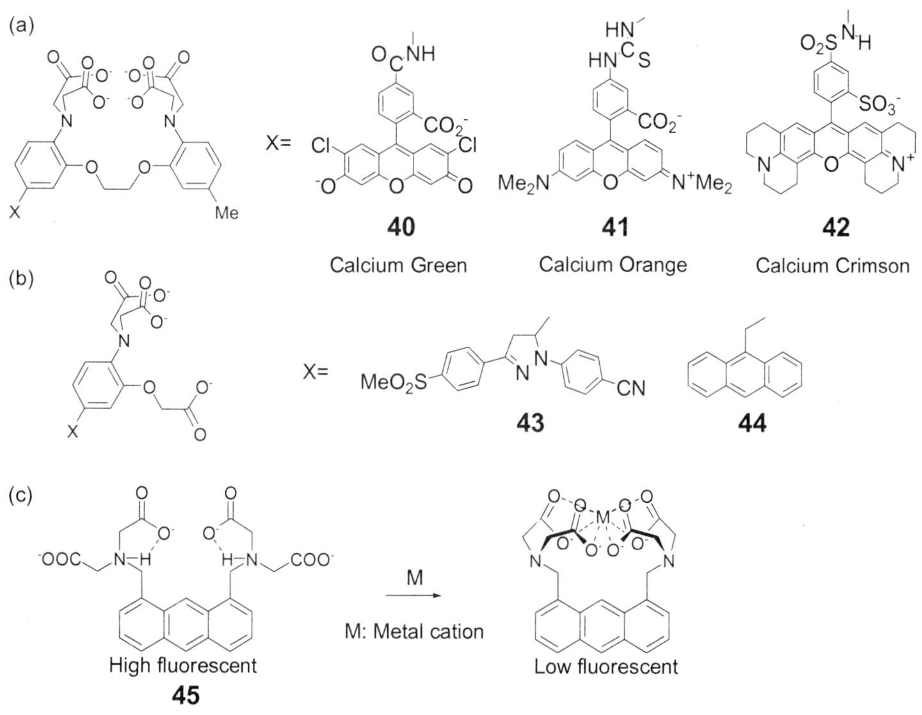

Figure 6.16. Chelator-based PET sensors described by (a) Kuhn, 1993; (b) de Silva, 1994; and (c) Choi, 2001.

Receptors containing diboronic acids can precisely recognize saccharide molecules.

In fact, one boronic acid can reversibly form a boronate ester with two OH groups (one diol group), and one diboronic acid can immobilize two suitably positioned diol units to form a saccharide-containing macrocycle. Selectivity can be achieved by controlling the relative spatial position of the two boronic acids in relation to the cis-diol moieties on the saccharide. A given monosaccharide possesses at least two binding sites that differ from other monosaccharides. Appropriate combinations of boronic acid and fluorophores lead to a remarkable class of PET sensors for saccharides. A boronic acid moiety is combined intramolecularly with an aminomethylfluorophore, and consequently, PET from the amine to the fluorophore causes fluorescence quenching of the latter. In the presence of a bound saccharide, the interaction between boronic acid and the amine is intensified, which inhibits the PET process. Boronic acid **51**[64] is an outstanding example of a selective sensor for glucose.

6.4.3. Cryptand-Based PET Sensors

Cryptands **52-56**[65-69] are examples of macrobicyclic structures (Figure 6.19). An interesting feature of **52**[65] is its ability to form exciplexes characterized by an additional band at higher wavelengths, thus allowing ratiometric measurements at two different observation wavelengths. The cavities of **53**[66] and **54**[67] fit the size of K^+ well. Coumarin **53** has been used successfully for monitoring the levels of potassium in blood and across biological membranes, but the pH must be controlled because of the pH sensitivity of this compound via protonation of the nitrogen atoms. This difficulty has been elegantly overcome in benzannelated cryptand **54**, in which the aromatic nitrogens have lower pKa than those of aliphatic amines. Cryptand **55**[68] shows a fluorescence enhancement with Zn^{2+}, Ni^{2+}, and Cu^{2+}. Its fluorescence quantum yield increases from 0.001 to 0.329 upon binding of Zn^{2+} in THF solution. Binding of pentane-1,4-diammonium ions by **56**[69] is supported by several lines of evidence, but fluorescence signaling is not one of them. Since protons are very successful in switching the emission "ON", the failure of pentane-1,4-diammonium ions must be attributable to their inability to block all four nitrogen lone electron pairs sufficiently strongly.

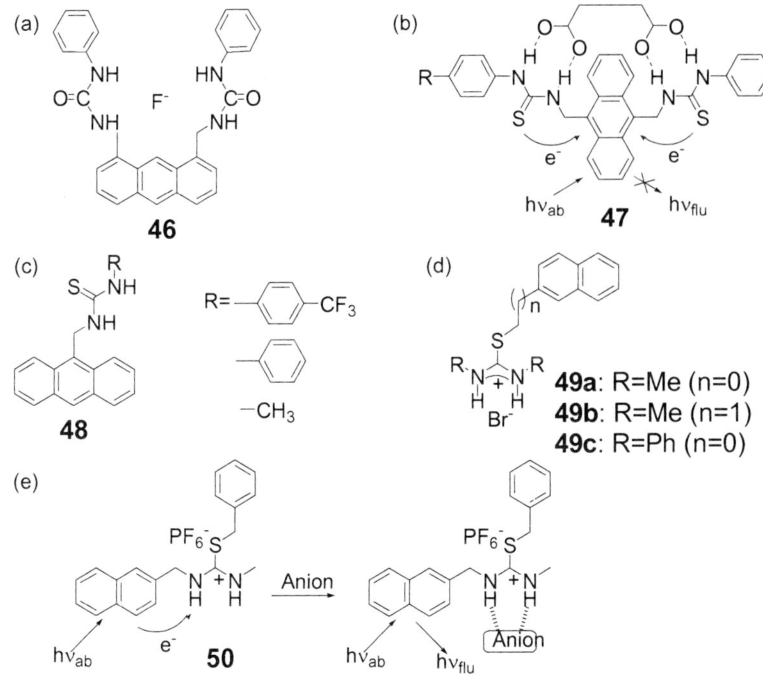

PET SENSORS

Figure 6.17. PET sensors containing urea or thiourea groups described by (a) Kim, 2002; (b) Gunnlaugsson, 2002; (c) Gunnlaugsson, 2001; (d) Nishizawa, 2002; and (e) Kubo, 2002.

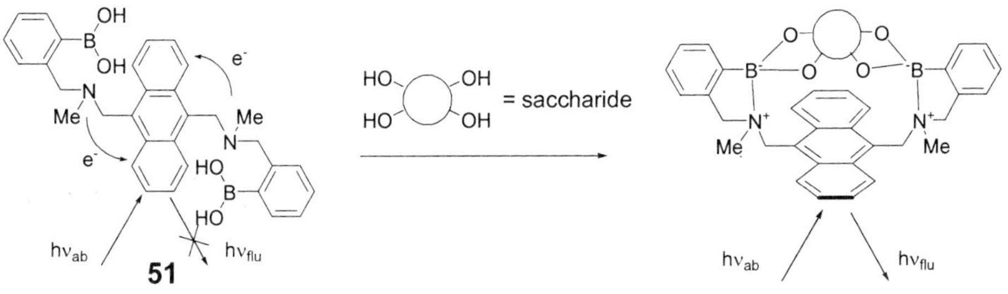

Figure 6.18. PET sensor containing boronic acid for saccharides described by James, 1995.

6.4.4. Cavitand-Based PET Sensors

Calixarene **57**[70] has been designed for selective recognition of sodium (Figure 6.20). It contains four carbonyl functional groups, two of which are linked to pyrene and nitrobenzene at opposite sites on the calixarene lower rim. Complexation with Na^+ prevents the close approach of pyrene and nitrobenzene and thus reduces the probability of PET. The fluorescence quantum yield increases from 0.0025 to 0.016. Calixarene **58**[71] is responsive to transition metal ions such as Zn^{2+} and Ni^{2+}. Calixarene **58** is an "OFF-ON" switch for Zn^{2+} and an "ON-OFF" switch for Ni^{2+}.

6.4.5. Polymer-Supported PET Sensors.

Polymers **59**[72] and **60**[73] are examples of simple PET systems which have been connected to inorganic and organic polymers, respectively. When one-fifth of the initial amino functional groups are grafted, the silica **59** behaves as a pH probe, the fluorescence of which switches on when the amino groups are protonated. On the contrary, if the silica loading is almost complete, pH probing is no longer possible. Compound **60** is interesting as it stands because of the sensitivity amplification due to the local concentration of K^+ induced by the polyanion.

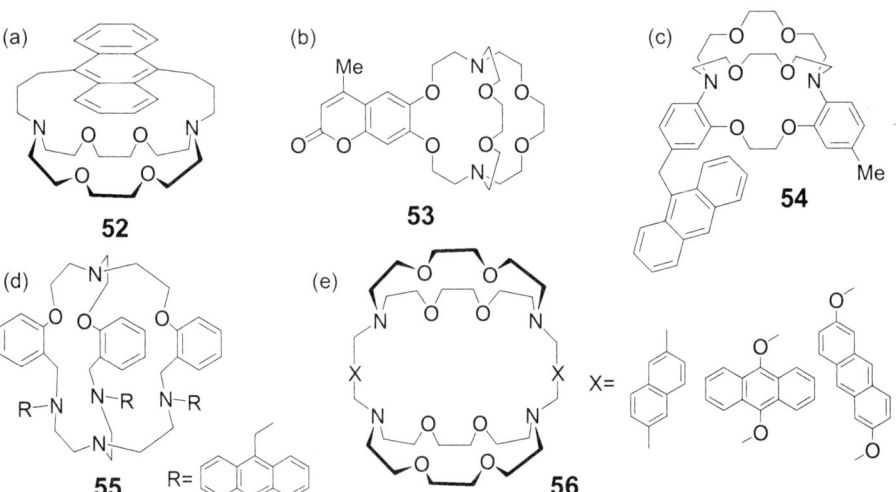

Figure 6.19. Cryptand-based PET sensors described by (a) Fages, 1989; (b) Golchini, 1990; (c) de Silva, 1993; (d) Ghosh, 1996; and (e) Fages, 1993.

6.4.6. PET Sensors Involving Excimer Formation

PET sensor **61** consists of a diazacrown ether with two pyrene pendants (Figure 6.22). Cation binding results in a large change in the monomer/excimer ratio. There is a concomitant increase in the overall fluorescence emission as a result of the reduction in PET from the nitrogen atom to the pyrenyl groups. The monomer/excimer ratio was found to be strongly dependent on the nature of the alkali metal ion. Among the investigated metal ions, larger stability constants for the complexes were obtained for K^+ and Ba^{2+}, in accordance with the size of these cations with respect to the crown diameter. In the same way, the emission spectrum of **62**[74], existing as the protonated form in acetonitrile, exhibits an excimer band whose intensity decreases upon binding of Cd^{2+} and Pb^{2+}.

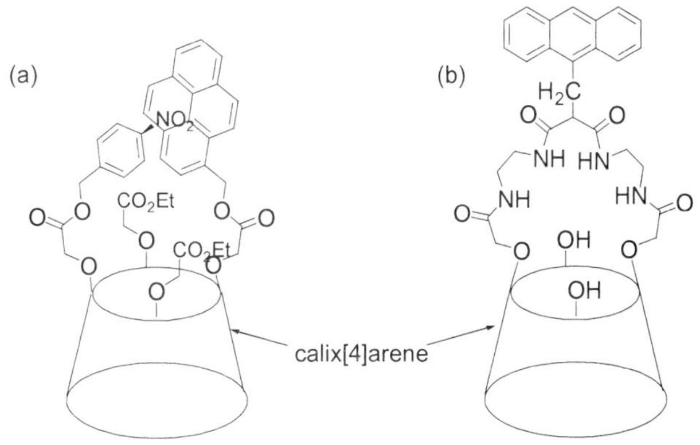

Figure 6.20. Cavitand-based PET sensors described by (a) Aoki, 1992 and (b) Unob, 1998.

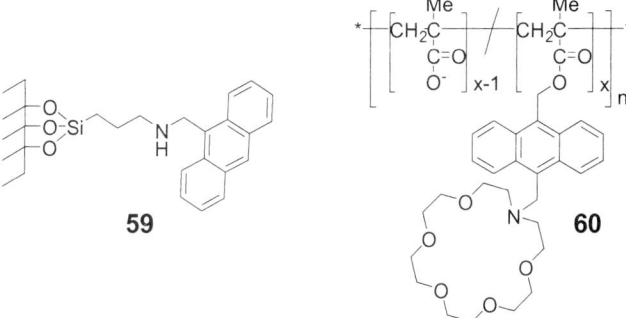

Figure 6.21. Polymer-supported PET sensors described by (a) Ayadim, 1996 and (b) Klok, 1996.

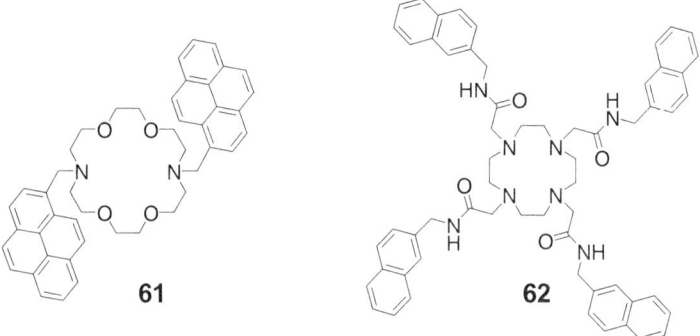

Figure 6.22. PET sensor involving excimer formation described by (a) Kubo, 1997 and Parker, 1995.

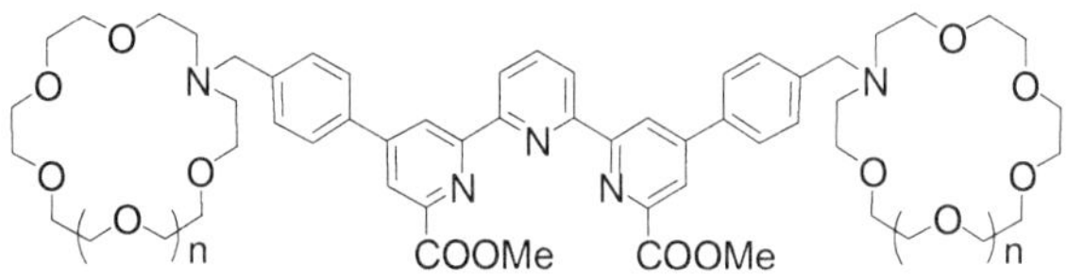

Figure 6.23. PET sensor involving energy transfer described by de Silva, 1997.

6.4.7. PET Sensors Involving Energy Transfer

Compounds **63a** and **63b**[75] (Figure 6.23) contain a terpyridyl diester that can strongly bind Eu^{3+} and a crown (15-crown-5 and 18-crown-6) to potentially bind Na^+ and K^+, respectively. Excitation energy transfer from this type of ligand to Eu^{3+} is known to occur via the triplet state and should result in luminescence from Eu^{3+}. However, when the crown is empty, only weak luminescence is detected because of quenching due to PET from the nitrogen atom of the crown. Binding of Na^+ and K^+ cause a very large enhancement of the luminescence quantum yield as expected from cation-induced eduction of the PET efficiency. **63a** and **63b** provide the first example of metal-triggered metal-centered emission. Due to the long lifetime of Eu^{3+} (hundreds of microseconds), time-delayed detection of the luminescence is possible which allows removal of the fast intrinsic fluorescence of biological samples.

PET SENSORS

6.5. PET SENSORS AS LOGIC GATES[76-78]

Fluorescent chemosensing is useful in biomedical research, and has very recently been developed into chemical logics. In the chemical logic system, the binding of a guest molecule to a host compound corresponds to the logic input and the resulting physical property changes such as absorption and/or fluorescence spectra correspond to the logic output. Whenever multiple chemical inputs independently produce one output, they can be regarded as OR logic gates, in other words, poorly chemoselective systems behave as OR functions. The OR gate **43**[57, 76] depends on the receptor unit being practically nonselective towards the two input ions as far as the fluorescence output level is concerned. In this case, the ion-induced FE values are 67 and 57 for Ca^{2+} and Mg^{2+}, respectively. Both cations cause similar conformational changes in the receptor unit and subsequently, similar increases in its oxidation potential. On the other hand, an AND logic gate recognizes multiple chemical inputs coincidentally and produces one output where high chemoselectivity is necessary. de Silva designed a cyanoanthracene system **64**[76] having both a benzocrown ring and a tertiary amino group which functions as an AND logic gate. The emission intensity of the cyanoanthracene increases in the presence of the sodium cation AND the proton. Two receptor sites, benzocrown ether for alkali metal ions and a tertiary amino group or protons, are characteristic of AND logic gates.

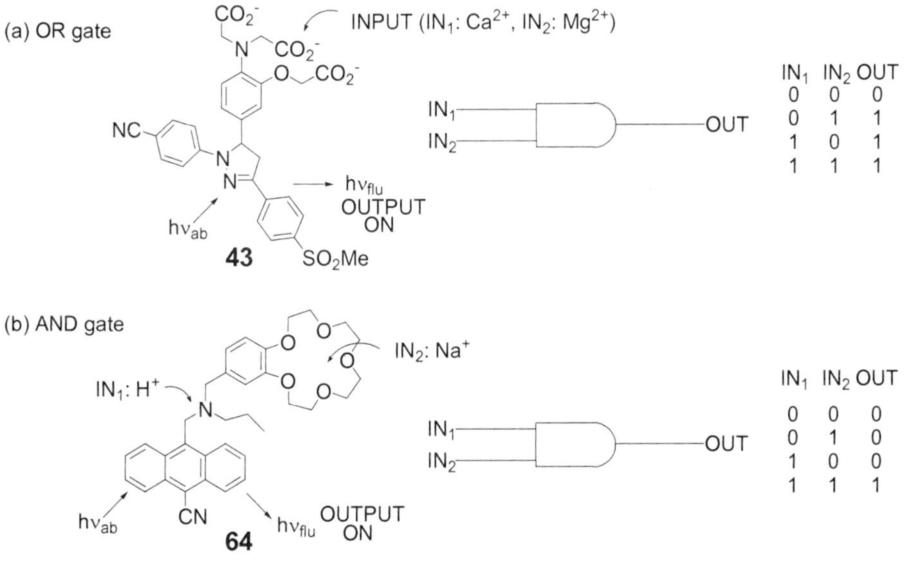

Figure 6.24. (a) OR and (b) AND logic gates. (Reproduced from Ref. 76. Copyright 1996 IUPAC.)

AND and OR logic gates each require a minimum of two inputs. Compound **65**[77] can now be shown to act as a NOR logic gate which integrates NOT and OR functions (Figure 6.25). As with **65**, the integration is achieved functionally without attempting the difficult task of physically linking molecular logic gates and other components. The bipyridyl receptor binds either H^+ or Zn^{2+}. Flattening and electrical charging of the receptor leads to rapid PET from the excited state of the anthracene fluorophore. Hence, fluorescence output from **65** is only observed if IN_1 (H^+) and IN_2 (Zn^{2+}) are both absent.

Simpler logic gates, i.e., NOT, only need one input. Creation of higher level logic systems requires integration of lower level devices. For instance, the INHIBIT gate,[77] as usually represented, requires a minimum of three inputs and a particular combination of AND and NOT logic. Its electronic symbols and truth table are shown in Figure 6.26. The challenge of integration has been taken up in INHIBIT gate **66**. In the language of Figure

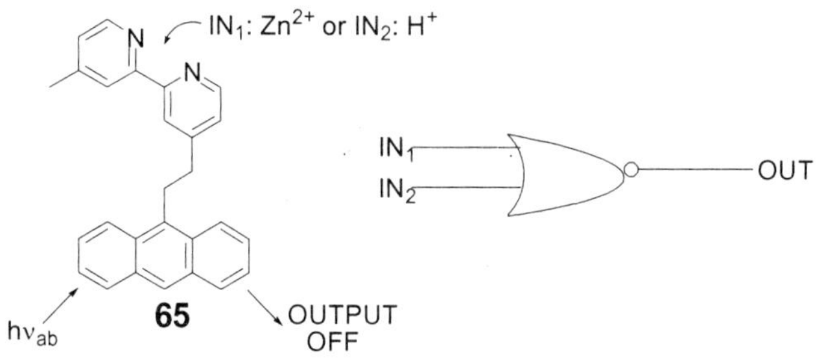

IN1	IN2	OUT
0	0	1
0	1	0
1	0	0
1	1	0

Figure 6.25. Electronic symbols and truth table for the NOR logic gate. (Reproduced from Ref. 77. Copyright 1999 American Chemical Society.)

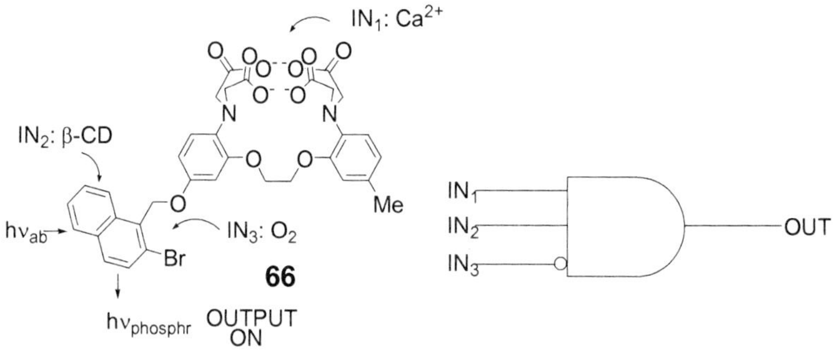

IN1	IN2	IN3	OUT
0	0	0	0
0	1	0	0
1	0	0	0
1	1	0	1
0	0	1	0
0	1	1	0
1	0	1	0
1	1	1	0

Figure 6.26. Electronic symbols and truth table for the INHIBIT logic gate. (Reproduced from Ref. 77. Copyright 1999 American Chemical Society.)

6.26, IN_1 = Ca^{2+}, IN_2 = β-cyclodextrin, IN_3 = O_2, OUT = phosphorescence intensity. Ca^{2+} binding to the amino acid moieties within Tsien's receptor stops PET involving the bromonaphthyl phosphor. Phosphorescence is also susceptible to PET-induced switching. However, phosphorescence emission cannot be observed without enveloping the phosphor in a transparent shield such as β-cyclodextrin. This prevents triplet-excited states centered on bromonaphthyl phosphors from encountering and deactivating each other. Intense phosphorescence emission still requires the absence of O_2. As a small paramagnetic molecule, O_2 wreaks havoc on magnetic triplet-excited states. In the present case of **66**, O_2 deactivates the triplet-excited state despite the reversible protection offered by the β-cyclodextrin shield. Thus, O_2 serves as a special disabling input (IN_3) for the INHIBIT gate.

6.6. CONCLUSIONS

It is hoped that this chapter will have given the reader a better understanding of the basic principles of PET sensors, switches, and logic gates. The present review illustrates the immense variety of PET sensors that have been designed for various analytes. There are many cations, anions and organic molecules in organisms (blood) and nature (e.g. seawater, rainwater, groundwater, river water etc.), and it is very difficult to detect a specific guest in complicated environments. It is necessary for the practical use of a PET sensor to improve the selectivity of guest capture and signal transduction in a specific condition. Progress in the relevant fields of supramolecular chemistry and photochemistry will certainly lead to new selective PET sensors in the future.

Finally, PET sensors can be used as switches with potential for information technology and sensors for monitoring species and properties in medicine, biology, environmental science, analytical chemistry, materials science, and even some branches of engineering. Further development of new PET sensors is expected.

6.7. REFERENCES

1. J. W. Steed and J. L. Atwood eds., *Supramolecular Chemistry*, John Wiley & Sons, Chichester, England (2000).
2. J. Mattay ed., *Topics in Current Chemistry 156, Photoinduced electron transfer I*, Springer-Verlag, Berlin, Tokyo (1990).
3. B. Valeur and I. Leray, Design principles of fluorescent molecular sensors for cation recognition, *Coord. Chem. Rev.* 205, 3-40 (2000).
4. L. Fabbrizzi and A. Poggi, Sensors and switches from supramolecular chemistry, *Chem. Soc. Rev.* 24, 197-202 (1995).
5. A. P. de Silva, H. Q. N. Gunaratne, T. Gunnlaugsson, A. J. M. Huxley, C. P. McCoy, J. T. Rademacher, and T. E. Rice, Signaling Recognition Events with Fluorescent Sensors and Switches, *Chem. Rev.* 97, 1515-1566 (1997).
6. A. P. de Silva and S. A. de Silva, Fluorescent signaling crown ethers: switching on of fluorescence by alkali metal ion recognition and binding in situ, *J. Chem. Soc., Chem. Commun.*, 1709-1710 (1986).
7. K. Kubo and T. Sakurai, Molecular recognition of PET fluoroionophores, *Heterocycles 52*, 945-975 (2000).
8. A. P. de Silva, S. A. de Silva, A. S. Dissanayake, and K. R. A. S. Sandanayake, Compartmental fluorescent pH indicators with nearly complete predictability of indicator parameters; molecular engineering of pH

sensors, *J. Chem. Soc., Chem. Commun.*, 1054-1056 (1989).
9. A. P. de Silva, H. Q. N. Gunaratne, and P. L. M. Lynch, Luminescence and charge transfer. Part 4. 'On-off' fluorescent PET (photoinduced electron transfer) sensors with pyridine receptors: 1,3-diaryl-5-pyridyl-4,5-dihydropyrazoles, *J. Chem. Soc., Perkin Trans. 2*, 685-690 (1995).
10. A. W. Czarnik, Chemical communication in water using fluorescent chemosensors, *Acc. Chem. Res. 27*, 302-308 (1994).
11. L. Fabbrizzi and A. W. Czarnik, Transition metals as switches, *Acc. Chem. Res. 32*, 846-853 (1999).
12. L. Fabbrizzi, M. Licchelli, P. Pallavicini, and A. Taglietti, A Zinc(II)-Driven Intramolecular Photoinduced Electron Transfer, *Inorg. Chem. 35*, 1733-1736 (1996).
13. J. Yoon and A. W. Czarnik, Fluorescent chemosensors of carbohydrates. A means of chemically communicating the binding of polyols in water based on chelation-enhanced quenching, J. Am. Chem. Soc. 114, 5874-5875 (1992).
14. A. P. de Silva, H. Q. N. Gunaratne, and C. P. McCoy, Direct visual indication of pH windows: 'off-on-off' fluorescent PET (photoinduced electron transfer) sensors/switches, *Chem. Comm.*, 2399-2400 (1996).
15. S. A. de Silva, B. Amorelli, D. C. Isidor, K. C. Loo, K. E. Crooker, and Y. E. Pena, A fluorescent 'off-on-off' proton switch with an overriding 'enable-disable' sodium ion switch, *Chem. Commun.*, 1360-1361 (2002).
16. S. L. Murov, I. Carmichael, and G. L. Hug eds., *Handbook of Photochemistry Second Edit., Revised and Expanded*, Marcel Dekker, Inc., New York (1993).
17. R. S. Davidson and K. R. Trethewey, The geometrical requirements for fluorescent intramolecular exciplex formation and fluorescence quenching, *J. Chem. Soc., Chem. Commun.*, 827-829 (1976).
18. K. Kubo, R. Ishige, J. Kubo, and T. Sakurai, Synthesis and complexation behavior of N-(1-naphthylmethyl)-1,4,7,10,13-pentaoxa-16-azacyclooctadecane, *Talanta 48*, 181-187 (1999).
19. K. Kubo, N. Kato, and T. Sakurai, Synthesis and complexation behavior of diaza-18-crown-6 carrying two pyrenylmethyl groups, *Bull. Chem. Soc. Jpn. 70*, 3041-3046 (1997).
20. S. Iwata and K. Tanaka, A novel cation 'AND' anion recognition host having pyrido[1',2':1,2]imidazo[4,5-b]pyrazine as the fluorophore, *J. Chem. Soc., Chem. Commun.*, 1491-1492 (1995).
21. K. Kubo, R. Ishige, and T. Sakurai, Synthesis and complexation behavior of N,N'-bis(9-anthrylmethyl)-1,4,10,13-tetraoxa-7,16-diazacyclooctadecane, *Heterocycles 48*, 347-351 (1998).
22. K. Kubo, R. Ishige, and T. Sakurai, Complexation and fluorescence behavior of diazacrown ether carrying two anthryl pendants, *Talanta 49*, 339-344 (1999).
23. K. Kubo, R. Ishige, N. Kato, E. Yamamoto, and T. Sakurai, Synthesis and complexation behavior of N,N'-bis(1-naphthylmethyl)-1,4,10,13-tetraoxa-7,16-diazacyclooctadecane, *Heterocycles 45*, 2365-2380 (1997).
24. K. Kubo, E. Yamamoto, and T. Sakurai, Complexation and fluorescence behavior of diaza-15-crown-5 carrying two naphthyl pendants, *Kyushu Daigaku Kino Busshitsu Kagaku Kenkyusho Hokoku 12*, 137-140 (1998).
25. K. Kubo, E. Yamamoto, and T. Sakurai, Synthesis, complexation and fluorescence behavior of diaza-12-crown-4 carrying two naphthyl pendants, *Heterocycles 48*, 2133-2139 (1998).
26. K. Kubo, E. Yamamoto, and T. Sakurai, Synthesis, complexation and emission behavior of N,N',N'',N'''-tetrakis(1-naphthylmethyl)-1,4,8,11-tetraazacyclotetradecane, *Heterocycles 48*, 1477-1481 (1998).
27. K. Kubo, S. Sakaguchi, and T. Sakurai, Synthesis, complexation, and fluorescence behavior of armed crown ethers carrying naphthyl group, *Talanta 49*, 735-744 (1999).
28. E. Yamamoto, K. Kubo, and A. Mori, Synthesis and complexation behavior of diaza-18-crown-6 having troponyl and anthryl groups, *Heterocycles 56*, 305-311 (2002).
29. T. Gunnlaugsson, B. Bichell, and C. Nolan, A novel fluorescent photoinduced electron transfer (PET) sensor for lithium, *Tetrahedron Lett. 43*, 4989-4992 (2002).
30. K. Yoshida, T. Mori, S. Watanabe, H. Kawai, and T. Nagamura, Synthesis and metal ion-sensing properties of fluorescent PET chemosensors based on the 2-phenylimidazo[5,4-a]anthraquinone chromophore, *J. Chem. Soc., Perkin Trans. 2*, 393-398 (1999).
31. J. R. Lakowicz ed., *Topics in Fluorescence Spectroscopy Vol. 4: Probe Design and Chemical Sensing*, Plenum Press, New York (1994).
32. T. Hirano, K. Kikichi, Y. Urano, T. Higuchi, and T. Nagano, Novel zinc fluorescent probes excitable with

33. A. P. de Silva and K. R. A. S. Sandanayake, Off-on fluorescence indicates the length and bonding of α,α-alkanediammonium ions of 9,10-bis[(1-aza-4,7,10,13,16-pentaoxacyclooctadecyl)methyl]- anthracenes, *Angew. Chem., Int. Ed. Engl. 29*, 1173-1175 (1990).
34. A. P. de Silva, H. Q. N. Gunaratne, C. McVeigh, G. E. M. Maguire, P. R. S. Maxwell, and E. O'Hanlon, Fluorescent signaling of the brain neurotransmitter γ-aminobutyric acid and related amino acid zwitterions, *Chem. Commun.*, 2191-2192 (1996).
35. S. Sasaki, A. Hashizume, D. Citterio, E. Fujii, and K. Suzuki, Fluororeceptor for zwitterionic form amino acids in aqueous methanol solution, *Tetrahedron Letters 43*, 7243-7245 (2002).
36. H. Kijima, M. Takeuchi, A. Robertson, S. Shinkai, C. Cooper, and T. D. James, Exploitation of a novel 'on-off' photoinduced electron-transfer (PET) sensor against conventional 'off-on' PET sensors, *Chem. Commun.*, 2011-2012 (1999).
37. F. D'Souza, Molecular Recognition via Hydroquinone-Quinone Pairing: Electrochemical and Singlet Emission Behavior of [5,10,15-Triphenyl-20-(2,5-dihydroxyphenyl)porphyrinato]zinc(II)-Quinone Complexes, *J. Am. Chem. Soc. 118*, 923-924 (1996).
38. G. F. Mes, H. J. V. Ramesdonk, and J. W. Verhoeven, Photoinduced electron transfer in polychromophoric systems. 2. Protonation directed switching between tri- and bichromophoric interaction, *J. Am. Chem. Soc. 106*, 1335-1340 (1984).
39. A. P. de Silva and R. A. D. D. Rupasinghe, A new class of fluorescent pH indicators based on photoinduced electron transfer, *J. Chem. Soc., Chem. Commun.*, 1669-70 (1985).
40. S. Draxler and M. E. Lippitsch, pH sensors using fluorescence decay time, *Sensors and Actuators, B: Chem. B29*, 199-203 (1995).
41. A. P. de Silva, H. Q. N. Gunaratne, P. L. M. Lynch, A. J. Patty, and G. L. Spence, Luminescence and charge transfer. Part 3. The use of chromophores with ICT (internal charge transfer) excited states in the construction of fluorescent PET (photoinduced electron transfer) pH sensors and related absorption pH sensors with aminoalkyl side chains, *J. Chem. Soc., Perkin Trans. 2*, 1611-1616 (1993).
42. L. M. Daffy, A. P. De Silva, H. Q. N. Gunaratne, C. Huber, P. L. M. Lynch, and W. O. S. Tobias, Arenedicarboximide building blocks for fluorescent photoinduced electron transfer pH sensors applicable with different media and communication wavelengths, *Chem. Eur. J. 4*, 1810-1815 (1998).
43. M. Sauer, Single-molecule-sensitive fluorescent sensors based on photoinduced intramolecular charge transfer, *Angew. Chem., Int. Ed. 42*, 1790-1793 (2003).
44. P. Nanjappan and A. W. Czarnik, Metal ion catalyzed reactions of acrylonitrile, acrylamide, and ethyl acrylate by way of their Diels-Alder cycloadducts, *J. Am. Chem. Soc. 109*, 1826-1833 (1987).
45. M. E. Huston, E. U. Akkaya, and A. W. Czarnik, Chelation enhanced fluorescence detection of non-metal ions, *J. Am. Chem. Soc. 111*, 8735-8737 (1989).
46. D. H. Vance and A. W. Czarnik, Real-Time Assay of Inorganic Pyrophosphatase Using a High-Affinity Chelation-Enhanced Fluorescence Chemosensor, *J. Am. Chem. Soc. 116*, 9397-9398 (1994).
47. S. Arimori and T. D. James, A competition assay for diols using 9-(N,N-diethanolaminomethyl)- anthracene and phenylboronic acid, *Tetrahedron Lett. 43*, 507-509 (2002).
48. W. Wang, G. Springsteen, S. Gao, and B. Wang, The first fluorescent sensor for boronic and boric acids with sensitivity at sub-micromolar concentrations, *Chem. Commun.*, 1283-1284 (2000).
49. G. Springsteen, C. E. Ballard, S. Gao, W. Wang, and B. Wang, The development of photometric sensors for boronic acids, *Bioorg. Chem. 29*, 259-270 (2001).
50. K. Kubo, T. Sakurai, and A. Mori, Complexation and fluorescence behavior of 9,10-bis[bis(β-hydroxyethyl)aminomethyl]anthracene, *Talanta 50*, 73-77 (1999).
51. A. P. de Silva, A. Zavaleta, D. E. Baron, O. Allam, E. V. Isidor, N. Kashimura, and J. M. Percarpio, A fluorescent photoinduced electron transfer sensor for cations with an off-on-off proton switch, *Tetrahedron Lett. 38*, 2237-2240 (1997).
52. K. Kubo and A. Mori, Crystal Structures of 9,10-Bis[bis(2-pyridylmethy)aminomethyl]anthracene and Its $ZnCl_2$ Complex: Intramolecular π–π Interaction Between Anthracene and $ZnCl_2$-Complexed Pyridine, *Chem. Lett.*, in pressing (2003).
53. A. Ojida, Y. Mitooka, M. Inoue, and I. Hamachi, First artificial receptors and chemosensors toward phosphorylated peptide in aqueous solution, *J. Am. Chem. Soc. 124*, 6256-6258 (2002).
54. S. C. Burdette, G. K. Walkup, B. Spingler, R. Y. Tsien, and S. J. Lippard, Fluorescent Sensors for Zn^{2+} Based on a Fluorescein Platform: Synthesis, Properties and Intracellular Distribution, *J. Am. Chem. Soc.*

123, 7831-7841 (2001).
55. S. C. Burdette, C. J. Frederickson, W. Bu, and S. J. Lippard, ZP4, an improved neuronal Zn^{2+} sensor of the zinpyr family, *J. Am. Chem. Soc. 125*, 1778-1787 (2003).
56. A. W. Czarnik, Fluorescent Chemosensors for Ion and Molecule Recognition. (Developed from a Symposium Sponsored by the Division of Organic Chemistry at the 204th National Meeting of the American Chemical Society, Washington, DC, August 23-28, 1992), ACS Symp. Ser., (1993)
57. A. P. de Silva,; H. Q. N. Gunaratne, and G. E. M. Maguire, 'Off-on" fluorescent sensors for physiological levels of magnesium ions based on photoinduced electron transfer (PET), which also behave as photoionic OR logic gates, *J. Chem. Soc., Chem. Commun.*, 1213-1214 (1994).
58. M. Choi, M. Kim, K. D. Lee, K. N. Han, I. A. Yoon, H. J. Chung, and J. Yoon, A New Reverse PET Chemosensor and Its Chelatoselective Aromatic Cadmiation, *Organic Lett. 2*, 3455-3457 (2001).
59. S. K. Kim and J. Yoon, A new fluorescent PET chemosensor for fluoride ions, *Chem. Commun.*, 770-771 (2002).
60. T. Gunnlaugsson, A. P. Davis, J. E. O'Brien, and M. Glynn, Fluorescent Sensing of Pyrophosphate and Bis-carboxylates with Charge Neutral PET Chemosensors, *Org. Lett. 4*, 2449-2452 (2002).
61. T. Gunnlaugsson, A. P. Davis, and M. Glynn, Fluorescent photoinduced electron transfer (PET) sensing of anions using charge neutral chemosensors, *Chem. Commun.*, 2556-2557 (2001).
62. S. Nishizawa, Y. Y. Cui, M. Minagawa, K. Morita, Y. Kato, S. Taniguchi, R. Kato, and N. Teramae, Conversion of thioureas to fluorescent isothiouronium-based photoinduced electron transfer sensors for oxoanion sensing, *J. Chem. Soc., Perkin Trans. 2*, 866-870 (2002).
63. Y. Kubo, S. Ishihara, M. Tsukahara, and S. Tokita, Isothiouronium-derived simple fluorescent chemosensors of anions. *J. Chem. Soc., Perkin Trans. 2*, 1455-1460 (2002).
64. T. D. James, K. R. A. S. Sandanayake, R. Iguchi, and S. Shinkai, Novel saccharide-photoinduced electron transfer sensors based on the interaction of boronic acid and amine, *J. Am. Chem. Soc. 117*, 8982-8987 (1995).
65. F. Fages, J. P. Desvergne, H. Bouas-Laurent, P. Marsau, J. M. Lehn, F. Kotzyba-Hibert, A. M. Albrecht-Gary, and M. Al-Joubbeh, Anthraceno-cryptands: a new class of cation-complexing macrobicyclic fluorophores, *J. Am. Chem. Soc. 111*, 8672-80 (1989).
66. K. Golchini, M. M.-Basic, S. A. Gharib, D. Masilamani, M. E. Lucas, and I. Kurtz, Synthesis and characterization of a new fluorescent probe for measuring potassium, *Am. J. Phys. 258*, F438-F443 (1990).
67. A. P. de Silva, H. Q. N. Gunaratne, and K. R. A. S. Sandanayake, A new benzo-annelated cryptand and a derivative with alkali cation-sensitive fluorescence, *Tetrahedron Lett. 31*, 5193-5196 (1990).
68. P. Ghosh, P. K. Bharadwaj, S. Mandal, and S. Ghosh, Ni(II), Cu(II), and Zn(II) Cryptate-Enhanced Fluorescence of a Trianthrylcryptand: A Potential Molecular Photonic OR Operator, *J. Am. Chem. Soc. 118*, 1553-1554 (1996).
69. F. Fages, J. P. Desvergne, K. Kampke, H. B. Laurent, J. M. Lehn, M. Meyer, and A. M. Albrecht-Gary, Linear molecular recognition: spectroscopic, photophysical, and complexation studies on α,α-alkanediyldiammonium ions binding to a bisanthracenyl macrotricyclic receptor, *J. Am. Chem. Soc. 115*, 3658-3664 (1993).
70. I. Aoki, T. Sakaki, and S. Shinkai, A new metal sensory system based on intramolecular fluorescence quenching on the ionophoric calix[4]arene ring, *J. Chem. Soc. Chem. Commun.*, 730-732 (1992).
71. F. Unob, Z. Asfari, and J. Vicens, An anthracene-based fluorescent sensor for transition metal ions derived from calix[4]arene, *Tetrahedron Lett. 39*, 2951-2954 (1998).
72. M. Ayadim, J. L. H. Jiwan, A. P. De Silva, and J. P. Soumillion, Photosensing by a fluorescing probe covalently attached to the silica, *Tetrahedron Lett. 37*, 7039-7042 (1996).
73. H. A. Klok and M. Moeller, Polymer bound chromogenic crown ethers for optical ion-detection devices, *Macromol. Chem. Phys. 197*, 1395-409 (1996).
74. D. Parker and J. A. G. Williams, Luminescence behavior of cadmium, lead, zinc, copper, nickel and lanthanide complexes of octadentate macrocyclic ligands bearing naphthyl chromophores, *J. Chem. Soc., Perkin Trans. 2*, 1305-1314 (1995).
75. A. P. de Silva, H. Q. N. Gunaratne, T. E. Rice, and S. Stewart, Switching "on" the luminescence of one metal ion with another: selectivity characteristics with respect to the emitting and triggering metal, *Chem. Commun.*, 1891-1892 (1997).
76. A. P. de Silva, H. Q. Nimal, N. Gunaratne, T. Gunnlaugsson, C. P. McCoy, P. R. S. Maxwell, and J. T. Rademacher, T. E. Rice, Photoionic devices with receptor-functionalized fluorophores, *Pure&Appl. Chem.*

68, 1443-1448 (1996).
77. A. P. de Silva, I. M. Dixon, H. Q. N. Gunaratne, T. Gunnlaugsson, P. R. S. Maxwell, and T. E. Rice, Integration of Logic Functions and Sequential Operation of Gates at the Molecular-Scale, *J. Am. Chem. Soc.* *121*, 1393-1394 (1999).

SIGMA-COUPLED CHARGE-TRANSFER PROBES OF THE FLUOROPROBE AND FLUOROTROPE TYPE

Jan W. Verhoeven[*]

7.1. INTRODUCTION

A plethora of optical probes based on charge-transfer (CT) is known. Evidently such probes must contain at least one moiety with electron donor properties (D) and at least one with electron acceptor properties (A). While in some cases intermolecular D/A complexes have been employed, in general D and A are either linked directly or via a conjugated pi-system (D–π–A). However, in this chapter the focus will be on some fluorescent probes in which D and A are linked by an extended sigma-system (D-σ-A) that prevents direct D/A contact but still allows electronic interaction via a through-bond mechanism.

In virtually all cases CT probes exploit the sensitivity of the spectral position or the intensity of a CT absorption band or a CT emission band for changes in the direct molecular environment of the D/A pair.

In the present chapter such changes will mainly be related to the solvating power and dynamics of that environment, but these are not the only properties which can be investigated with CT probes. Thus many CT probes have been reported which allow e.g. specific complexation of metal ions at the D site or protonation at that site. This typically reduces its electron donating properties thereby causing a significant blue-shift of the CT transition of the probe or even leads to a state inversion with the CT state no longer being the lowest excited state of the probe. The latter not only leads to dramatic changes in the absorption spectrum but especially also in the fluorescent behavior (e.g. switching from essentially non-fluorescent to strongly fluorescent). Examples of such probes are reviewed in other chapters of this series, but closer to the main stream of this chapter are the very well known polarity probes[1,2] based on the spectral shift ('solvatochromism') of a CT absorption or emission. Some examples of these have been compiled in Figure 7.1. The most prominent examples of solvatochromic CT absorption probes are the $E_T(30)$-probe and Kosower's Z-probe (see Figure 7.1).

---

[*] Jan W. Verhoeven, Laboratory of Organic Chemistry, University of Amsterdam, Nieuwe Achtergracht 129, 1018WS Amsterdam, The Netherlands.

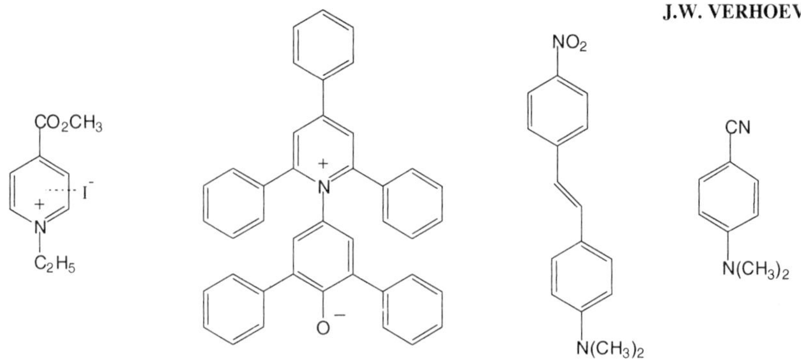

Figure 7.1. Typical charge-transfer probes with solvatochromic absorption and/or emission.

These probes have been used to define[1,2] the $E_T(30)$ and the Z-value solvent polarity scales, which simply reflect the position of the CT absorption maximum expressed in the somewhat odd units of kcal/mole. These CT absorption probes can be applied in virtually any solvent or solvent mixture (this is especially true for the closely related family of E_T probes) but of course suffer from the relatively low sensitivity of absorption measurements. This limits application in very small volumes and in highly scattering or absorbing media and makes remote monitoring quite difficult.

Evidently fluorescent CT probes could circumvent these limitations and in fact a plethora of such molecules has been reported, of which DMANS and DMABN are examples (see Figure 7.1). The transition responsible for their fluorescence has significant charge-transfer character and thereby makes the spectral position of that fluorescence sensitive to various changes in the molecular environment, which for the moment we will loosely indicate as polarity changes. Some of such molecules, such as DMABN, display dual fluorescence[3] from two states that differ in CT character and in conformation.

Before proceeding to the type of CT fluorescent molecules that are the subject of this chapter, a brief discussion will be given of the factors which govern the polarity dependence of CT probes in general. In doing so we will especially concentrate on the question how a maximum sensitivity (i.e. the largest spectral shift as a function of 'polarity') can be achieved.

7.1.1. Maximizing the Solvatochromism of CT Probes

The theory of solvatochromism has been treated extensively in many texts.[4] A crucial factor is always the difference in dipole moment between the ground and excited state of the probe molecule, which should be maximized. Furthermore strong solvatochromism in absorption requires that the ground state has a large dipole moment and the excited state a small dipole moment while strong solvatochromism in fluorescence requires the opposite. Thus the solvatochromic $E_T(30)$ and Z absorption probes shown in Figure 7.1 are essentially zwitterionic in the ground state and excitation in their CT absorption band leads to (partial) cancellation of charges. Solvatochromic fluorescence probes contain D and A moieties which (ideally) are neutral in the ground state (D-A) and present a lowest CT excited state with zwitterionic character (D$^+$- A$^-$) from which the solvatochromic fluorescence stems.

Using a continuum dielectric model for the solvent and under the assumption that the dipole moment of the initial state (i.e. the ground state dipole moment μ_g for absorption probes and the excited state dipole moment μ_e for emission probes) is much larger than that of the final state the solvatochromism of absorption and emission probes can be expressed[4,5] via the relatively simple Eq. (1) resp. Eq. (2).

$$v_{abs} = v_0 + 10070\,(\mu_g^2/\rho^3)(f\text{-}f'/2) = v_0 + 10070\,(\mu_g^2/\rho^3)\Delta f = v_0 + S\Delta f \tag{1}$$

$$v_{fl} = v_0 - 10070\,(\mu_e^2/\rho^3)(f\text{-}f'/2) = v_0 - 10070\,(\mu_e^2/\rho^3)\Delta f = v_0 - S\Delta f \tag{2}$$

In these equations spectral positions are expressed in cm^{-1} and v_0 stands for the (extrapolated) position of the absorption or emission maximum in the gas phase. The dipole moments (μ) are in Debye units and ρ (in Å) indicates the effective radius of the (spherical) solvent cavity surrounding the probe molecule. The solvent parameter Δf is defined on the basis of the bulk dielectric permittivity (ε) and the refractive index (n) as:

$$\Delta f = (\varepsilon\text{-}1)/(2\varepsilon+1) - (n^2\text{-}1)/(4n^2+2) \tag{3}$$

Because of the definition of Δf it should of course be clear that the Eqs. (1) and (2) cannot be expected to describe correctly the influence of eventual specific solvation mechanisms such as hydrogen bonding. Nevertheless, and as we will see later on, the linear dependence predicted by Eq. (2) is often confirmed in the so called Lippert-Mataga plots of v versus Δf over a wide range of solvent polarities where Δf varies between ca. 0.1 for saturated hydrocarbon solvents to ca. 0.4 for polar solvents like acetonitrile and the lower alcohols. The slope of such plots (S) is thus an objective measure for the polarity sensitivity of the probe employed at least as long as solvents giving specific interactions such as strongly hydrogen bonding solvents are excluded. Furthermore, for fluorescent probes S in principle carries quantitative information about the dipole moment of the excited state of the probe. Concerning the latter it should, however, be stressed that much uncertainty arises in determining an appropriate value for ρ, which is quite critical because S relates to the inverse cube of ρ. Probe molecules often have an elongated shape, which is thought to make the spherical cavity model underlying Eqs. (1) and (2) unsatisfactory. As an alternative Lippert proposed to set ρ at 40% of the long axis of an ellipsoidal cavity in which the probe fits. This may seem a small modification as compared to the spherical model where ρ is 50% of that axis, but for the same value of μ it predicts S to be almost twice as large and thus for the same value of S it leads to a calculated μ which is 1.4 times smaller than that derived via the spherical model! Nevertheless it is interesting to use the expressions given above for estimation of the maximum value that S might attain. In doing so we note that the length of a probe molecule (L) limits the dipole length it may contain to L-$2d$ where d is the shortest distance between a center of charge in the dipole and the Van der Waals surface of the molecule as schematically indicated in Figure 7.2.

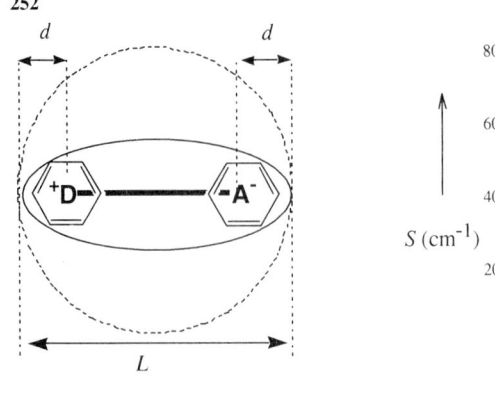 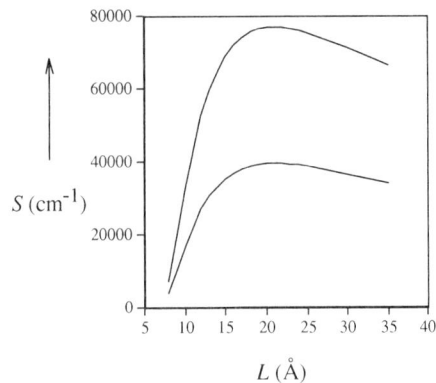

Figure 7.2. Spherical and ellipsoidal solvent cavity models for D-bridge-A molecules and the maximum solvent sensitivity (S) of their solvatochromic CT transition estimated therefrom as a function of the overall length (L) for $d = 3.5$ Å.

In most probes these charges are located in aromatic donor and acceptor moieties for which the Van der Waals radius $d \geq 3.5$ Å. In Figure 7.2 we now plot the slope S predicted by Eq. (2) as a function of the length of the probe assuming $d = 3.5$ Å and applying either the spherical cavity model where $\rho = 0.5L$ or the ellipsoidal cavity model with $\rho = 0.4L$.

While, as already noted above, the ellipsoidal model predicts S values that are twice those of the spherical model, both models indicate that S maximizes for probe molecules around 20Å in length. Although further elongation of course allows for larger μ values, its effect on the solvent sensitivity is offset and even overcompensated by the concomitant increase of ρ.

Quantitatively the graphs in Figure 7.2 should be taken with a 'grain of salt', but they demonstrate why extension of probe length may increase μ but not necessarily enhances the solvent sensitivity.

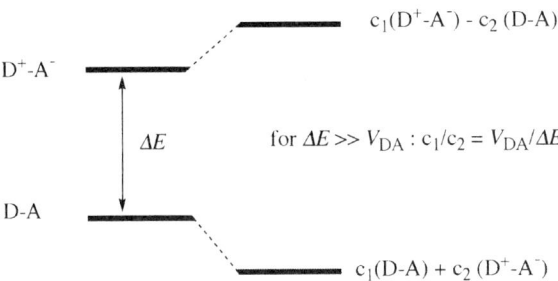

Figure 7.3. Mulliken two-state description of CT interaction.

$$V_{DA} = \frac{2.06 \times 10^{-2} \sqrt{\varepsilon_{max} \upsilon_{max} \Delta \upsilon_{1/2}}}{R_c} \qquad (4)$$

$$V_{DA} = \sqrt{\frac{1.4 \times 10^5 k_r}{n^3 R_c^2 \upsilon_{max}}} \qquad (5)$$

In the discussion above it has tacitly been assumed that for a CT probe the CT transition switches the molecule between a state devoid of charge transfer (DA) and a state with full charge transfer character (D$^+$A$^-$). In most CT probes used (like the ones in Figure 7.1), however, these states mix to a certain degree and in fact the CT transition often derives a large part of its oscillator strength from such state mixing, which results from a non-zero electronic coupling (V_{DA}) between D and A. Thus the molar extinction coefficient (ε_{max} in M^{-1}cm^{-1}) of a CT absorption can be connected to V_{DA} via the well known Hush expression Eq. (4) and to the radiative rate constant (k_r) of CT fluorescence via the related Eq. (5).[6,7]

In these expressions V_{DA}, $\Delta v_{1/2}$ (the full bandwidth at half height), and v_{max} are in cm^{-1}, k_r is in s^{-1}, while R_c represents the center-to-center distance of D and A in Å. Both expressions are based on the Mulliken two-state model for CT interaction[8] in which the 'no-bond' DA configuration and the 'ionic' D$^+$A$^-$ configuration mix to give the actual ground state and CT excited state as schematized in Figure 7.3.

This mixing reduces the difference in dipole moment between the two states and thereby the solvent sensitivity of CT probes. Especially for absorption probes this forms a problem because a strong CT absorption band is a desirable feature, but inevitably requires (see Eq. 4) a high value of V_{DA} and thus a relatively strong contribution (see Figure 7.3) of the CT configuration to the ground state wave function.

For fluorescent CT probes V_{DA} and thereby the degree of mixing can in principle be made much smaller. First of all because a high fluorescence quantum yield (Φ) may be achieved even when k_r is small if at the same time the radiationless decay rate (k_d) can be minimized because $\Phi = k_r/(k_r+k_d)$. And – as we shall see later – especially because the radiative rate can also be enhanced by mechanisms that do not require mixing of the DA configuration into the wavefunction of the CT radiative state.

7.2. SIGMA-COUPLED CHARGE TRANSFER PROBES; D-σ-A

7.2.1. Fluoroprobe (FP) an Extremely Solvatochromic Fluorescent D-σ-A Probe.

As we have seen above, the length of a CT probe and thus the distance between the D and A moieties should be limited for maximum sensitivity (see Figure 7.2) but at the same time also the electronic coupling V_{DA} between D and A should be limited.

For D-bridge-A probes in which the bridge is a conjugated pi-electron moiety (i.e. D–π–A) these two criteria are difficult to combine. A well studied example of a CT absorbing and fluorescent D-π-A probe is 4-dimethylamino-4'-nitrostilbene (DMANS, see Figure 7.1). The centers of the aniline donor moiety and the nitrobenzene acceptor moiety are about 6.7 Å apart, which suggests that in the CT excited state the dipole moment could increase by up to ≈ 32 D. However, D and A are strongly coupled via the ethylene pi–bridge and as a result the ground-state dipole moment of DMANS already has a high value (μ_g = 7.5 D) which increases with only ≈ 15.9 D in the excited state to μ_e = 23.4 D.[2] For longer pi–bridges the charges furthermore tend to become delocalised into the pi–bridge, which further offsets the change in dipole moment between ground and excited states.

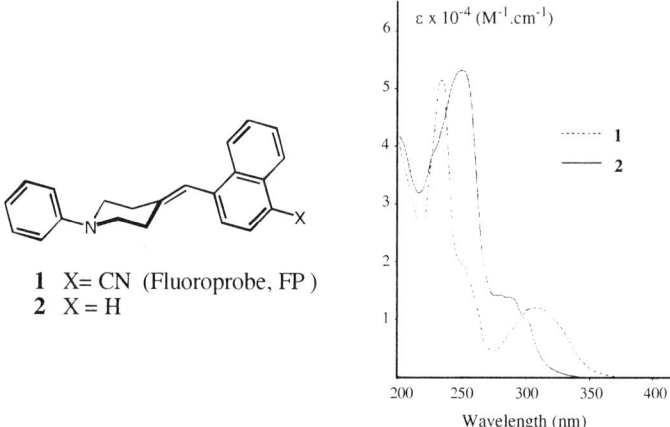

1 X= CN (Fluoroprobe, FP)
2 X = H

Figure 7.4. Structure of Fluoroprobe (**1**) and of its analogue **2** and their absorption spectra (in cyclohexane).

Early in our research on long-range electron transfer we found that saturated hydrocarbon bridges often allow remarkably fast photoinduced electron transfer to occur between D and A moieties connected by such bridges (i.e. D–σ–A) even if these bridges have a rigidly extended conformation which precludes direct D/A contact. Some of these systems even display discrete intramolecular CT absorption bands and some show CT fluorescence from the dipolar state thus populated upon excitation into that CT absorption or into local transitions of D or A.[9-11]

When D and A are neutral moieties - as in most systems we studied - their CT absorption shows very little solvatochromism but their CT fluorescence displays very pronounced solvatochromism, which is fully in line with the expected change from an essentially non-dipolar ground state (D–σ–A) to a dipolar CT excited state (D$^+$–σ–A$^-$). Regrettably the fluorescence quantum yield was in general quite low and often the CT fluorescence - if any - is only observable in a limited solvent polarity range. Evidently the through-sigma-bond interaction (TBI) responsible for the electronic coupling between D and A in extended D–σ–A systems is relatively weak thereby making the radiative rate (k_r) of the CT fluorescence in general too small to be of much practical importance for application of such systems as fluorescent probes.

This situation changed dramatically with the more or less serendipitous finding[12] of the molecule that was later dubbed 'Fluoroprobe' (FP) (**1**) followed by a series of analogues[13,14] such as **2** (see Figure 7.4).

Fluoroprobe contains an N,N-dialkylanilino electron donor moiety connected to a 1-vinyl-4-cyanonaphthalene electron acceptor moiety via a saturated piperidine bridge in an extended conformation. This structure implies that the D and A pi–systems are separated by at least three sigma-bonds and, as classically expected, the absorption spectrum of FP (see Figure 7.4) and its analogues thus provides little indication for any D/A interaction except perhaps for some broadening and tailing of the absorption spectrum towards the red, which might indicate the presence of a weak intramolecular CT absorption. In sharp contrast, however, the fluorescence spectrum of FP (as well as that of many of its analogues such as **2**) does neither display emission attributable to D nor to A but instead displays a strong structureless fluorescence with extreme solvatochromism attributed to intramolecular CT fluorescence.

Table 7.1. CT fluorescence maxima (v_{ct} in cm^{-1}) of FP (**1**) and absorption maxima of the $E_T(30)$ probe in a series of 14 solvents. For comparison the $E_T(30)$ values given have been converted from kcal/mole to cm^{-1}.

Solvent	Δf	$E_T(30)$ (cm^{-1})	v_{ct} (**1**) (cm^{-1})
1) n-hexane	0.092	10876	24600
2) cyclohexane	0.1	10841	24400
3) benzene	0.116	12035	20920
4) 1,4-dioxane	0.122	12631	19380
5) di-n-butylether	0.194	11578	21500
6) diisopropylether	0.237	11964	20400
7) diethylether	0.251	12105	19500
8) chloroform	0.251	13719	18830
9) ethylacetate	0.292	13368	17500
10) tetrahydrofuran	0.292	13122	17500
11) 1,2-dimethoxyethane	0.309	13403	16810
12) dichloromethane	0.319	14280	17300
13) pyridine	0.326	14210	15950
14) acetonitrile	0.393	15999	14400

Table 7.1 compiles data about the fluorescence maximum of FP in a series of 14 solvents and Figure 7.5 shows a Lippert-Mataga plot based on these data.

The most prominent feature is of course the extreme solvatochromism evidenced by a slope $S = -35,000$ cm^{-1} for FP, which brings it close to the theoretical maximum predicted via the spherical model although the ellipsoidal model suggests that there is still room for improvement (see Figure 7.2).

In Figure 7.5 we have also plotted as a function of Δf the $E_T(30)$ values of the solvents converted to a wavenumber scale. Evidently – at least in the nonhydroxylic solvents employed here – the sensitivity of the $E_T(30)$ probe is less than half that of FP. As predicted by Eqs. (1) and (2) the slope of the Lippert-Mataga plot is negative for FP and positive for $E_T(30)$. However, while for FP the correlation (with two exceptions, see Figure 7.5) is excellent, this is less so for $E_T(30)$. The smaller sensitivity of $E_T(30)$ and its weaker correlation with Δf can probably both be related to the fact that in this probe strong mixing of the type indicated in Figure 7.3 occurs. Thus, while the ground-state dipole moment of the $E_T(30)$ probe is large ($\mu_g = 14.7$ D), it remains rather large in the excited state ($\mu_e = 6$ D)^2 and the assumption $\mu_g >> \mu_e$ underlying Eq. (1) therefore does not apply.

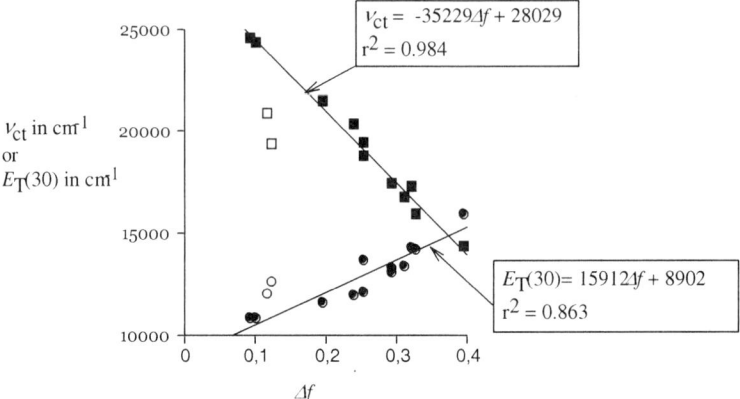

Figure 7.5. Lippert-Mataga plot (filled squares) for FP in a series of solvents. The correlation between Δf and the $E_T(30)$ values (filled circles) of the solvents is also shown. Benzene and 1,4-dioxane (open symbols) were omitted in the regressions.

By contrast the increase in dipole moment of FP in the excited state has been estimated from the Lippert-Mataga plot[12] as well as via other, independent techniques such as time resolved microwave conductivity (TRMC)[12] and electrochromism[15] to be as large as 27 ± 2 Debye. This huge increase in dipole moment corresponds to separation of two elementary charges at 5.6 ± 0.4 Å, which closely matches the distance between e.g. the C-1 atoms of the donor benzene and the acceptor naphthalene rings thereby substantiating the full CT character of the emitting state in FP.

As evident from Figure 7.5 the list of solvents (see Table 7.1) used contains two entries that do not fit the Lippert-Mataga plot. These are benzene and 1,4-dioxane, which both behave much more 'polar' than predicted by their Δf values. The anomaly of aromatic solvents and of 1,4-dioxane has been observed with many other polarity probes. It is generally agreed[4] that this anomaly is not related to specific solvent-solute interactions but stems from the fact that for close range interaction the nonspherical polarizability tensor of benzene and the quadrupolar nature of 1,4-dioxane become important to a degree not expressed in their bulk dielectric properties.

On the other hand specific interaction in the form of hydrogen bonding is probably involved in the strong quenching action of hydroxylic solvents on the fluorescence of FP. It is well known[4] that hydroxylic solvents also have a specific influence on other probes such as the $E_T(30)$ probe, which in alcohols shows a much stronger blue shift than would be predicted on the basis of bulk dielectric properties like Δf.

The extreme solvatochromism of FP implies a bathochromic shift of its fluorescence of $\Delta\upsilon > 10,000$ wavenumbers upon transfer from cyclohexane to acetonitrile. Importantly this shift occurs within the visible region (maxima ranging between 400 and 700 nm) which makes FP a very useful visible probe as may be clear from the photograph shown in Figure 7.6A, which displays the colors observed visually for various solutions of FP under UV illumination.

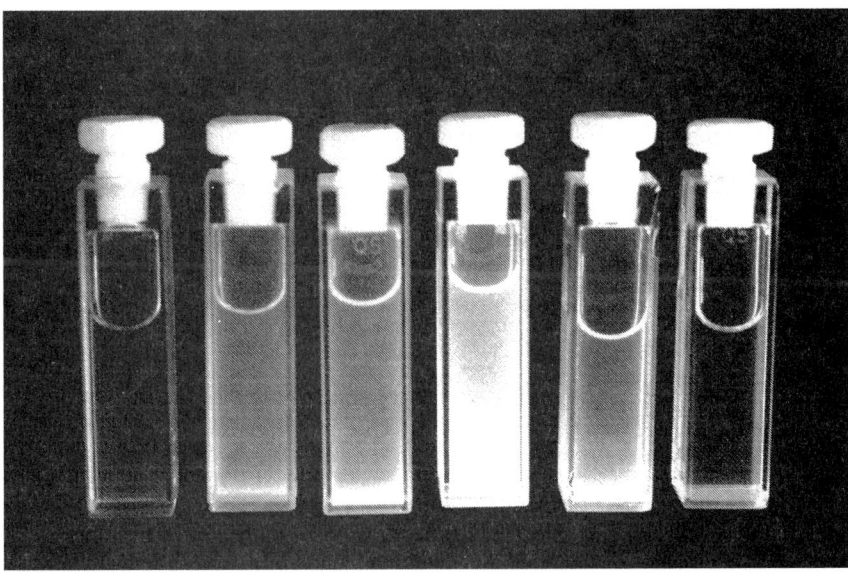

Figure 7.6A. (Please see Color Inserts Section) Fluorescence colors shown at room temperature by Fluoroprobe (FP) under UV illumination in a series of solvents with increasing polarity. From left to right : cyclohexane, benzene, toluene, 1,4-dioxane, chloroform, and pyridine.

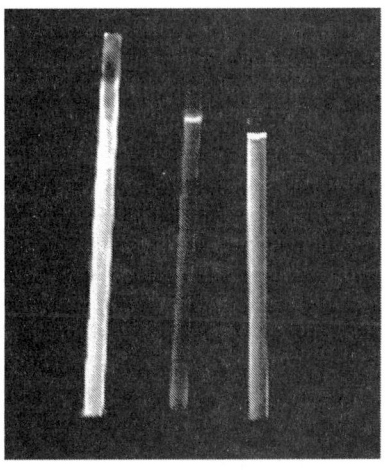

77 K   140 K   293 K

Figure 7.6B. Fluorescence of FP in MTHF at three temperatures. (Please see Color Inserts Section).

Figure 7.6C. Blue light emitted by a spin coated PLED based on **9** in PVK. (Please see Color Inserts Section).

Table 7.2. Fluorescence data of FP **1** and (in brackets) for **2** in various solvents at room temperature.

solvent	Δf	v_{ct} (cm^{-1})	Φ	τ (ns)	k_r (10^7 s^{-1})	k_d (10^7 s^{-1})
n-hexane	0.092	24600	0.2	0.84	23.8	95.2
		(27300)	(0.06)	(0.17)	(35.3)	(553)
cyclohexane	0.1	24400	0.21	1.2	17.5	65.8
		(27300)	(0.06)	(0.2)	(30)	(470)
benzene	0.116	20920				
1,4-dioxane	0.122	19380				
di-n-butylether	0.194	21500	0.85	11.4	7.46	1.32
		(25800)	(0.19)	(1)	(19)	(81)
diisopropylether	0.237	20400	0.78	11.6	6.72	1.90
		(25100)	(0.28)	(2.1)	(13.3)	(34.3)
diethylether	0.251	19500	0.58	13.4	4.33	3.13
		(24300)	(0.46)	(4.5)	(10.2)	(12.0)
chloroform	0.251	18830				
ethylacetate	0.292	17500	0.19	7.3	2.60	11.1
		(22300)	(0.85)	(13.5)	(6.3)	(1.1)
tetrahydrofuran	0.292	17500	0.16	8.7	1.84	9.66
		(22400)	(0.74)	(13.9)	(5.32)	(1.87)
1,2-diMeO-ethane	0.309	16810				
dichloromethane	0.319	17300	0.21	8.3	2.53	9.52
		(21600)	(0.68)	(14)	(4.86)	(2.29)
pyridine	0.326	15950				
acetonitrile	0.393	14400	≈ 0.01			
		(19000)	(0.28)	(12.7)	(2.20)	(5.67)

7.2.2. Quantum Yields and Fluorescence Lifetimes of FP and Related Probes as a Function of Solvent

In Table 7.2 we have collected for FP and for the related probe **2**, which contains the weaker naphthalene instead of cyanonaphthalene acceptor (see Figure 7.4), some more fluorescence data.

These now include quantum yields (Φ), fluorescence lifetimes (τ) and the radiative (k_r) and dark (k_d) decay rates calculated from Φ and τ as: $k_r = \Phi/\tau$ and $k_d = (1/\tau)-k_r$. A collection of fluorescence spectra in various solvents is shown in Figure 7.7.

The absorption spectra of FP and **2** (see Figure 7.4) display only very minor solvent dependence but just like for FP also for **2** the fluorescence is CT in nature and displays extreme solvatochromism ($S = -27,400$).

As might be expected, in each solvent the CT fluorescence of **2** occurs at significantly higher energy than that of FP which simply relates to the higher energy of the CT excited state in the former. These observations have been made with many FP related probes[13,14] in which both the acceptor and the donor site were varied.

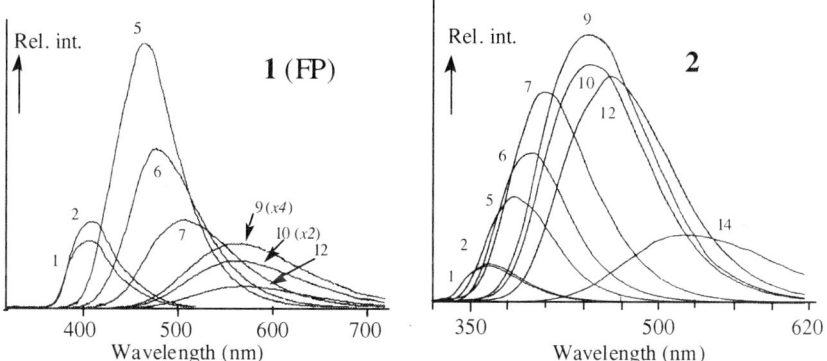

Figure 7.7. Fluorescence spectra of **1** (FP) and **2** in various solvents (see Table 7.1 for the numbering of solvents).

Another general observation, which deserves some more attention in the present context, is the solvent dependence of the quantum yield. This is plotted in Figure 7.8A as a function of Δf and in Figure 7.8B as a function of ν_{ct}. From Figure 7.8A it may be concluded that FP-like probes can be tuned to display their maximum fluorescence quantum yield in a certain polarity range, which has been substantiated by experiments with many such probes.

As already indicated by the data plotted in Figure 7.8B such maximum quantum yields always occur[13] in a polarity range where the combination of D and A incorporated in the probe leads to a fluorescence maximum around 22,000 cm^{-1} (≈ 450 nm).

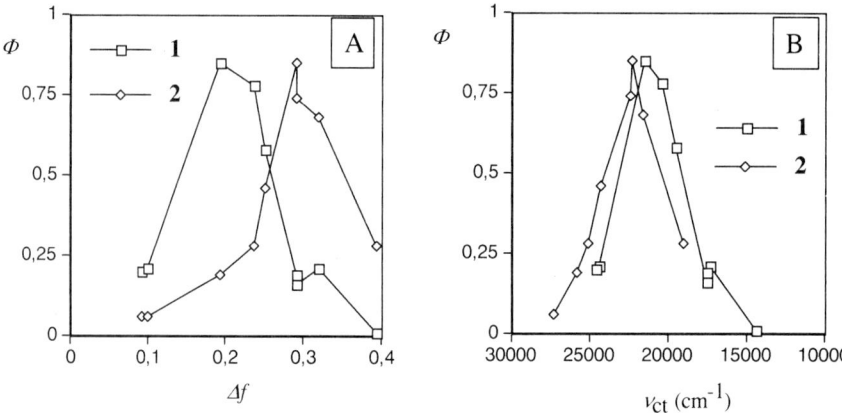

Figure 7.8. Fluorescence quantum yield as a function of solvent polarity Δf (A) and of the position of the fluorescence maximum ν_{ct} (B) for **1** (FP) and **2**. Lines are drawn between data points.

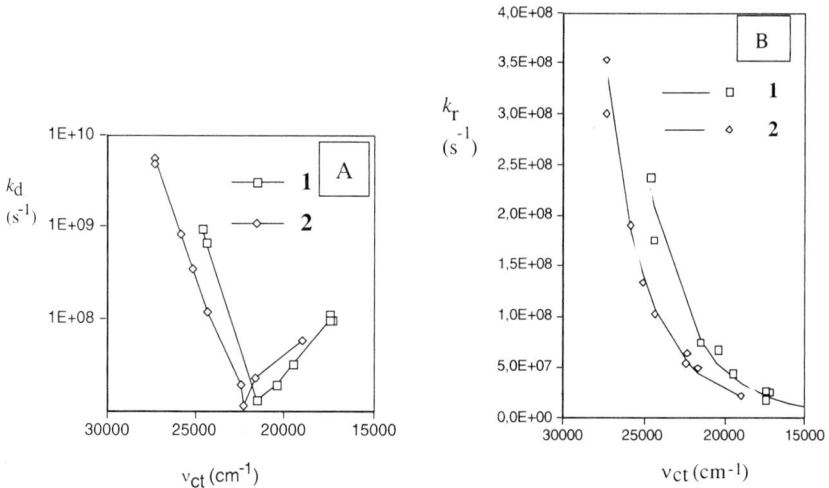

Figure 7.9. Dark decay (k_d) and radiative decay (k_r) rates for **1** (FP) and **2** as a function of the fluorescence maximum. Lines in Fig. 9A are drawn between data points, in Fig. 9B lines represent a best fit to Eq. (6), see text for details.

Closer inspection of the data in Table 7.2 reveals that this remarkable behavior is related to a strong dependence of both the radiative (k_r) and the dark (k_d) decay rates of the

CT excited state on the position of the CT fluorescence maximum. This is plotted for **1** and **2** (as representative examples of a large series of such systems investigated) in Figure 7.9A for k_d and in Figure 7.9B for k_r. As evident from Figure 7.9A k_d displays a minimum in the same range where the optimal quantum yield occurs and rises sharply (note the logarithmic scale) both in solvents where the fluorescence occurs at longer and at shorter wavelengths.

The increase of k_d in polar solvents, which induce a red shift, is not unexpected. This may simply be related to the energy-gap law, which implies that the vibrational coupling between an excited state and the ground state increases for smaller energy gaps thereby enhancing radiationless decay.

The sharp rise of k_d at higher energies, however, requires the contribution of an additional dark decay channel not directly leading to the ground-state. Originally we proposed[13] that this channel might be intersystem crossing from the CT singlet state to a local triplet state of the acceptor that is enhanced when the energy of the CT state rises to close the gap between these states. Later studies[16,17] have, however, made clear that instead a thermally activated[17] internal conversion to a local singlet state of the acceptor must be involved in which twisting of the exocyclic double bond of the acceptor occurs. This is a mechanism we suggested[11] earlier for other D–σ–A systems displaying a bell shaped dependence of the quantum yield of their (much weaker) CT fluorescence on solvent polarity. Interestingly, the nature of this mechanism implies that it can be suppressed by lowering the temperature as well as by increasing the viscosity of the medium, factors that thus increase the fluorescence quantum yield of FP-type probes especially under circumstances where it is blue shifted above 22,000 cm^{-1} (i.e. below ≈ 450 nm).

An equally remarkable solvent dependence is displayed by k_r (see Figure 7.9B), which increases strongly as the solvent is changed to induce a blue shift. It should be noted that this is the opposite of what is predicted by Eq. (5) based on the original Mulliken two-state model of D/A interaction (Figure 7.3), because upon increasing the energy gap between the ground and CT state their mixing should decrease leading to a concomitant decrease of k_r. Again, this non-Mulliken behavior of k_r has been observed by us for many D–σ–A systems. While a full discussion can be found in the literature[11,18,19] it suffices here to point out that instead of the simple two state model of Figure 7.3, a three (or more) state model adding one (or more) locally excited states of D or A seems required to describe the spectroscopic behavior of D–σ–A systems. In Figure 7.10 this is sketched for addition of a locally in A excited state (DA*).

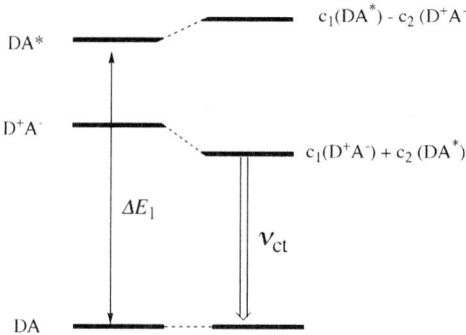

Figure 7.10. Three state model for D/A interaction leading to intensity borrowing by the CT transition from a local transition.

While we have analyzed this model allowing mixing between all three states[18,19] it has turned out that for D–σ–A systems the Mulliken-type mixing between DA and D^+A^- can often be neglected as compared to that between D^+A^- and DA*. This is not only due to the smaller energy gap between the two latter states, but probably also is related to the nature of the through-bond-interaction (TBI) in D–σ–A systems, which allows stronger interaction between higher energy states.

If – as assumed in Figure 7.10 – only mixing between excited configurations is taken into account the result is that the CT transition borrows most of its intensity from the local A→A* transition. If furthermore the mixing is relatively weak, the radiative rate of the CT fluorescence can be expressed by Eq. (6). Here ΔE_1 is the energy gap between the ground state and the locally excited state of A involved (see Figure 7.10), μ^* the transition dipole moment of the unperturbed A→A* transition and V_{DA*} the electronic coupling matrix element between the D^+A^- and DA* configurations.

$$k_r = \frac{64\pi^4 n^3}{3h}(V_{DA*}\mu^*)^2 \frac{v_{ct}^3}{(\Delta E_1 - v_{ct})^2} = Const.\frac{v_{ct}^3}{(\Delta E_1 - v_{ct})^2} \quad (6)$$

In contrast to Eq. (5), Eq. (6) predicts that the radiative rate will rise sharply when the CT emission frequency shifts to the blue thereby reducing the energy gap between the CT state and the locally excited state. As can be seen in Figure 7.9B the radiative rates for FP and **2** as a function of the solvent induced shift can satisfactorily be fitted via Eq. (6) using $\Delta E_1 = 32,300$ cm^{-1} for FP and $\Delta E_1 = 33,300$ cm^{-1} for **2**. This corresponds to transitions at 310 nm and at 300 nm, which is close to the actual position of the first A→A* π–π* transition of the acceptor in FP and **2** (see Figure 7.4).

Further analysis[19] revealed that for FP the actual degree of mixing is small over the whole polarity range so that the emissive state retains its essentially complete CT character even in the most non-polar solvents where it was estimated that DA* contributes less than 10% to the emissive CT state, which nevertheless allows sufficient intensity borrowing from the strong A→A* transition to give the CT fluorescence a radiative rate $\geq 10^8$ s$^-$

[1] (see Table 7.2). Even in the gasphase[20] the fluorescence of FP is strongly CT in nature with ν_{ct} = 26,880 cm^{-1}, which is close to the position extrapolated from solution via the Lippert-Mataga plot shown in Figure 7.5. Interestingly studies under supersonic jet cooled conditions revealed[20,21] that under such conditions the lowest Frank-Condon excited state is local A* in nature and that a very small barrier ($\approx$ 100 cm^{-1}) separates it from the CT state, a barrier that disappears when the donor strength is enhanced by substitution with a *p*-methyl or *p*-methoxy substituent.[22]

The fact that little mixing between the ground-state and the CT state occurs is of course a very beneficial factor contributing to the extreme fluorescence solvatochromism of FP (and other D–σ–A probes) as compared to most D–π–A probes where the ground-state dipole moment is enhanced and the excited state dipole moment reduced by such mixing.

7.2.3. Improving the Absorption Characteristics of Fluoroprobe, from FP (1) to FT (3)

As evident from their absorption spectra (see Figure 7.4) efficient excitation of probes like FP and **2** requires light sources emitting rather deep in the UV. Although FP can still be excited in its absorption tail via the common 365 nm Hg line or with the frequency tripled output (355 nm) of a Nd-YAG laser, shorter excitation wavelengths such as produced by a nitrogen laser (337 nm) are more appropriate. This constitutes a drawback in some applications. Substitution of the naphthalene acceptor chromophore for larger aromatics such as anthracene and pyrene has been investigated[14, 21, 23] and has been found to lead to useful FP analogues absorbing at longer wavelength. However, the lowering of the DA* state resulting from such substitution also implies that it comes closer to the CT state and may even be lower in energy which implies that the CT fluorescence is substituted by a fluorescence with strong A*→A character. This was indeed observed[21,23] for such FP analogues in non-polar solvents and these analogues therefore display strongly solvatochromic CT fluorescence only in more polar media.

It should be borne in mind that when the CT state is the fluorescent state of a D/A system - like it is in FP - this also implies that excitation into the corresponding CT absorption should in principle constitute the longest wavelength excitation path of such a system unless extensive conformational reorganization occurs between the locally excited and CT state. Indeed the first absorption band of D–π–A systems, of directly coupled D–A systems, and of intermolecular D/A complexes (such as DMANS, $E_T(30)$, and the Z-probe respectively, see Figure 7.1) is CT in nature. So why does FP not display a discrete intramolecular CT absorption band?

The expected position of the CT absorption maximum of FP can be estimated by assuming that the absorption band should show an approximate mirror relation with the fluorescence of FP in saturated alkane solvents (see Figure 7.7), which places it around 350 nm. Furthermore the oscillator strength of an absorption is in principle related to the radiative rate constant of the corresponding fluorescence, a relation which also connects Eqs. (4) and (5). In a simplified form this implies[24] that in the visible and near UV region the molar extinction of an absorption band and the radiative rate of the corresponding fluorescence are approximately related via Eq. (7):

$$k_r \approx 10^4 \, \varepsilon_{max} \tag{7}$$

Since (see Table 7.2) $k_r \geq 10^8$ s^{-1} for FP in non-polar solvents, this would imply that the CT absorption of FP should have $\varepsilon_{max} \approx 10^4$ (M^{-1}cm^{-1}). Although FP displays a distinct long wavelength absorption tail in the 350 nm region (see Figure 7.4) it is clear that this is much weaker than predicted.

The only way to explain this discrepancy is that the electronic coupling between D and A (i.e. V_{DA} in the two state model and V_{DA^*} in the three state model) is much weaker in the ground state than in the (relaxed) emissive state due to some conformational relaxation following or accompanying the intramolecular electron transfer.

In more flexibly linked systems this is often the result of conformational changes that reduce the D to A distance eventually driven by electrostatic attraction forces.[20,22,23,25-29] However, the very large and solvent independent dipole moment of the emissive state in FP makes it unlikely that this type of conformational processes occurs. Instead a more subtle conformational process seems to be involved, which selectively modulates the through-bond interaction between D and A. We assume that this occurs mainly via a change in the pyramidalization of the anilino nitrogen of the donor moiety. As substantiated by NMR and X-ray diffraction data[30] in the ground state of FP the central saturated piperidine ring adopts a chair conformation in which the donor nitrogen is pyramidalized with its phenyl ring in an equatorial orientation as expected from sterical considerations. This implies that the 'lone pair' orbital on that nitrogen occupies an axial position, which disfavors through-bond interaction with the acceptor pi-orbitals via the piperidine C-C sigma bonds. This is indicated in Figure 7.11 by the minor delocalization calculated for an axial lone pair in a model system (piperidone-4) where it can be seen that for an axial lone pair (hyperconjugative) delocalization of the lone pair occurs mainly into the properly aligned axial C-H bonds on the vicinal C2 and C6 atoms.

Enforcing an axial orientation of the N-substituent brings the lone pair in an equatorial position and thereby allows strong delocalization into the C2-C3 and C5-C6 bonds leading to significant TBI with the carbonyl group (which acts as the acceptor) via a path that closely fulfills the *all-trans* rule[31,32] for optimal TBI. Also planarization of the nitrogen coordination (see central pictures in Figure 7.11) is predicted to lead to a strongly increased TBI.[32] Interestingly, such planarization is exactly what should occur in the CT excited state of FP because amine radical cations adopt a planar structure.[33-35]

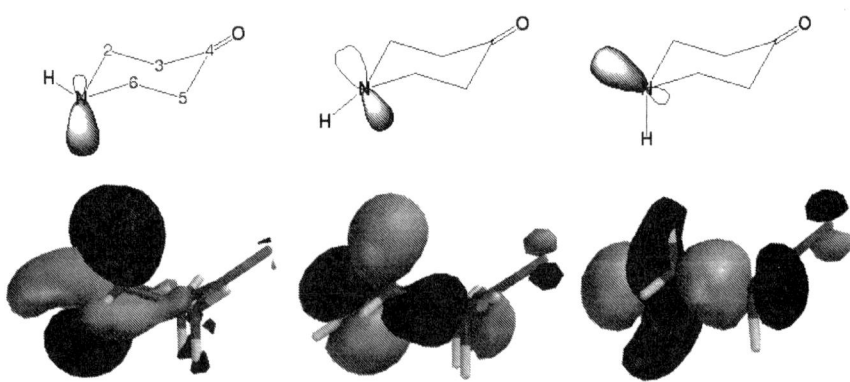

Figure 7.11. Three conformations of piperidone-4 and the calculated (AM1) effect on the delocalization of its (nitrogen 'lone pair') HOMO.

These qualitative considerations thus already provide a simple explanation for the apparent discrepancy between the high k_r of FP and its weak CT absorption. In the ground state the pyramidal structure around the anilino nitrogen and the equatorial orientation of the phenyl substituent allow only for weak TBI between D and A, but in the relaxed CT excited state planarization around the anilino nitrogen strongly enhances the TBI and thereby allows for a high k_r. Evidently this also implies that the CT absorption could be enhanced if the preferred ground state orientation of the phenyl substituent is changed from equatorial to axial. This has been achieved by expanding the Piperidine moiety to the bicyclic Tropane unit, leading to a system (see Figure 7.12) we named FluoroTrope (FT).[36] The additional ethylidene bridge in FT (**3**) sterically destabilizes the equatorial orientation of the phenyl ring and – as we have shown before[37-39] in systems containing stronger acceptor moieties – together with the additional stabilization of the axial orientation by TBI this suffices to make the axial orientation the preferred one in the ground state.

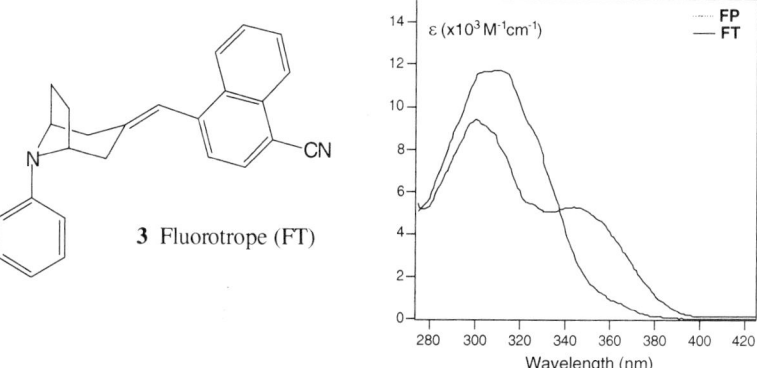

Figure 7.12. Structure of Fluorotrope (FT) and comparison of its absorption spectrum with that of FP (both in cyclohexane).

Comparison of the absorption spectra of FT and FP (Figure 7.12) confirms the dramatic increase of TBI in the former leading to the appearance of a discrete CT absorption band at the predicted position (350 nm) and with an intensity ($\varepsilon_{max} \approx 5000$ M^{-1}cm^{-1}) compatible with the k_r values found for FP. It is important to note that at the same time the first A→A* absorption of FT is weakened and blue shifted with respect to that in FP. This phenomenon has been observed[11] in many through-bond coupled D–σ–A systems and may be taken as strong evidence for dominant intensity borrowing of the CT transition from the local A→A* transition via the mechanism indicated in Figure 7.10. Again this implies that also in FT the degree of CT in the ground state can be very minor and that therefore its solvatochromism should be comparable to that of FP.

Table 7.3. Comparison of CT fluorescence maxima (v_{ct} in 10^3 cm^{-1}), quantum yields (Φ), and fluorescence lifetimes (τ in ns) and radiative rates (k_r in 10^7 s^{-1}) for FT and FP.

Solvent	FT				FP			
	v_{ct}	Φ	τ	k_r	v_{ct}	Φ	τ	k_r
n-hexane	23.8	0.55	3.7	14.9	24.6	0.20	0.84	23.8
cyclohexane	23.7	0.61	3.9	15.6	24.4	0.21	1.2	17.5
di-n-pentylether	21.3	0.77	11.2	6.9				
di-n-butylether	21.0	0.74	11.5	6.4	21.5	0.81	11.4	7.1
di-n-propylether	20.7	0.72	12	6.0	20.4	0.78	11.6	6.7
diethylether	19.7	0.71	15.7	4.5	19.5	0.58	13.4	4.3
ethylacetate	17.6	0.25	6.7	3.7	17.5	0.19	7.3	2.6
tetrahydrofuran	17.5				17.5	0.16	8.7	1.8
dichloromethane	17.1				17.3	0.21	8.3	2.5

This is confirmed by the experimental results compiled in Table 7.3. These data demonstrate that FT and FP show a close similarity with respect to fluorescence position and quantum yield as a function of solvent. There are, however, a few differences that deserve comment. Both FP and FT display their strongest fluorescence (Φ around 80%!) in solvents of moderate polarity, but the decrease in quantum yield at lower polarity is less pronounced for FT than for FP. Thus FT retains $\Phi \geq 50\%$ in saturated alkane solvents where for FP it drops to ca. 20%. As explained above the rather sharp drop of Φ in non-polar solvents must be attributed to the opening of a non-radiative path related to twisting of the exocyclic double bond of the acceptor when the CT state comes closer to the locally in A excited state. Fluorescence maxima of FT and FP in non-polar media indicate that in FT the CT state is at slightly lower energy and thus further removed from locally excited states than for FP in such non-polar environment. A very important conclusion to be drawn from the data in Table 7.3 is that – as visualized in Figure 7.13 – for any given position of the CT fluorescence maximum the radiative rate constants of FT and FP are virtually identical and both display the dependence on the energy of the emission maximum expected for a situation in which intensity borrowing from the local A→A* transition is dominant (see Eq. 6).

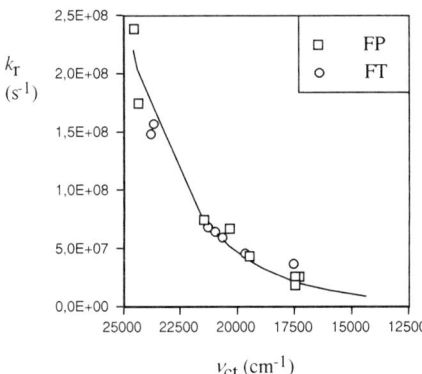

Figure 7.13. Comparison of the radiative rate constants of FP and FT as a function of their fluorescence maxima in a series of solvents (see Table 7.3 for data).

It may seem contradictory that FT and FP display similar radiative CT emission probabilities although the CT absorption is much stronger for FT than for FP. However, as already indicated above and as visualized in Figure 7.14, while the difference in CT absorption is caused by the difference in ground-state conformation, the conformation of the relaxed emissive CT state in FT and FP will be similar because the radical cation character of the donor in this state induces planarization of the anilino nitrogen.

In conclusion the results for FT show that it is possible to enhance significantly the long-wavelength absorption characteristics of FP without losing its uniquely strong solvatochromic properties.

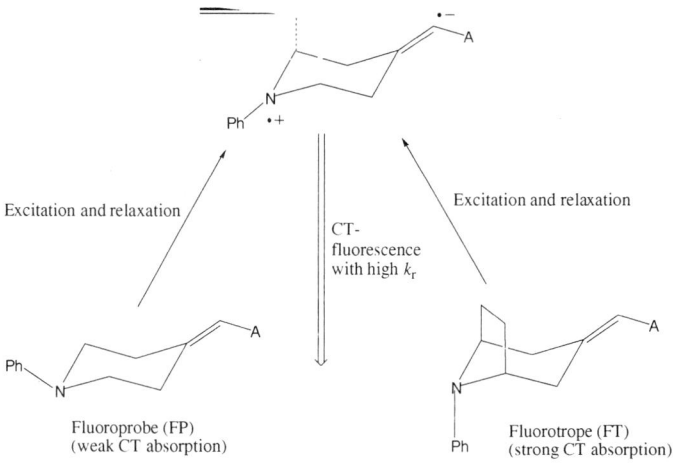

Figure 7.14. The conformation of FP and FT differs in the ground-state but is similar in the CT excited state resulting in very different CT absorption properties but nearly identical radiative CT fluorescence rates.

7.2.4. Fluorogenic Derivatives of FP and FT which Allow Covalent Attachment of the Probes; from FP and FT to MFP (4) and MFT (5)

In many applications it is desirable that optical probes can be attached covalently in a well defined location and especially for the investigation of biological systems many fluorescent probes have been extended with functional groups that allow them to form a covalent bond or eventually a very strong complex by reaction with specific functionalities in bio(macro)molecules.[40] Fluorogenic probes constitute a special class of such reactive probes. Fluorogenic probes are defined as being essentially non-fluorescent before reaction and showing strongly enhanced fluorescence after reaction. This property implies that the removal of unreacted probe molecules, which otherwise lead to nonspecific background fluorescence, can be omitted.

We decided to attach a maleimido group at the donor site of FP and FT because this group can be attached covalently to various nucleophilic centers and can also be copolymerized to become part of e.g. acrylic polymers as sketched in Figure 7.15.

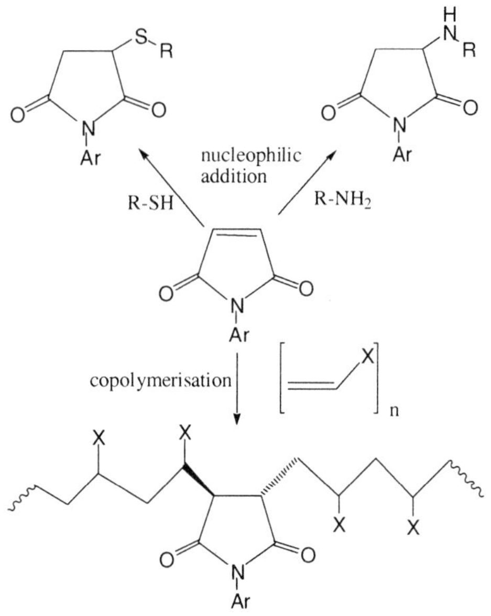

Figure 7.15. Reactivity of (N-aryl)maleimides converting the maleimide ring to a succinimide ring.

Furthermore it has been reported[40] that the maleimido group quenches the fluorescence of various fluorophores and that this fluorescence is restored upon reaction of the maleimido group.

This behaviour is dramatically demonstrated by the maleimido derivatives of FP and FT, named MFP[41] and MFT[36] depicted in Figure 7.16.

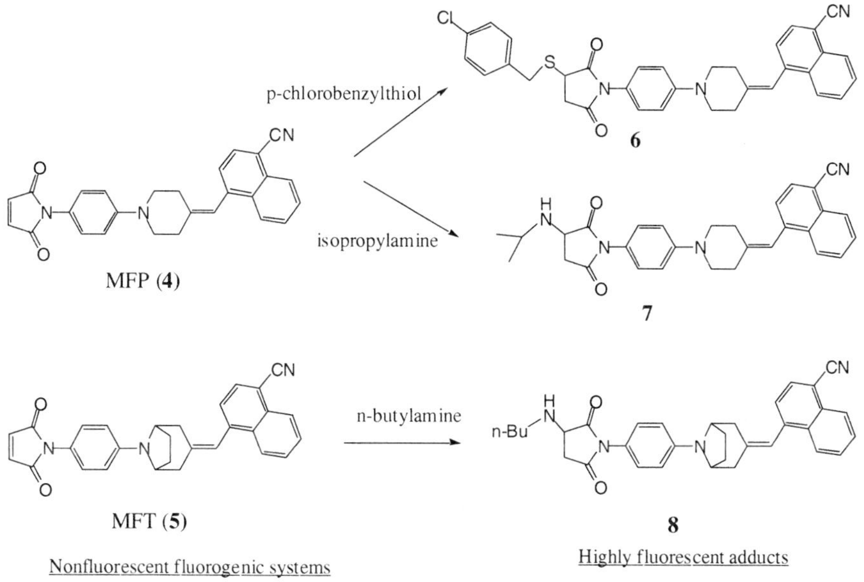

Figure 7.16. Structures of fluorogenic systems MFP and MFT and of their fluorescent adducts (**6**, **7** and **8**).

While as described above both FP and FT are highly fluorescent, MFP and MFT show no detectable fluorescence in any solvent. But after reactions of the type depicted in Figures 15 and 16 the strongly solvatochromic fluorescence of the FP or FT moiety is restored as shown by the data in Table 7.4, which compares fluorescence maxima of the adducts **6**, **7**, and **8** (see Figure 7.16 for structures) with those of FP and FT in a number of solvents.[36,41]

Table 7.4. CT fluorescence maxima (nm) of FP and FT compared to the adducts **6**, **7**, and **8** in various solvents.

solvent	FP (1)	6	7	FT (3)	8
			λ_{max}(nm)		
cyclohexane	410	411	411	420	420
di-n-butylether	465	466	461	476	a)
diethylether	513	518	516	508	503
ethylacetate	571	568	560	568	559
dichloromethane	578	564	559	585	575

a) Not determined

While the maleimido group thus exerts a strong quenching action, saturation of its double bond to yield a succinimide derivative removes this quenching and furthermore leads to a situation in which the donor capacities of the now para-substituted anilino moiety differ little from those in the parent FP and FT systems.

The latter finding is not unexpected because an N-succinimide substituent is estimated to exert little electronic influence since e.g. its Hammett sigma value (σ_p) is probably close to zero in analogy to other N-amido substituents.[42] The mechanism of the quenching action that the N-maleimide substituent exerts on FP and FT as well as on other fluorophores, such as pyrene[40] and electron rich stilbenes[40], seems to be not fully established yet. It has been noted[41] that N-phenylmaleimides show a very weak long wavelength absorption which shifts hypsochromically upon increasing solvent polarity and bathochromically upon introduction of electron releasing substituents on the phenyl ring. From the solvent effect this absorption was assigned to an n-π* transition and the quenching effect was therefore tentatively attributed[41] to internal conversion populating such a low lying, non-emissive n-π* state. On the other hand it has been noted[36,43] that maleimides have strong electron accepting properties. Semiempirical calculations[36,43] in fact suggested that in MFP and MFT the lowest unoccupied molecular orbital (LUMO) is not localized on the vinyl-cyanonaphthalene acceptor, as it is in FP and FT, but on the maleimido moiety. This implies that in MFT and MFP the lowest excited state is charge transfer in nature but with the maleimide acting as the acceptor. This is now further supported by the results of calculations on the model compound N-(4-dimethylamino phenyl)maleimide.

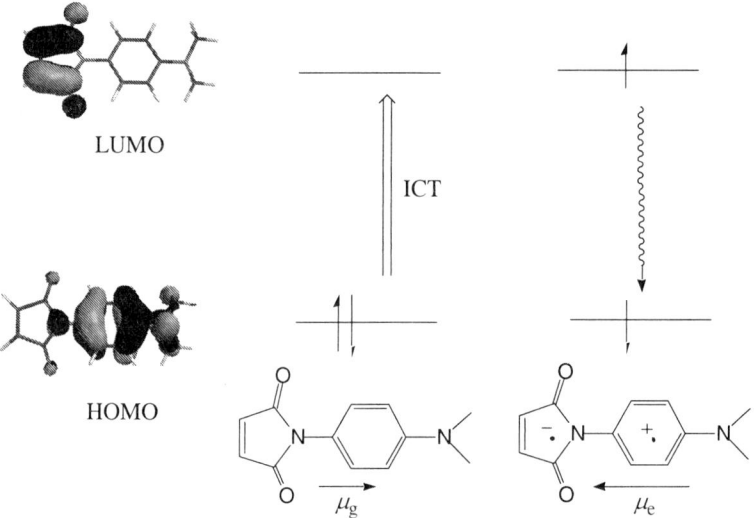

Figure 7.17. Frontier orbitals (AM1) of N-(4-dimethylaminophenyl)maleimide and the reversal of its dipole moment predicted to result from an ICT transition between these.

As shown in Figure 7.17 these calculations suggest that in this compound a weak internal charge transfer (ICT) transition should be available (weak because of the nodal properties of the HOMO and LUMO involved) leading to an apparently non-emissive CT state with a dipole moment oriented in a direction opposite to that of the (smaller) ground-state dipole moment. The inversion of the dipole moment is predicted to cause a negative solvatochromism of the ICT absorption. This is because although more polar solvents stabilize the weakly dipolar ground state (μ_g calculated to be ca. 1.3 D) the orientation polarization of the solvent induced by the ground state dipole moment destabilizes to a larger extent the highly dipolar Franck Condon excited state. We therefore now propose that the weak long-wavelength absorption of some N-aryl maleimides, attributed earlier[41] to an n-π* transition because of its negative solvatochromism, may in fact stem from such an ICT transition. For a model N-(4-dialkylaminophenyl)maleimide this absorption was found[41] at 380 nm (ε_{max} 300 M^{-1}cm^{-1}) in dichloromethane (390 nm in cyclohexane, 360 nm in methanol), which implies that the corresponding excited state is below the Franck-Condon CT state of FT (and FP) for which the CT absorption (which displays negligible solvatochromism) was found to occur around 350 nm (see e.g. Figure 7.12). Saturation of the double bond in the maleimide destroys its electron acceptor properties and thereby cancels the dark ICT state of the maleimide-donor moiety restoring the fluorescent CT state of the FT or FP probe as the lowest excited state.

7.3. APPLICATIONS OF D-σ-A PROBES

As already discussed above, FP, FT, derivatives of the fluorogenic MFP and MFT, as well as closely related systems in which the aromatic parts of D or A are modified or substituted can be employed as extremely sensitive polarity probes in a variety of liquid solvents. Also the behavior in solvent mixtures has been studied[21] revealing e.g. preferential solvation of the dipolar CT state by the most polar solvent component. The scope of application is, however, much wider than probing of micropolarity in liquid solution. Thus FP and some of its derivatives have been investigated isolated in a supersonically cooled molecular beam[20-22] and on the other hand have been attached covalently at a solid-liquid interface.[44] In this section, however, we will mainly concentrate on the behavior in complex media such molecular glasses, high molecular weight melts, (semi)crystalline solids, and polymers. We select here a number of examples of such studies to demonstrate the scope of FP and FT type fluorescent probes as well as their recently discovered electroluminescent properties.

7.3.1. Probing Solvation and its Dynamics as a Function of Temperature in Glass Forming Solvents

Until now we have tacitly assumed that the fluorescence observed from our probes stems from a relaxed CT state, which is at equilibrium with its surroundings. For (CT) absorption probes the assumption that their absorption occurs from a relaxed ground state is allowed, but with fluorescent probes after excitation unavoidably some emission already occurs before the nuclear reorganization of the surroundings as well as the internal reorganization of the probe itself have been completed. In low viscosity media at room temperature reorganization typically occurs on a (sub)picosecond time scale and thus for probes with nanosecond excited state lifetimes like FP and FT the contribution of fluorescence from non-relaxed states to the total emission is almost zero. Nevertheless, femtosecond time-resolved fluorescence studies of FP in liquid solvents have clearly shown the occurrence of a dynamic Stokes-shift over the first few picoseconds.[45,46]

At lower temperatures and especially when the solvent solidifies the reorganization is bound to slow down and thus fluorescence from non-relaxed states should become more important.

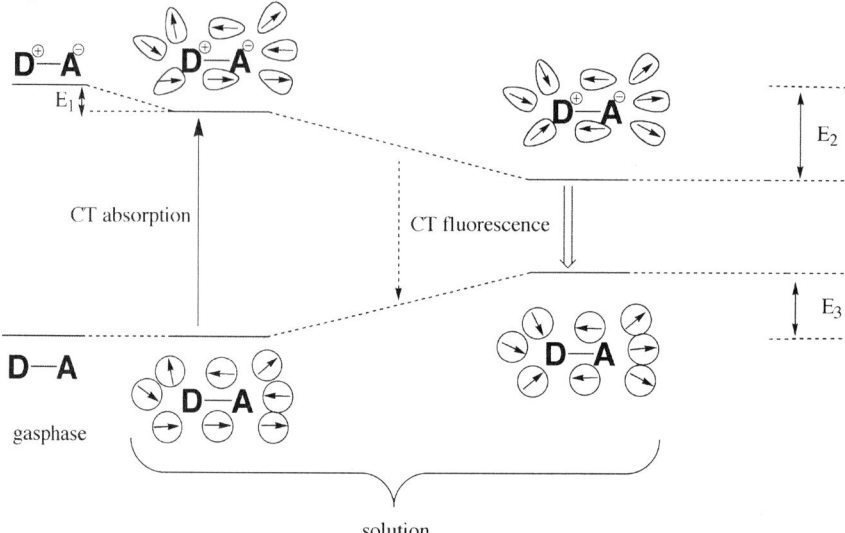

Figure 7.18. Solvation dynamics of a CT probe with zero ground-state dipole moment. Permanent solvent dipoles are indicated by small arrows, induced electronic polarization of solvent molecules is schematically indicated by a contour distortion.

Qualitatively this can easily be demonstrated by visual observation of solutions of FP or FT in dipolar solvents upon cooling down in e.g. liquid nitrogen. While under UV illumination at room temperature these solutions show a variety of colors depending on solvent polarity, as e.g. shown in Figure 7.6A, they all turn blue upon solidification as if solvent polarity is lost.

Figure 7.18 sketches the underlying dynamics of this process for a D-A system with zero ground state dipole moment and excited state dipole moment μ_e. In the ground state dipolar solvent molecules surround such a system randomly. Upon CT excitation in solution a Franck-Condon CT state is reached, which is (as compared to the situation for the isolated molecule) only stabilized by the rapid induced electronic polarization of the solvent molecules indicated in Figure 7.18 by a contour distortion. This stabilization amounts to:

$$E_1 = - (\mu_e^2/\rho^3)\{(n^2-1)/(2n^2+1)\} = - (\mu_e^2/\rho^3)f \qquad (8)$$

It is interesting to note that according to Eq. (8) D-A systems with zero ground state dipole moment should show a significant red shift of their CT absorption upon transfer from the gasphase ($n^2 = 1$) to solution but that variation of the solvent should lead to minor further solvatochromism because the refractive index of common solvents varies little from $n^2 \approx 2$.

As sketched in Figure 7.18, orientation polarization of the solvent dipoles leads to further stabilization of the dipolar CT state, which after full relaxation increases to:

$$E_2 = -(\mu_e^2/\rho^3)\{(\varepsilon-1)/(2\varepsilon+1)\} = -(\mu_e^2/\rho^3)f \qquad (9)$$

Fluorescence from a fully relaxed situation then leads to a Franck-Condon ground state in which the electronic polarization of the solvent is lost but its orientation polarization is still retained. This causes it to be destabilized by:

$$E_3 = +(\mu_e^2/\rho^3)\{(\varepsilon-1)/(2\varepsilon+1) - (n^2-1)/(2n^2+1)\} = +(\mu_e^2/\rho^3)(f-f') \qquad (10)$$

Combination of Eqs. (9) and (10) directly leads to the Lippert-Mataga equation (Eq. 2) mentioned earlier, in which it is thus assumed that the fluorescence observed stems from a fully relaxed CT state, which is a valid assumption for low viscosity media at ambient temperature. As indicated by a dashed arrow in Figure 7.18, fluorescence from an incompletely relaxed situation should be blue shifted to a degree depending on the degree of relaxation and thus its time dependence (i.e. a dynamic Stokes shift) provides direct information about the solvation dynamics.

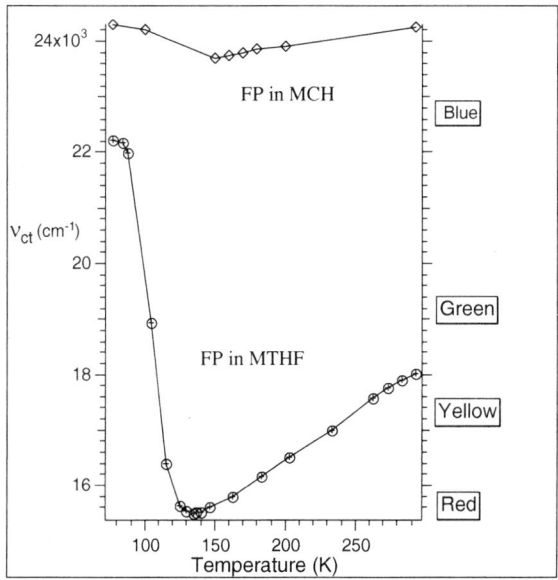

Figure 7.19. Continuous fluorescence maximum as a function of temperature for FP in 2-methyltetrahydrofuran (MTHF) and in methylcyclohexane (MCH).

The temperature dependence (see Figure 7.19) of the fluorescence of FP in the glass forming polar solvent 2-methyltetrahydrofuran (MTHF) and in the non-polar solvent methylcyclohexane (MCH) provides a striking demonstration of the solvation dynamics sketched in Figure 7.18.[47] While in MCH the fluorescence maximum of FP was found to stay within the narrow range of 24,000 ± 200 cm^{-1} over the temperature range 77-300K, in MTHF (see Figure 7.19) the FP fluorescence traverses a range of about 7000 cm^{-1} spanning the visible spectrum as also demonstrated by the color photographs shown in Figure 7.6B.

At 293 K an MTHF solution of FP shows strong green-yellow fluorescence with a maximum located at 562 nm (17,800 cm^{-1}) and a fluorescence lifetime of τ = 12.6 ns. Upon cooling the fluorescence gradually shifts to the red (over more than 100 nm!) with an average of 15 cm^{-1}/K until a bright red color is attained around 140 K. It turns out that this red-shift can be reproduced quite well by the Lippert-Mataga equation (Eq. 2) taking into account the increase of the dielectric constant (ε) and the refractive index (n) at lower temperatures. Microscopically this can be understood from the increased density of the solvent and from the increased dipolar orientation of the solvent molecules with respect to the solute dipole as the counteracting thermal motions diminish upon cooling.[4]

The fact that the Lippert-Mataga equation applies down to ca. 140 K implies that over this temperature range the solvent reorganization remains fast as compared to the fluorescence lifetime of FP.

Below 140 K the red-shift first levels off and below 130 K changes to a very dramatic blue-shift that in the region 120-90 K amounts to -153 cm^{-1}/K and levels off again at lower temperatures. It is known[48] that the solvent relaxation time of MTHF increases to nanoseconds upon cooling into the supercooled liquid region (i.e., below the melting point of T_M = 135 K) and rapidly increases further to seconds upon cooling to the glass transition point $T_G \approx$ 91 K. That as a result below 140 K a sizable fraction of the overall fluorescence occurs from not fully relaxed states can easily be shown from the fact that in this region the fluorescence displays a time dependent Stokes shift (see Figure 7.20), which extends over periods ranging from a few nanoseconds at 130 K to way over 40 ns (the maximum time window in which fluorescence can be detected) below 110 K.[47]

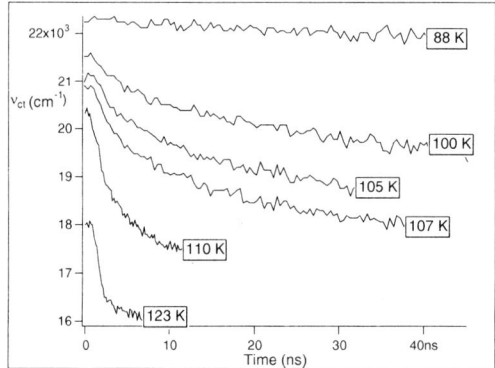

Figure 7.20. Dynamic Stokes shift of the CT fluorescence of FP in MTHF at various temperatures (excitation at 337 nm with 600 ps (fwhm) nitrogen laser pulses).

The measurements described above for FP in MTHF and MCH were also performed[49] for a solution of FP in sucrose octaacetate (SOA). When molten (melting point ca. 360 K) SOA behaves as a viscous liquid which solidifies to a clear glass upon cooling. In fact (see Figure 7.21) the thermochromic shift of FP in SOA is qualitatively similar to that in MTHF, but occurs at much higher temperatures. A noticeable difference between the situation in SOA and in MTHF is that in SOA in the 300-400 K region the shift of the maximum only amounts to ca. -25 cm^{-1}/K whereas in the 'undercooled liquid' region in MTHF this shift was over -150 cm^{-1}/K. This difference is probably brought about by the higher thermal energy in the SOA matrix as compared to the undercooled MTHF, making it harder to freeze nuclear relaxation processes in SOA. This is also indicated by the results of time resolved measurements (see Figure 7.21) on the shift of the FP emission maximum after pulsed laser excitation. It turns out that over most of the temperature region investigated including the viscous liquid region above the melting point and even down to 300 K, where SOA forms a mechanically very hard glass, a significant dynamic Stokes shift occurs over a time window spanning more than the fluorescence lifetime. This implies that the continuous fluorescence spectra, of which the maxima are plotted in Figure 7.21, are always a weighed average over the time domain in which emission occurs. Furthermore these results show that at room temperature significant nuclear reorganization processes can still take place in a very hard matrix on a nanosecond time scale. We will come back to this phenomenon later in this chapter.

The observations described in this section demonstrate that, in contrast to CT absorption probes, fluorescent CT probes cannot only detect changes in the solvating power of the medium but are also able to provide direct information about the local mobility of the medium. For FP like probes the resulting thermochromism in a suitably chosen medium easily spans the whole visible range of the spectrum and can act as a 'molecular-scale thermometer' of which the temperature range and sensitivity can be tuned by changing the medium directly surrounding the probe.

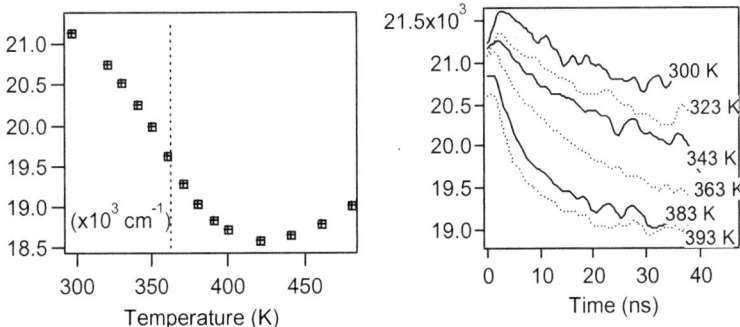

Figure 7.21. Continuous fluorescence maximum of FP in sucrose octaacetate (SOA, melting point indicated by vertical dashed line) as a function of temperature (left) and the dynamic Stokes shift observed after laser excitation (337 nm, fwhm 600 ps) at various temperatures.

7.3.2. Probing of Mobility Changes and Phase Transitions in (Semi)Crystalline Materials

The experiments in the previous section related to the transition between liquid solutions and an amorphous, glassy matrix. As expected from the decreased molecular mobility such a transition is always reflected by a blue shift of the fluorescence of our CT probes. Furthermore, in the solid glassy matrix cooling always leads to a further blue shift, the extent and steepness of which depends on the type of medium involved and the temperature domain over which the glassy state exists.

Striking deviations from this behavior have been observed for a number of transitions between the liquid and (semi) crystalline state as well as for the temperature dependence in that crystalline state. A few examples of such observations will now be described.

7.3.2.1. Octadecyl octadecanoate (Stearylstearate), $CH_3\text{-}(CH_2)_{17}\text{-}O\text{-}(CO)\text{-}(CH_2)_{16}\text{-}CH_3$

This is a crystalline solid at room temperature and in its melt (mp 334 K) FP can readily be dissolved to give a strongly fluorescent solution.[50] Further heating of this solution leads to a gradual blue shift in analogy with the thermochromism in other solvents (see above). However, upon crystallization instead of the expected blue shift a significant red shift occurs from $\approx$ 460 nm at 334 K to $\approx$ 495 nm at 310 K. Further cooling then induces a gradual blue shift (to e.g. 440 nm at 220 K). We suspected that this behavior is related to a solid state phase transition, but DSC (differential scanning calorimetry) did not reveal the existence of such a transition and only showed a sharp melting at 334 K. However, temperature dependent wide angle X-ray scattering (WAXD) revealed that a phase transition indeed takes place and occurs about 20 K below the melting point. While this explains why below this point FP shows the expected blue shift upon decreasing temperature it remains to be explained why the transition from the liquid to the solid state is accompanied by a red shift. Clearly the crystalline state initially formed is metastable and only exists over a small temperature domain of ca. 20 K. It must be concluded that during its formation from the melt the FP molecules are incorporated in such way that the medium dipoles are fortuitously already pre-oriented so as to stabilize the CT excited state better than can be achieved via the dipolar field induced orientation in the liquid state. That this pre-orientation is far from optimal can be concluded from the very significant dynamic Stokes shift that FP undergoes in solidified octadecyloctadecanoate.[50,51]
Similar observations were made[51] with ethylstearate ($C_2H_5\text{-}O\text{-}(CO)\text{-}(CH_2)_{16}\text{-}CH_3$).

7.3.2.2. α,ω-Diacetyl polyethyleneglycols, $CH_3CO\text{-}(OCH_2CH_2)_m\text{-}OCOCH_3$

The occurrence of a dynamic Stokes shift in solid matrices is a recurring theme and while a discussion of the underlying mechanism will be postponed to a later point it is clear that this must be related to a certain mobility of the matrix and/or the probe. This mobility cannot only be varied by temperature - as shown above - but also by e.g. varying the molecular weight of the matrix molecules such as was done in a series of homologous α,ω-diacetyl polyethylene glycols $CH_3CO\text{-}(OCH_2CH_2)_m\text{-}OCOCH_3$ with m = 2-225.[51-55] At room temperature the lower members in this series are viscous liquids, the higher members are waxy to crystalline solids. Over the whole range of m values investigated, including the solid regime, a distinct blue shift of the fluorescence maximum of FP was observed with increasing molecular weight. Comparison with the behavior of FP in low viscosity binary mixtures of ethylacetate and diglyme containing increasing mole fractions of the latter suggests that in the absence of mobility restrictions the fluorescence in fact should shift to the red for higher m. Coupled with the observation[51,53,54] of significant dynamic Stokes shifts in a time window up to 60 ns, this implies that the blue shift is governed by a decreasing mobility over the whole molecular weight range investigated including that of the solid members.

7.3.2.3. Thermal and Mechanical Effects in Copolymers and Block-copolymers

Copolymers and especially block-copolymers have a complex morphology related to the presence of microdomains that vary from completely amorphous to highly crystalline. The behavior was investigated[51,56] in some depth of FP as a function of temperature and of mechanical stress in a number of such block-copolymers having molecular weights in the range 30,000 – 60,000 composed of alternating blocks of polybutylene terephthalate (PBT) and polyethylene glycol (PEG). In mechanical stretching experiments it was found that the FP fluorescence maximum undergoes significant and partly reversible spectral shifts and that the magnitude as well as the sign of these shifts depends strongly on the mechanical and thermal pretreatment of the samples. In thermographic (DSC) measurements the PBT/PEG copolymers display various phase transitions which can e.g. be related to melting of the PEG resp. PBT regions. Interestingly, all of these phase transitions are reflected by significant changes in the continuous fluorescence maximum of FP incorporated in these copolymers.

7.3.3. Probing of Polymerization Processes

As shown above, the fluorescence of CT probes is not only sensitive to the polarity but also to the mobility of the medium. It was realized that this should enable us to detect the progress of a polymerization process and as early as 1987 this was indeed demonstrated[57] by monitoring the fluorescence of FP dissolved in the monomer methylmethacrylate (MMA) during thermal polymerization to give PMMA. In MMA at room temperature the fluorescence of FP occurs at 565 nm (17,700 cm^{-1}), which compares well with that observed (see Table 7.1) in e.g ethylacetate. After polymerization to PMMA the fluorescence maximum has shifted dramatically to the blue (ca. 450 nm at 293 K) which can mainly be attributed to the reduced mobility.

Similar observations were made during polymerization of several other monomers.[51] In all cases, however, the majority of the shift occurs over a small part of the overall polymerization process since it sets in only when the polymerization has progressed sufficiently to increase significantly the viscosity (i.e. during the gelation and vitrification stages). This limitation has more recently been removed by the use of the fluorogenic derivatives MFT and MFP. As sketched in Figure 7.15 the maleimide group of MFT and MFP can enter into copolymerization with acrylic monomers. This reaction both triggers the fluorescence and makes the fluorescent probes a part of the growing polymer chain. Recent studies by us[58,59] and by others[60] of radiation induced polymerization of MMA have demonstrated that this allows monitoring of the full polymerization process. In the early stages only the fluorescence intensity rises from zero while the position remains constant as the probe is incorporated in the growing chain without significant viscosity increase. In later stages (just like with FP) also a sharp blue shift upon gelation and vitrification occurs.

Already during the earliest polymerization experiments[57] with FP in (P)MMA it was noted that in PMMA the FP fluorescence undergoes a dynamic Stokes shift extending over a time domain up to 50 ns. This implies that most of the fluorescence stems from non-relaxed states and that the spectrum observed by continuous spectroscopy is a weighed average of those emitted during relaxation. As a result the continuous fluorescence maximum of FP in PMMA was found to be rather temperature dependent shifting to the red at higher temperatures where relaxation is faster (ca. 430 nm at 110 K, 450 nm at 293 K, 490 nm at 433 K, the latter being well above the glass transition point $T_g \approx 393$ K). This latter finding already makes it unlikely that the dynamic Stokes shift observed stems from an inhomogenenous distribution of sites in the PMMA homopolymer occupied by FP molecules in such a way that those emitting more to the red have a longer fluorescence lifetime. Recent studies by Frahn et al.[60] further support that the dynamic Stokes shift in PMMA stems from relaxation processes and not from inhomogeneity. Both for FP in PMMA and for MFP incorporated covalently in PMMA very similar dynamic Stokes shifts and shift kinetics were found. Furthermore it was observed that at the red side of the spectrum the emission shows a distinct growth during the first nanoseconds after excitation, which occurs over the same time domain as a fast component of the decay at the blue side (see Figure 7.22).

This is fully in line with a real dynamic Stokes shift from a homogeneous collection of sites just like in the amorphous glassy matrices like MTHF, MCH and SOA discussed earlier. An interpretation problem still arises from the fact that the time scale of the relaxations indicated by these Stokes shifts is much shorter than that of the fastest relaxation processes detected by e.g. dielectric relaxation measurements[61] on polymers like PMMA. Initially it was suggested[57] that internal relaxation processes of the probe might be involved rather than reorganization processes in the matrix. However, the lack of a significant shift upon cooling a non-polar matrix like MCH from the liquid to the solid state (see Figure 7.19) makes it unlikely that this plays an important role and/or implies that internal reorganizations such as the planarization of the donor nitrogen (see Figure 7.14) remain very fast even in solid matrices.

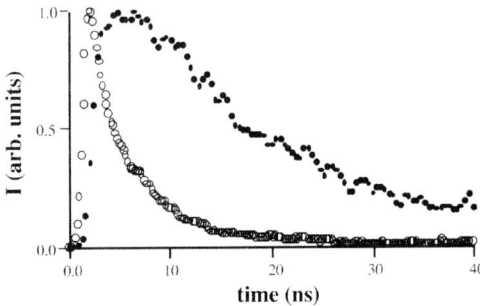

Figure 7.22. Fluorescence decay curves monitored[60] at 380 nm (circles) and at 580 nm (dots) after pulsed laser excitation (337 nm, fwhm 500 ps) of FP in PMMA at room temperature. Note the fast decay at the short wavelength extreme of the emission and the growth followed by a much slower decay at the long wavelength extreme.

Alternatively it was suggested[55] that the dynamic Stokes shift of FP like probes in polymers might be related to reorientation of the probe with respect to the surrounding polymer in the free volume occupied by the probe. While fluorescence depolarization measurements[57] show that such reorientation cannot be excluded, the recent observation by Frahn[60] that for FP in PMMA and for covalently attached MFP in PMMA the magnitude and kinetics of the dynamic Stokes shift are virtually identical also seems to exclude such reorientation as a major source of the dynamic Stokes shift observed. As a way-out it has been proposed by Frahn et al.[60] that the strong dipolar field created by the CT excited state of the probe could accelerate reorientation of polar side groups of the surrounding polymer to occur on a nanosecond time scale rather than on the much longer time scales concluded from e.g. dielectric relaxation measurements. It remains to be seen whether this probe-effect is operative or whether certain rapid microscopic motions in polymers detectable with CT fluorescent probes have until now just escaped observation by other techniques.

7.3.4. Penetration of Solvents and Vapors in Polymers

Small molecules can penetrate many polymers from the liquid or gaseous phase. An example of this is the swelling of polymer particles dispersed in an aqueous solution ('latices') upon addition of certain organic solvents, which can be followed by monitoring the particle size via quasi-elastic light scattering (QELS) or scanning electron microscopy (SEM). For a monodisperse ($\varnothing \approx 400$ nm) polystyrene latex such QELS data are shown in Figure 7.23 with dichloromethane (DCM) as the swelling agent.[51,62]

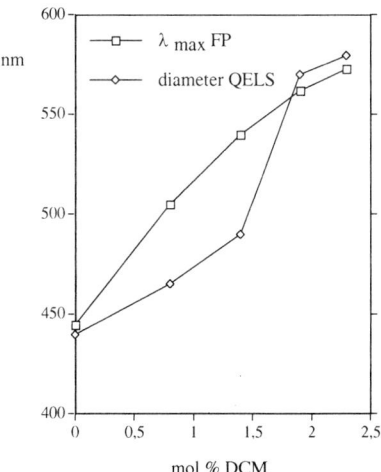

Figure 7.23. Solvent penetration and swelling of a 400 nm polystyrene latex by dichloromethane (DCM) as detected via the fluorescence shift of incorporated FP and by QELS.

Strong swelling sets in above 1.5 mol% DCM added to the aqueous dispersion medium. When the polymer particles are sparsely doped with FP they become brightly fluorescent and in the absence of DCM such a doped latex fluoresces at 445 nm, which is identical to the wavelength of FP in bulk polystyrene. This implies that the FP probes are embedded in the particles so as to be shielded from the aqueous medium. Upon addition of DCM the fluorescence shifts to the red as far as to 573 nm at 2.4 mol% signaling penetration of the solvent to the embedded FP molecules. Similar observations were made with other low molecular weight solvents. With DCM as well as with some other solvents the fluorescence of the embedded and non-extractable FP molecules is shifted to a position nearly identical to that for FP in the pure solvent.[51,62] This not only provides an optical probe for the solvent penetrability but, as can be seen from Figure 7.23, also allows detection of such solvent penetration at early stages where the size of the particles has not yet changed significantly. In later investigations these phenomena were e.g. exploited to study the crosslinking and drying of latex based coatings employing a crosslinker covalently tagged with MFP.[63] Also the kinetics of solvent penetration into latex particles was examined in more detail. While for the polystyrene latices used above penetration reaches a steady state within the mixing time, other latices show much slower solvent penetration especially when their T_g is high.[64]

When the FP labeled polystyrene latex used in Figure 7.23 is spread to a thin film and dried, a material is obtained of which the fluorescence is not only highly solvatochromic but also vapochromic. Thus contacting it with e.g. dichloromethane vapor rapidly and reversibly shifts its fluorescence from blue (445 nm) to orange green (573 nm) providing a nice class-room demonstration of a 'fluorescent nose'.[65]

7.3.5. Thin Film Electroluminescence, OLEDs and PLEDs

At present much interest is focussed on the development of light emitting diodes (LEDs) based on the electroluminescence of semiconducting thin organic films when sandwiched between a cathode consisting of a low work function metal (e.g. Al, Ca or Ba) and a transparent anode (mostly indium-tinoxide, ITO). Electroluminescence in such devices occurs if the energy, set free when holes and electrons injected at the anode and cathode respectively recombine, is sufficient to excite an embedded fluorophore or a fluorophoric polymer segment. We argued that the CT fluorescence of our FP and FT type probes in fact constitutes a hole-electron recombination, which is inherently radiative by it self. Therefore thin films of these probes in pure form or mixed with an appropriate semiconducting (but not necessarily electroluminescent) polymer might lead to small-molecule based LED's (OLEDs) or polymer based LEDs (PLEDs) respectively. The latter have the advantage that the thin film can be produced by spin coating, while for the former vapor phase deposition is required. Both types of devices have indeed been realised.[49,66,67] Vapor phase deposition of a homogeneous thin film, as required for the production of an OLED, turned out to be possible with FT as well as with its analogues **9** and **10** in which the cyanonaphthalene is substituted by the weaker acceptor pyrene.

9 : R = H
10 : R = CH$_3$

PLEDs with polyvinylcarbazole (PVK) as the polymer matrix were produced from various CT probes including FT, **9** and **10** by spin coating solutions of PVK and a few weight % of the probe on an ITO anode and covering with a low work function metal anode.

The PLEDs as well as the OLEDs based on FT, **9**, and **10** show bright electroluminescence, as e.g. exemplified in Figure 7.6C, which is easily tuned in color by variation of the D and A moieties as well as of the embedding medium. Regrettably the operational lifetime of the LEDs reported until now is rather limited, which may be related to the chemical reactivity of the donor radical cation and acceptor radical anion species involved. Especially the successful construction of OLEDs based on pure thin layers of these D-σ-A probes, however, nicely supports the mode of operation proposed for creation of their CT excited state via an electrochemical pathway.

7.4. CONCLUDING REMARKS

The data discussed above constitute the main part of the results obtained until now with Fluoroprobe and Fluorotrope type D-σ-A probes since their discovery in 1984. These results demonstrated the extreme solvatochromism, thermochromism, and vapochromism of these probes and also their ability to detect subtle mobility changes in the surrounding matrix brought about by chemical, thermal or mechanical factors. It appears that these properties as well as their recently established electroluminescent properties have as yet not been exploited fully. The author hopes that this review will stimulate interest in further application and development of this type of probes.

7.5. ACKNOWLEDGEMENTS

The author thanks Dr. J.M. Warman and Dr. M.S. Frahn for making available the data plotted in Figure 7.22 and Dr. A.M. Brouwer for communicating unpublished results about the kinetics of solvent penetration in polymer latex particles.

7.6. REFERENCES

1. C. Reichardt, Solvatochromic Dyes as Solvent Polarity Indicators, *Chem. Rev.* **94**, 2319-2358 (1994).
2. C. Reichardt, *Solvents and Solvent Effects in Organic Chemistry*, chap. 6 (Wiley-VCH,Weinheim, 2003).
3. Z.R. Grabowski, K. Rotkiewicz, and W. Rettig, Structural Changes Accompanying Intramolecular Electron Transfer: Focus on Twisted Intramolecular Charge-Transfer States, *Chem. Rev.* **103**, 3899-4031 (2003).

4. P. Suppan, Solvatochromic Shifts: The Influence of the Medium on the Energy of Electronic States, *J. Photochem. Photobiol, A* **50**, 293-330 (1990).
5. H. Knibbe, *Charge-Transfer Complex Formation in the Excited State*, PhD dissertation (Free University, Amsterdam, 1969).
6. H. Oevering, J.W. Verhoeven, M.N. Paddon-Row, and J.M. Warman, Charge-transfer absorption and emission resulting from long-range through-bond interaction; Exploring the relation between electronic coupling and electron-transfer in bridged donor-acceptor systems, *Tetrahedron* **45**, 4751-4766 (1989).
7. J.W. Verhoeven, J. Kroon, M.N. Paddon-Row, and J.M. Warman, Kinetic and spectroscopic investigation of the influence of conformation and orbital-symmetry on long-range intramolecular donor-acceptor interaction, in: *Supramolecular Chemistry, NATO ASI Series Vol. 371*, edited by V. Balzani and L. De Cola, (Kluwer Academic Publishers, Dordrecht 1992) pp. 181-200.
8. G. Briegleb, *Elektronen-Donator-Acceptor-Komplexe* (Springer Verlag, Berlin, 1961).
9. A.W.J.D. Dekkers, J.W. Verhoeven, and W.N. Speckamp, On the nature of sigma-coupled transitions; through-bond interactions in 1-aza-adamantane derivatives, *Tetrahedron* **29**, 1691-1696 (1973).
10. P. Pasman, J.W. Verhoeven and Th.J. de Boer, Fluorescence of intramolecular electron donor-acceptor systems; the importance of through-bond interaction, *Chem. Phys. Lett.* **59**, 381-385 (1978).
11. P. Pasman, F. Rob and J.W. Verhoeven, Intramolecular charge-transfer absorption and emission resulting from through-bond interaction in bichromophoric molecules, *J. Am. Chem. Soc.* **104**, 5127-5133 (1982).
12. G.F. Mes, B. de Jong, H.J. van Ramesdonk, J.W. Verhoeven, J.M. Warman, M.P. de Haas, and L.E.W. Horsman- van de Dool, Excited-state dipole moment and solvatochromism of highly fluorescent rod-shaped bichromophoric molecules, *J. Am. Chem. Soc.* **106**, 6524-6528 (1984).
13. R.M. Hermant, N.A.C. Bakker, T. Scherer, B. Krijnen, and J.W. Verhoeven, Systematic Study of a Series of Highly Fluorescent Rod-Shaped Donor-Acceptor Systems, *J. Am. Chem. Soc.* **112**, 1214-1221 (1990).
14. T. Scherer, W. Hielkema, B. Krijnen, R.M. Hermant, C. Eijckelhoff, F. Kerkhof, A.K.F. Ng, R. Verleg, E.B. van der Tol, A.M. Brouwer and J.W. Verhoeven, Synthesis and exploratory photophysical investigation of donor-bridge-acceptor systems derived from N-substituted 4-piperidones, *Recl. Trav. Chim. Pays-Bas* **112**, 535-548 (1993).
15. S.V. Rodrigues, A.K. Maiti, H. Reis, and W. Baumann, Electro optical emission measurements on a non-conjugated bichromophoric donor-acceptor molecule, *Mol. Phys.* **75**, 953-960 (1992).
16. S. Depaemelaere, L. Viaene, M. van der Auweraer, F.C. De Schryver, R.M. Hermant, and J.W. Verhoeven, Non-radiative decay processes of the intramolecular charge transfer state in a rigid bichromophoric system, *Chem. Phys. Letters* **215**, 649-655 (1993).
17. R.J. Willemse, D. Theodori, J.W. Verhoeven, and A.M. Brouwer, Decay pathways of charge separated states in strongly fluoresecnt electron donor-acceptor compounds, *Photochem. Photobiol. Sci.* **2**, 1134-1139 (2003).
18. M. Bixon, J. Jortner, and J.W. Verhoeven, Lifetimes for radiative charge recombination in donor-acceptor molecules, *J. Am. Chem. Soc.* **116**, 7349-7355 (1994).
19. J.W. Verhoeven, T. Scherer, B. Wegewijs, R.M. Hermant, J. Jortner, M. Bixon, S. Depaemelaere, and F.C. De Schryver, Electronic coupling in inter- and intra-molecular donor acceptor systems as revealed by their solvent dependent charge-transfer fluorescence, *Recl. Trav. Chim. Pays-Bas* **114**, 443-448 (1995).
20. B. Wegewijs and J.W. Verhoeven, On long-range charge separation in solvent free donor-bridge-acceptor systems, in: *Advances in Chemical Physics, Vol. 106, Electron Transfer- From Isolated Molecules to Biomolecules, Part One*, Edited by J. Jortner and M. Bixon, (John Wiley & Sons, Inc., 1999), pp. 221-264.
21. R.M. Hermant, *Highly Fluorescent Donor-Acceptor Systems: Fundamentals and Applications*, PhD dissertation (University of Amsterdam, 1990).
22. B. Wegewijs, *Long-range charge separation in solvent-free Donor-bridge-Acceptor systems: Donor-bridge-Acceptor molecules in splendid isolation*, PhD dissertation (University of Amsterdam, 1994).
23. T. Scherer, *Conformational Dynamics of Fluorescent Exciplexes*, Phd dissertation (University of Amsterdam, 1994).
24. N.J. Turro, *Modern Molecular Photochemistry*, p. 90 (The Benjamin/Cummings Publ. Co., Menlo Park CA, 1978).
25. B. Wegewijs, T. Scherer, R.P.H. Rettschnick, and J.W. Verhoeven, Exciplex formation in jet-cooled donor-bridge-acceptor compounds incorporating bridges with three degrees of flexibility, *Chem. Phys.* **176**, 349-357 (1993).
26. W. Schuddeboom, T. Scherer, J.M. Warman, and J.W. Verhoeven, The formation of extended and folded charge separated states of donor-spacer-acceptor molecules with flexible and semi-rigid sigma-bond spacers, *J. Phys. Chem.* **97**, 13092-13098 (1993).

27. B. Wegewijs, A.K.F. Ng, and J.W. Verhoeven, Coulomb-induced intramolecular exciplex formation in semiflexible Donor-bridge-Acceptor compounds in nonpolar solvents as a function of temperature, *Recl. Trav. Chim. Pays-Bas*, **114**, 6-12 (1995)
28. X.Y. Lauteslager, M.J. Bartels, J. J. Piet, J. M. Warman, J.W. Verhoeven, and A.M. Brouwer, Exploring the limits of the electrostatically induced conformational folding process in charge-separated excited states: retarding effect of long alkyl tails attached to the chromophores, *Eur. I. Org. Chem.*, 2467-2481 (1988).
29. X.Y. Lauteslager, I.H.M. van Stokkum, H.J. van Ramesdonk, A.M. Brouwer, and J.W. Verhoeven, Conformational dynamics of semiflexibly bridged D-A systems studied with a streak camera and spectrotemporal parametrization of fluorescence, *J. Phys. Chem. A*, **103**, 653-659 (1999).
30. G.F. Mes, *Photo-induced Electron-transport in Multichromophoric Systems*, PhD dissertation (University of Amsterdam, 1985).
31. R. Hoffmann, A. Imamura, and W.J. Hehre, Benzynes, dehydroconjugated molecules, and the interaction of orbitals separated by a number of intervening sigma-bonds, *J. Am. Chem. Soc.* **90**, 1499-1509 (1968).
32. J.W. Verhoeven, From close-contact to long-range intramolecular electron transfer, in: *Advances in Chemical Physics, Vol. 106, Electron Transfer- From Isolated Molecules to Biomolecules, Part One*, Edited by J. Jortner and M. Bixon (John Wiley & Sons, Inc., 1999), pp. 603-644.
33. A.M. Brouwer and R. Wilbrandt , Vibrational spectra of N,N-dimethylaniline and its radical cation. An interpretation based on quantum chemical calculations, *J. Phys. Chem.* **100**, 9678-9688 (1996).
34. A.M. Brouwer, Ab initio study of the structures and vibrational spectra of some diamine radical cations, *J. Phys. Chem. A* **101**, 3626-3633 (1997).
35. A.M. Brouwer, J.M Zwier, C. Svendsen, O.S. Mortensen, F.W. Langkilde, and R. Wilbrandt, The radical cation of N,N-dimethylpiperazine: dramatic structural effects of orbital interactions through bonds, *J. Am. Chem. Soc.* **120**, 3748-3757 (1998).
36. M. Goes, X.Y. Lauteslager, J.W. Verhoeven, and J.W. Hofstraat, A blue excitable Charge-Transfer fluorescent probe and its fluorogenic derivative, *Eur. J. Org. Chem.*, 2373-2377 (1998).
37. B. Krijnen, H.B. Beverloo, and J.W. Verhoeven , Conformational Effects of Through-Bond Interaction in N-Aryl-Piperidine Derivatives, *Recl. Trav. Chim. Pays-Bas* **106**, 135-136 (1987).
38. B. Krijnen, H.B. Beverloo, J.W. Verhoeven, C.A. Reiss, K.Goubitz, and D. Heijdenrijk, Effect of through-bond interaction on conformation and structure of some N-arylpiperidone and N-aryltropanone derivatives, *J. Am. Chem. Soc.* **111**, 4433-4440 (1989).
39. D.J.A. de Ridder, K. Goubitz, H. Schenk, B. Krijnen, and J.W. Verhoeven, Effect of Through-bond Interaction on Conformation and Structure in Rod-shaped Donor-Acceptor Systems; Part 2, Crystal Structures of Seven N-Aryltropan-3-one Derivatives, *Helv. Chim. Acta* **86**, 812-826 (2003).
40. R.P. Haugland, *Handbook of Fluorescent Probes and Research Chemicals*, 5th edn. (Molecular Probes Inc., Eugene OR, 1994), pp. 14-16, see also: http//www.probes.com/handbook.
41. H.J. Verhey, J.W. Hofstraat, C.H.W. Bekker, and J.W. Verhoeven, A fluorogenic charge-transfer polarity probe for the derivatisation of thiols and amines, *New J. Chem.* **20**, 809-814 (1996).
42. E.S. Gould, *Mechanism and Structure in Organic Chemistry* (Holt, Rinehart and Winston Inc., New York, 1959) p. 221.
43. M. Frahn, personal communication.
44. M.J. Bartels, M. Koeberg, and J.W. Verhoeven, Fluorescence Probing of Solvent Accessibility and Micropolartity on Silica and Alkylated Silica Surfaces, *Eur. J. Org. Chem.* 2391-2395 (1999).
45. E.R. Middelhoek, P. Vermeulen, J.W. Verhoeven, and M. Glasbeek, Picosecond time-dependent Stokes shift studies of fluoroprobe in liquid solution, *Chem. Phys.* **198**, 373-380 (1995).
46. E.R. Middelhoek, H. Zhang, J.W. Verhoeven, and M. Glasbeek, Subpicosecond studies of the solvation dynamics of Fluoroprobe in liquid solution, *Chem. Phys.* **211**, 489-497 (1996).
47. M. Goes, M. de Groot, M. Koeberg, J.W. Verhoeven, N.R. Lokan, M.J. Shephard, and M.N. Paddon-Row, Temperature dependence of charge-transfer fluorescence from extended and U-shaped donor-bridge-acceptor systems in glass-forming solvents, *J. Phys. Chem. A* **106**, 2129-2134 (2002).
48. E. Görlach, H. Gygax, P. Lubini, and U. Wild, Solvent relaxation of oxazine-4 in 2-methyltetrahydrofuran, *Chem. Phys.* **194**, 185-193 (1995).
49. M. Goes, *Photo- and Electroluminescence Generated by Intramolecular Charge Transfer*, PhD dissertation (University of Amsterdam, 2002).
50. L.W. Jenneskens, H.J. Verhey, H.J. van Ramesdonk, J.W. Verhoeven, K.F. van Malssen, and H. Schenk, Intramolecular Charge Transfer Fluorescence of 1-phenyl-4-(4-cyano-1-naphthyl methylene) piperidine as a sensor for phase transitions in the solid state, *Recl. Trav. Chim. Pays-Bas* **111**, 507-510 (1992).
51. H.J. Verheij, *Fluorescence Probing of Polymers*, PhD dissertation (University of Amsterdam, 1997).

52. L.W. Jenneskens, H.J. van Ramesdonk, H.J. Verhey, G.D.B. van Houwelingen, and J.W. Verhoeven, Discrimination between polarity and mobility effects on the charge-transfer fluorescence of Fluoroprobe in organic matrices, *Recl. Trav. Chim. Pays-Bas* **108**, 453-454 (1989).
53. L.W. Jenneskens, H.J. Verhey, H.J. van Ramesdonk, A.J. Witteveen, and J.W. Verhoeven, Intramolecular charge-transfer fluorescence of 1-phenyl-4-[(4-cyano-1-naphthyl)methylene]piperidine as a morphology probe in α,ω–diacetylpoly(ethylene glycol) matrices, *Macromolecules* **24**, 4038-4040 (1991).
54. L.W. Jenneskens, G.D.B. van Houwelingen, H.J. van Ramesdonk, H.J. Verhey, and J.W. Verhoeven, Intramolecular charge transfer fluorescence of 1-phenyl-4-(4-cyano-1-naphthylmethylene)piperidine as a mobility probe in α,ω-diacetyl poly(ethyleneglycols), in: *Integration of Fundamental Polymer Science and Technology-5*, edited by P.J. Lemstra and L.A. Kleintjes (Elsevier Applied Science, London and New-York, 1991), pp. 291-295.
55. L.W. Jenneskens, J.P.B. van Deursen, H.J. Verhey, H.J. van Ramesdonk, and J.W. Verhoeven, Molecular inhomogeneity of polymeric matrices, as detected by site directed fluorescent probes, *Appl. Fluor. Techn.* **3**, 11-15 (1991).
56. J.W. Hofstraat, J. Veurink, B. Gebben, H.J. Verheij, and J.W. Verhoeven, Charge-Transfer Fluorescent Probes Applied to the Characterization of Thermal and Mechanical Properties of Polymers, *J. Fluor.* **8**, 335-342 (1998).
57. H.J. van Ramesdonk, M. Vos, J.W. Verhoeven, G.R. Möhlman, N.A. Tissink, and A.W. Meesen, Intramolecular charge-transfer fluorescence as a mobility probe in (poly)methylmethacrylate, *Polymer* **28**, 951-956 (1987).
58. J.M. Warman, R.D. Abellon, H.J. Verhey, J.W. Verhoeven, and J.W. Hofstraat, Maleimido-fluoroprobe: A dual-purpose Fluorogenic Probe of Polymerization Dynamics, *J. Phys. Chem. B*, **25**, 4913-4916 (1997).
59. J.M. Warman, R.D. Abellon, L.H. Luthjens, J.S. Suykerbuyk, H.J. Verheij, and J.W. Verhoeven, In-situ monitoring of radiation-induced polymerisation of methylmethacrylate using fluorogenic molecular probes, *Nuclear Instruments and Methods in Physics Research B* **151**, 361-366 (1999).
60. M.S. Frahn, *Radiation-Induced Polymerization Monitored with Fluorogenic Molecular Probes*, PhD dissertation (Delft University of Technology, 2003).
61. C.P. Smyth in: *Physics and Chemistry of the Organic Solid State*, edited by D. Fox, M.M. Labes, and A. Weissberger (Interscience, New York, 1963), Vol. 1, p. 697.
62. H.J. Verhey, B. Gebben, J.W. Hofstraat, and J.W. Verhoeven, Detection of nanoscale events in dispersed polymer particles containing a charge-transfer fluorescence probe, *J. Polym. Sci.: Part A: Polym. Chem.* **33**, 399-405 (1995).
63. H.J. Verhey, L.G.J. van der Ven, C.H.W. Bekker, J.W. Hofstraat, and J.W. Verhoeven, Cross-linking and drying of a two-component waterborne coating monitored by a functionalized CT fluorescence probe, *Polymer* **38**, 4491-4497 (1997).
64. A.M. Brouwer, personal communication.
65. J.W. Verhoeven, unpublished results.
66. J.W. Verhoeven, M. Goes, J.W. Hofstraat, and K. Brunner, Electroluminescence of Charge-Transfer Fluorescent Donor-bridge-Acceptor Systems, *The Spectrum* **15-2**, 1-13 (2002).
67. M. Goes, J.W. Verhoeven, J.W. Hofstraat, and K. Brunner, OLED and PLED Devices Employing Electrogenerated, Intramolecular Charge-Transfer Fluorescence, *ChemPhysChem.* **4**, 349–358 (2003).

A DUAL LUMINOPHORE PRESSURE SENSITIVE PAINT: ELIMINATING THE TEMPERATURE INTERFERENCE IN THE MEASUREMENT OF OXYGEN PARTIAL PRESSURE

Muhammet E. Kose, Joanne M. Bedlek-Anslow, James P. Hubner, Bruce F. Carroll, Kirk S. Schanze[*]

8.1. BACKGROUND

Luminescence imaging, combined with pressure sensitive paint (PSP) is an optical method for measuring surface air pressure distributions on aerodynamic models in wind tunnels.[1, 2] This method is of particular of interest to the aerospace industry for determining loads in aerodynamic prototype testing. When compared to conventional methods such as pressure tap measurements, luminescence imaging has the advantage of providing a non-invasive method to obtain full field surface air pressure distributions with high spatial resolution.

A typical formulation of a PSP is shown in Figure 8.1.[3] The first layer above the model surface is a primer layer. The primer is usually white, in order to reflect the luminescence intensity and hence increase the luminosity of the active layer. The white primer layer also provides optical uniformity on the model surface. The second active layer must have a very high gas permeability to enable rapid diffusion of oxygen. Typically, this layer consists of a high gas permeability silicone or fluorinated acrylate polymer.[4-8] The luminophore is dispersed or dissolved in the polymer binder and it is usually a long-lifetime luminescent molecule, such as a polypyridine ruthenium(II) complex or a platinum(II) porphyrin.[9-14] The reason for using a long-lifetime luminophore is that oxygen diffusion in the polymer binder is slow compared to the emission lifetime of typical fluorescent molecules ($\tau \sim 10$ ns). Typical luminophores used in PSPs have lifetimes in 5 – 50 µs range.[1, 5, 15]

Pressure sensitive paint measurements are based on the quenching of

---

[*] Muhammet E. Kose and Kirk S. Schanze, Department of Chemistry, University of Florida, Gainesville, FL, 32611. Joanne M. Bedlek-Anslow, DuPont Nylon, South Carolina. John P. Hubner and Bruce F. Carroll, Department of Mechanical and Aerospace Engineering, University of Florida, Gainesville, FL, 32611.

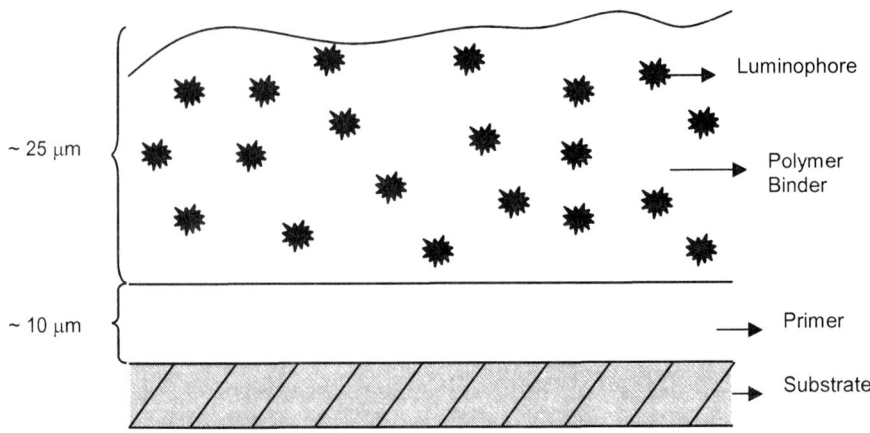

Figure 8.1. Scheme showing cross section of a typical pressure sensitive paint.

luminescence by oxygen in the air. The luminescence intensity of a PSP is inversely proportional to the partial pressure of oxygen over the model surface. This situation follows because the concentration of oxygen in the PSP film is proportional to the total air pressure, $[O_2]_{poly} = 0.21\, c\, P_{air}$, where P_{air} is the total air pressure above the sensor film, c is Henry's law coefficient, and the factor 0.21 accounts for the fact that $P_{O2} = 0.21\, P_{air}$. The luminescence quenching by oxygen follows bimolecular kinetics, which was early described by Stern and Volmer.[16] Initially, a luminophore (L) is photoexcited which results in an excited state L^* with a finite lifetime:

$$L + h\nu_{exc} \rightarrow L^* + heat \qquad (1)$$

There are several routes for the decay of the photoexcited state:

$$L^* \rightarrow L + h\nu_{em} \ (rate\ k_p) \qquad (2)$$

$$L^* \rightarrow L + heat \ (rate\ k_{nr}) \qquad (3)$$

$$L^* + {}^3O_2 \rightarrow L + {}^1O_2 + heat \ (rate\ k_q[O_2]) \qquad (4)$$

At constant temperature, the radiative and non-radiative decay rates (k_p and k_{nr}) are constant and the process in Eq. (4) reduces the luminescence intensity. This is the basis for the use of luminescence in wind tunnel studies. The total decay rate for L^* is then,

$$k = k_p + k_{nr} + k_q[O_2]_{poly} \qquad (5)$$

where k_p and k_{nr} are unimolecular decay rate constants (units s^{-1}), k_q is a bimolecular rate constant (unit $M^{-1}s^{-1}$), and $[O_2]_{poly}$ is the molar concentration of oxygen in the polymer. By using Henry's Law, we can also write:

$$[O_2]_{poly} = C_o (p/p_o) \tag{6}$$

p: air pressure over the polymer, p_o : reference air pressure
C_o: oxygen concentration in the polymer when $p = p_o$

The quantum yield for emission is then:

$$\Phi = \frac{k_p}{k_p + k_{nr} + k_q C_o (p/p_o)} \tag{7}$$

The response of a PSP is usually monitored by measuring the quantum yield in Eq. (7) through an intensity ratio, one of which is the reference intensity obtained at a known pressure.[1]

$$\frac{I_o}{I} = \frac{\Phi_o}{\Phi} = \frac{k_p + k_{nr} + k_q C_o (p/p_o)}{k_p + k_{nr} + k_q C_o} = A + B(p/p_o)$$

$$A = \frac{k_p + k_{nr}}{k_p + k_{nr} + k_q C_o} \tag{8}$$

$$B = \frac{k_q C_o}{k_p + k_{nr} + k_q C_o}$$

In wind tunnel studies Eq. (8) is applied as

$$\frac{I^o_{xy}}{I_{xy}} = A + B(p_{xy}/p_o) \tag{9}$$

where the subscripts x and y refer to the spatial coordinates of the point in an image matrix. In the laboratory the luminescence intensity can be measured at vacuum, then the Stern-Volmer equation takes the following form,

$$\frac{I_{vac}}{I} = 1 + K_{sv}[O_2]$$

$$K_{sv} = \frac{B}{A} = \frac{k_q C_o}{k_p + k_{nr}} = \tau_{vac} k_q C_o \tag{10}$$

$$\tau_{vac} = (k_p + k_{nr})^{-1}$$

τ_{vac}: lifetime of luminophore at vacuum conditions

Stern-Volmer (SV) plots for a typical PSP system is shown in Figure 8.2. Contrary to our theoretical expectations, the SV plot is non-linear. The non-linearity is attributed to inhomogeneity of the luminophore distribution in the binder.[17, 18]

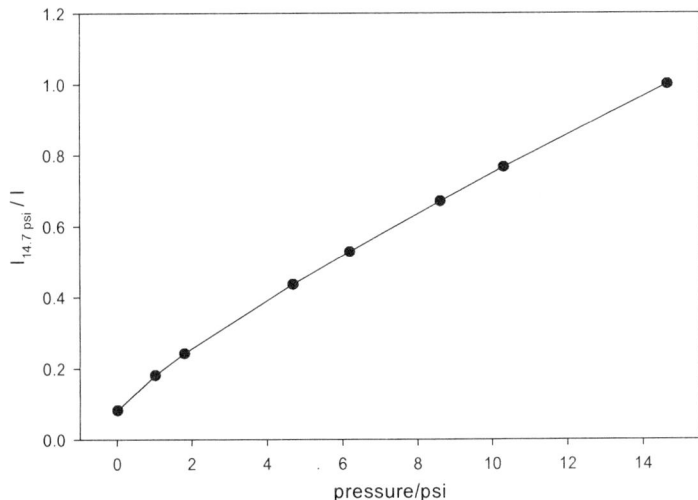

Figure 8.2. Stern-Volmer plot of $I_{14.7\,psi}/I$ versus P_{air} (psi) for PtTFPP in FIB (Fluoro/Isopropyl/Butyl acrylic polymer). (Data from our laboratory.)

8.2. TEMPERATURE EFFECTS

Under ideal conditions, the photoluminescence intensity of a PSP would respond only to changes in pO_2. Other environmental parameters such as temperature (T) would not interfere with the measurement. Unfortunately, it is well known that at constant air pressure, the photoluminescence intensity of most oxygen sensor coatings varies inversely with T as does the Stern-Volmer coefficient (K_{SV}).[19, 20] For many sensor coatings, K_{SV} varies linearly with T over a narrow range of temperatures as modeled by Eq. (11).

$$K_{SV}(T) = K_{SV}^{ref} + bT \qquad (11)$$

Because of the temperature dependence of K_{SV}, when a PSP is applied under non-isothermal conditions (most often the case in a wind-tunnel), the global map of P_{air} that is created will be in error. On the other hand, since K_{SV} is generally a well-behaved function of temperature as in Eq. (11), if one knows the temperature distribution over the surface of the aerodynamic model, it is possible to apply computer algorithms to correct for the temperature dependence of K_{SV} and obtain an accurate global surface pressure distribution.[19-21]

A convenient method for obtaining the necessary information regarding the temperature distribution over the aerodynamic model is to incorporate a second "temperature-sensitive" luminophore into the PSP. Such a dual-luminophore PSP allows one to construct a global temperature map for the surface of an aerodynamic model by imaging the photoluminescence from the temperature-sensitive luminophore. The global temperature map is then used to correct for the temperature dependence of the photoluminescence intensity image-map generated from the pressure sensitive luminophore. This process affords a global surface pressure distribution that is corrected for the non-uniform temperature distributions. When incorporated separately (i.e., one at a time, but not together) into a polymer binder, many photoluminescent probe molecules display desirable emission characteristics for oxygen or temperature sensing. However, when two or more probe molecules are mixed into the same binder, they typically do not display the desired oxygen and/or temperature sensing photoluminescence properties. This non-ideal behavior arises from molecular level physical and chemical interactions, and often leads to unpredictable results with respect to the overall photoluminescence properties of the sensor coating.[21] One solution to this problem is to design nanometer to micrometer sized capsules that separate the luminophores in the polymer binder at the molecular level, yet provide a coating that is spatially homogeneous on the millimeter scale (i.e., the spatial resolution of the CCD imager). In this manner, it is possible to produce a PSP incorporating two or more photoluminescent probe molecules, which display separate and well-defined photoluminescence intensity variations due to temperature and oxygen sensing.

8.3. A DUAL-LUMINOPHORE PRESSURE SENSITIVE PAINT

In the work presented in this chapter a dual-luminophore PSP was developed and characterized. This coating contains a temperature-sensitive luminophore adsorbed onto crosslinked polystyrene microspheres and an oxygen-sensitive luminophore that is dispersed in the PSP binder. The dyed-microspheres and the oxygen sensing luminophore were distributed in a gas-permeable polymer binder. Detailed schemes for the preparation of the dyed-microspheres and resulting dual-luminophore coatings are outlined in Figure 8.3.

The coating consists of Pt(II) *meso*-tetrakis(pentafluorophenyl)porphine (PtTFPP) as the oxygen sensing luminophore and DOCI-adsorbed microspheres produced via precipitation polymerization (DOCIpμsp, 3,3'-diethyloxacarbocyanine iodide = DOCI) as the temperature sensing luminophore dispersed in a vinyl terminated polydimethylsiloxane (VPDMS, Gelest Inc, Cat. # DMS-V25) polymer binder. The preparation of the coating is schematically represented in Figure 8.3. The components, PtTFPP, DOCIpμsp, and VPDMS are stirred in methylene chloride solution until all components well dispersed. Prior to application of the coating onto the substrate, a Pt catalyst solution (platinum-divinyl tetramethyl-disiloxane complex 3-3-5% platinum concentration in vinyl terminated polydimethylsiloxane, Gelest Cat. # SIP830) is added to the mixture in order to crosslink the VPDMS binder. Immediately afterwards, the coating is applied using a commercially available air-brush at 15 psi onto clean primer-coated borosilicate glass slides. The coating is allowed to cure to the touch at room temperature for 1 h. The PSP layer thickness is typically 25 μm.

Emission spectra of the coating at various pressures and temperatures are shown in Figure 8.4. Two emission bands are detected when the coating is excited with 450 nm light. The PtTFPP emission consists of two bands: an intense band, T(0,0), centered at 645 nm and a weaker band, T(0,1), centered at 710 nm. The emission bands are due to phosphorescence from the $^3T_1(\pi,\pi^*)$ state of the porphyrin macrocycle.[1] Back bonding between the d_{xz} and d_{yz} orbitals of the Pt with the empty $e_g(\pi^*)$ orbitals of the porphyrin

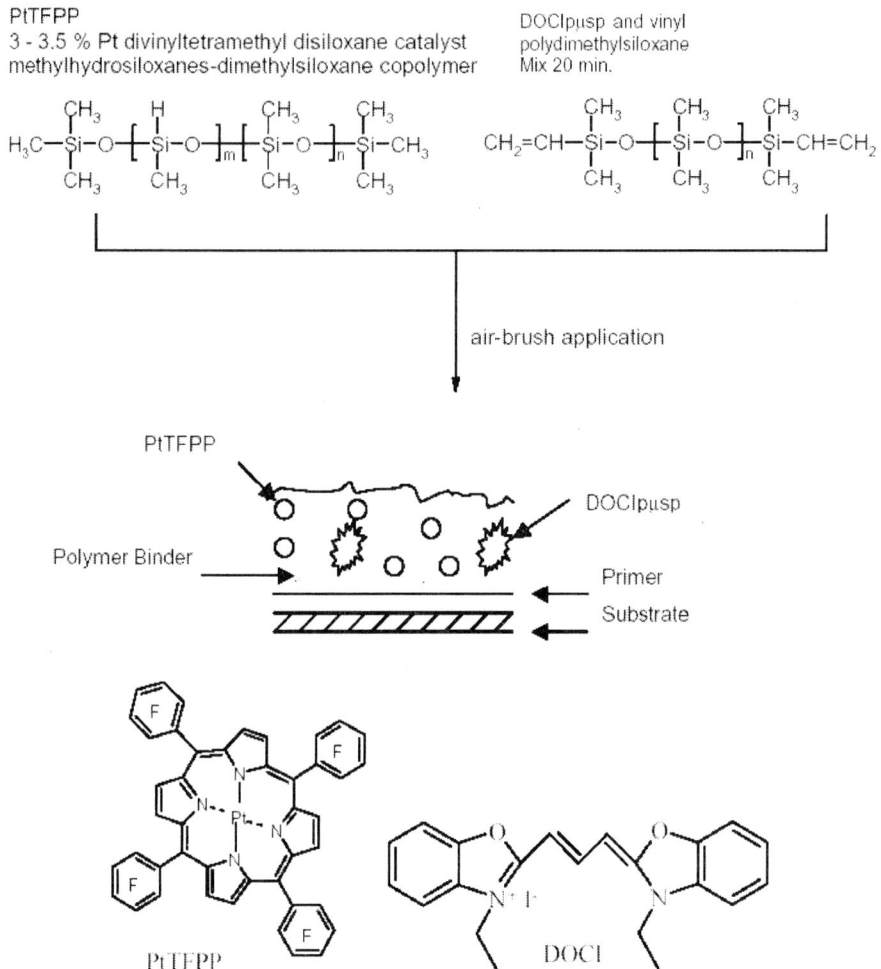

Figure 8.3. Scheme for preparation of dual-luminophore PSP formulation using PtDOCIpµsp/VPDMS.[22]

produces strong spin-orbit coupling. This leads to singlet-triplet mixing which increases the radiative decay rate from $^3T_1(\pi,\pi^*) \rightarrow {}^1S_0$. The long lifetime of the PtTFPP phosphorescence ($\tau \approx 50$ µs) facilitates efficient oxygen quenching.[23] Therefore, a decrease in emission intensity is observed as the air pressure surrounding the coating sample increases. The DOCIpµsp temperature-sensing luminophore exhibits a broad emission ranging from 500 to 600 nm. The DOCI fluorescence band exhibits a *dip* at 540 nm due to the Q-band absorption by PtTFPP that is co-located in the PSP film (i.e. an inner filter effect). Due to the short lifetime of the DOCI emission, the luminescence intensity does not vary with air pressure.

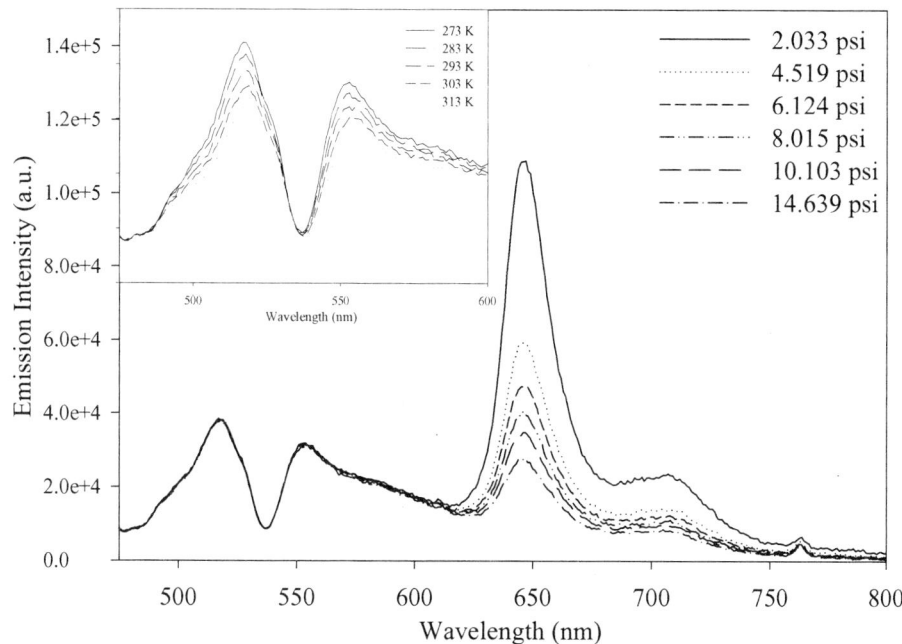

Figure 8.4. Emission spectra of PSP formulation consisting of PtTFPP and DOClpμsp dispersed in VPDMS polymer at various air pressures and temperatures. Inset - Emission intensity spectra for DOClpμsp in PtDOClpμsp/VPDMS PSP at 0.1 psi for five temperatures between 273 – 313 K. Excitation wavelength for all scans was 450 nm.[22]

The temperature dependence of PtTFPP photoluminescence from the dual luminophore PSP was analyzed. The luminescence intensity decreases moderately with increasing temperature at all pressures from 0.1 - 14.7 psi. This is a typical behavior for many pressure probes.[1, 24] Stern-Volmer analysis of the solid coatings were conducted utilizing Eq. (13),

$$I(\lambda^P_{em}, P_{air} = 1\ atm)/I(\lambda^P_{em}, P_{air}) = A + BP_{air} \quad (13)$$

and $(B/A) = K_{SV}$. The pressure probe's response to oxygen quenching at different temperatures yields linear SV plots with excellent correlation and large K_{SV} values. The coatings are strongly pressure sensitive, as the total light intensity decreases by nearly a factor of 10 when the pressure above the film increases from 0.1 – 14.7 psi. The SV plot for each temperature is displayed in Figure 8.5 (left). The temperature dependence of the coating's emission is evidenced by the *fanout* of the regression lines at higher pressure. In particular, the K_{SV} values for the PtTFPP emission increase with increasing temperature, in a manner consistent with Eq. (11).

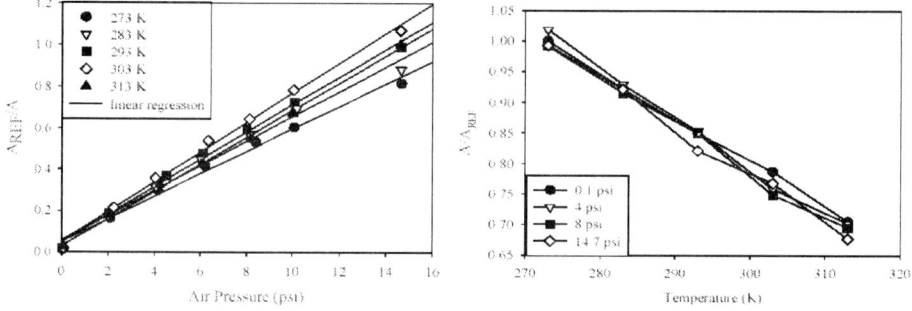

Figure 8.5. Left – Stern-Volmer plots for PtDOCIpμsp/VPDMS PSP for temperatures between 273 – 313 K. A$_{REF}$: area between 630 - 670 nm at 14.7 psi and 313 K. Right - Temperature dependence of emission for DOCIpμsp in PtDOCIpμsp/VPDMS PSP for a series of pressures between 0.1 - 14.7 psi. A$_{REF}$: area between 530 - 570 nm at 273 K and 0.1 psi.[22]

The temperature dependence of the photoluminescence from the DOCIpμsp temperature probe is shown in Figure 8.5 (right). The luminescence intensity of the temperature probe does not vary significantly with change in pressure at a constant temperature, however, the intensity decreases with increasing temperature. The temperature dependence is nearly linear over the temperature range 273 – 310 K with a slope of the linear correlation of approximately -0.80 %-K^{-1}. The thermal-stability of the temperature probes' emission was evaluated by subjecting the samples to cyclic temperature changes from 273 → 313 → 273 → 313 → 273 K. The DOCIpμsp exhibit good temperature dependence when the coating is heated and then again when it is cooled. The probe's emission intensity response is appreciable with changes in temperature, and it is reproducible throughout repeated cycling from low temperature to high temperature.

8.4. SCANNING ELECTRON MICROSCOPY: STRUCTURE OF THE DOCI LOADED MICROSPHERES

Scanning electron microscopy (SEM) was used to image the DOCIpμsp microspheres used in this work. Images were obtained on the undyed samples as well as on the dyed samples that were used in the coating preparations. The poly(divinylbenzene) microspheres were prepared by precipitation polymerization. A typical SEM image of the microspheres prior to DOCI (3,3'-diethyloxacarbocyanine iodide) adsorption is illustrated in Figure 8.6 (top). All images show the microspheres as aggregates of large and small particles. The larger particles are 3 to 5 μm in size while the smaller particles are as small as 0.5 μm. The polymer particles exhibit a non-spherical shape and no evidence of pore formation. It appears that in some instances the particles are fused together. This fusion most likely occurs during an early stage in their growth.[25] Placing carbon tape on an aluminum stud and pressing it against the prepared sample fractured

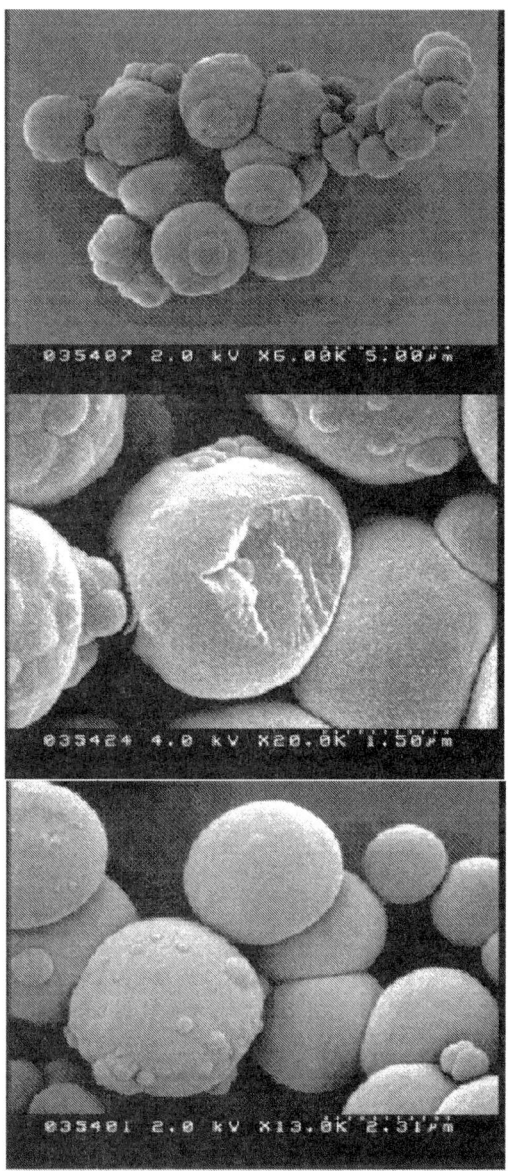

Figure 8.6. Top - Scanning electron micrograph of precipitation microspheres. The scale bar consists of 11 white vertical lines and is 5 μm long from the first line to the last line. Middle - Scanning electron micrograph of fractured precipitation microspheres. The scale bar consists of 11 white vertical lines and is 1.5 μm long from the first line to the last line. Bottom - Scanning electron micrograph of precipitation microspheres with 7.27 wt. % adsorbed DOCI (DOCIpμsp). The scale bar consists of 11 white vertical lines and is 2.31 μm long from the first line to the last line.[22]

the polymer particles. The fractured microspheres were transferred to the carbon tape and sputter-coated with 15 nm of gold. As a result of fracturing the polymer particles, a closer look of their interior morphology was obtained and imaged in Figure 8.6 (middle).

Magnification of the microspheres depicts the fusion of the smaller particles to the larger particles as seen in the upper half and left side of the image. Closer analysis of the exterior and interior morphologies reveals a cauliflower-like surface. This effect is most likely due to pore formation at the sub-micron level.[25] When 3,3'-diethyloxacarbocyanine iodide (DOCI) was adsorbed onto the microspheres, the particles feature a more spherical swollen shape with no evidence of pore formation. Figure 8.6 (bottom) displays a typical cluster of swollen 7.27 wt. % DOCI adsorbed microspheres (DOCIpµsp).

8.5. FLUORESCENCE MICROSCOPY OF DUAL LUMINOPHORE PRESSURE SENSITIVE PAINT

Fluorescence microscopy techniques were employed to analyze the size and distribution of PtTFPP and the DOCIpµsp in the VPDMS polymer binder. Microscopy was carried out on a typical PSP film that was applied by using an air-brush. DOCIpµsp were analyzed for size (length of particle major axis) and distribution (number of particles per region analyzed). The DOCIpµsp (0.6 wt. %, relative to VPDMS) were distributed in the VPDMS polymer binder. The DOCIpµsp/VPDMS film was approximately 10 µm thick, as measured by profilometry. For this experiment, an infrared blocking filter, a 50% neutral density filter, and a 425 nm 40 nm bandpass filter were placed in front of the high pressure 100-W mercury excitation source. The DOCIpµsp/VPDMS film was imaged with a 40X objective, and the emission light was imaged through a 475 nm long-pass filter. Five regions of each coating were analyzed using a CCD exposure time of 150 msec.

A typical fluorescence microscope image of the DOCIpµsp/VPDMS film is displayed in Figure 8.7 (image size is 218 µm x 173 µm). Within the image the white clusters and isolated particles are well-defined due to the DOCI fluorescence emission. Five regions were imaged for the film and the number and size of the particles in the images were analyzed and subjected to statistical analysis. White clusters were considered as one microsphere object rather than counting the individual microsphere components. The number average and the standard deviation for the number of microsphere objects for a field size of 0.0377 mm^2 are estimated to be 69.6 ± 13.7 which corresponds to approximately 1840 particles per mm^2. For all of the microsphere objects counted, the major axis lengths were determined using a statistical imaging program, Sigma Scan Pro (SPSS Inc.). The lengths of the objects' major axes from the five regions analyzed for each film are represented by histogram in Figure 8.8. The number of objects for a given size range are plotted versus the length of the objects (0.5 µm bin increments). As can be seen, the major fraction of the particles has sizes < 5 µm, but a small fraction of the particles consist of large clusters which are > 10 µm in size.

In summary, fluorescence microscope analysis of the dual luminophore PSP indicates that the DOCI emission emanates primarily from the microspheres. This is consistent with the notion that the dye is compartmentalized in the particles. Analysis of the images reveals that the microspheres are distributed uniformly in the film and the spatial density is small enough such that on the length scale of the CCD imager used for PSP application, the particles will not be resolved (i.e. the DOCI fluorescence will be spatially homogenous). This is borne out by the imaging tests described below.

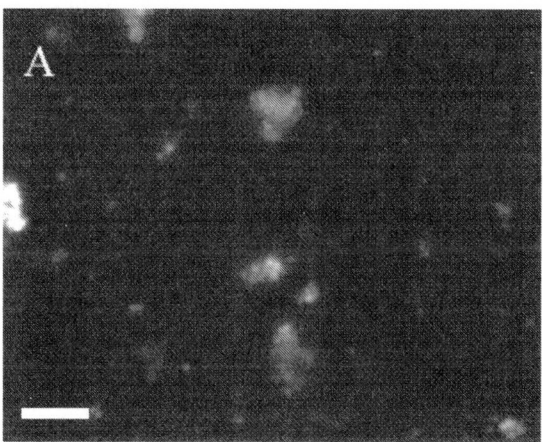

Figure 8.7. Fluorescence microscope image of DOCIpμsp/VPDMS film obtained with a CCD camera through a 40X objective. Excitation at 425 nm and emission imaged through 475 nm long-pass filter. White scale bar is 26.5 μm long.[22]

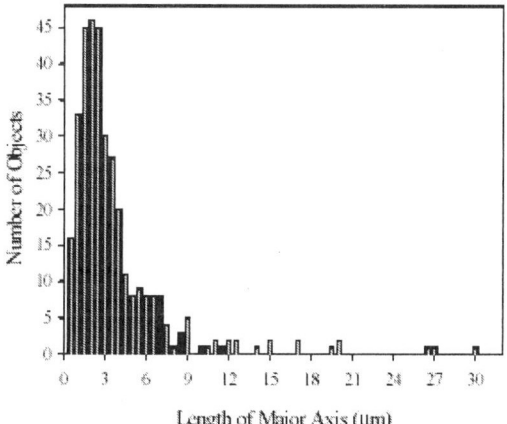

Figure 8.8. Histogram of the microsphere major axis lengths for DOCIpμsp/VPDMS film. Vertical bars equal a 0.5 μm length increment.[22]

8.6. DEMONSTRATION OF THE DUAL LUMINOPHORE PRESSURE SENSITIVE PAINT

A calibration test was conducted with the PtDOCIpμsp/VPDMS coating to assess its pressure measurement accuracy with a test specimen that has a spatial temperature gradient imposed across its surface. An aluminum coupon (4 in. x 2 in. x 1/8 in.) was coated with a white primer consisting of silanol terminated PDMS (Gelest Inc, Cat. # DMS-S27) and TiO_2 (CR-800, Tronox) which were pre-mixed in methylene chloride. After curing the primer, the dual luminophore PSP was applied in four successive coats by using an air brush. The coupon was clamped between a heat source (an aluminum block with an etched-foil heating element) and a heat sink (an aluminum block with a water channel). The heating apparatus generated a near one-dimensional temperature gradient across the coupon of approximately 20 K. Surface temperatures were monitored using an array of five T-type thermocouples mounted on the backside of the coupon. The coupon and heating apparatus was then placed into a pressure evacuation chamber which was used to control the global surface pressure. Figure 8.9 shows a schematic of the coupon, heating apparatus, and evacuation chamber. The surface pressure over the test coupon was varied between 2.0 to 14.7 psi and the surface temperature gradient was 300 to 321 K. The coupon was illuminated using two blue LED lamps with a center wavelength of 465 nm. Images of the coupon subjected to the temperature gradient were acquired at pressure levels of 2.0, 4.0, 6.0, 8.0, 10.0, 12.0 and 14.7 psi using a 16-bit digital camera fitted with two different interference bandpass filters: 550 nm center wavelength corresponding the DOCIpμsp temperature emission, and 650 nm center wavelength corresponding to the PtTFPP pressure-temperature emission. The optical bandwidth for each filter was 40 nm full-width-half-maximum. A final reference image was acquired at the pressure of 14.7 psi and temperature (no gradient) of 300 K.

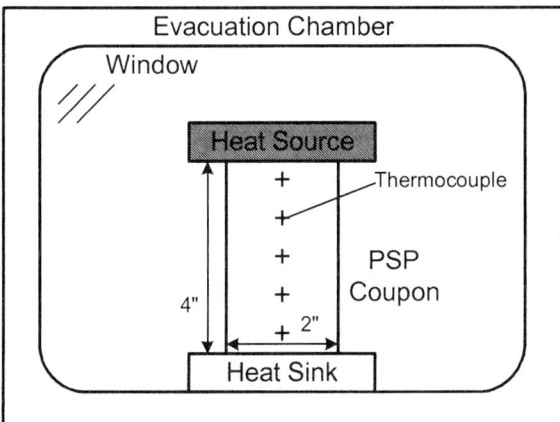

Figure 8.9. Front-view schematic of the PSP coated coupon, heating apparatus, and evacuation chamber used in the imaging test. (+): Indicates the position of thermocouples used to measure temperature at specific points on the specimen.

To correct for spatial variations in the excitation lamps and nonuniformity of the coating thickness, images at each pressure were ratioed with the reference image (I_{ref}/I). Because the emission intensity is inversely related to pressure and temperature, it is common to analyze and plot the inverse of the emission results (I_{ref}/I). The intensity-ratio (I_{ref}/I) results at the five thermocouple positions and seven pressure levels for PtDOCIpμsp/VPDMS are shown in Figure 8.10. (The plotted data correspond to the average of five 10 x 10 pixel matrices that correspond to the regions where the five thermocouples are positioned). Plots A and B correspond to the 550 nm emission, and plots C and D correspond to the 650 emission. As desired, the PtTFPP pressure probe

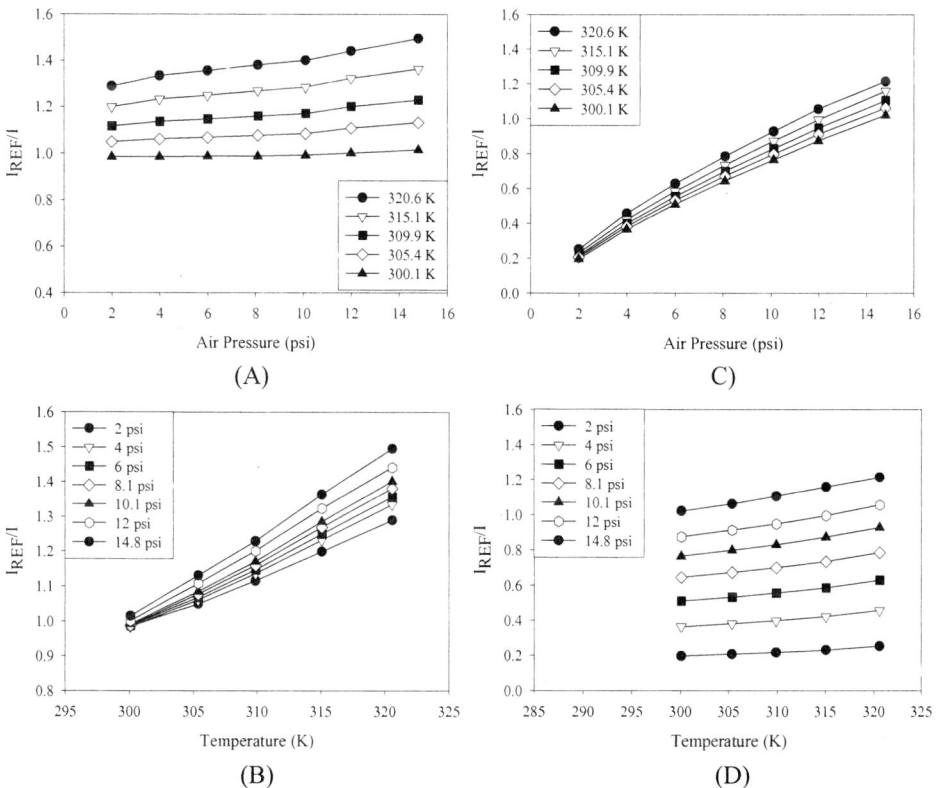

Figure 8.10. Ratioed emission intensity plots versus pressure and temperature for PtDOCIpμsp/VPDMS PSP imaged using a CCD camera. The plotted data correspond to the average of five 10 x 10 pixel matrices that correspond to the regions where the five thermocouples are positioned. Ratios are relative to the same 10 x 10 matrices obtained from the reference image which was acquired at a known temperature and pressure. A) Isothermal $I_{ref}/I(P,T)$ curves for the 550 nm filter, B) Isobaric $I_{ref}/I(P,T)$ curves for the 550 nm filter, C) Isothermal $I_{ref}/I(P,T)$ curves for the 650 nm filter, D) Isobaric $I_{ref}/I(P,T)$ curves for the 650 nm filter. The reference condition for all plots is 14.7 psi and 300 K. The solid lines are the fits calculated using the empirical equations as described in the text.

shows good pressure dependence (plot C and D for isotherms and isobars, respectively). The probe also exhibits is a moderate temperature dependence. If no temperature dependence existed in the 650 nm emission results, all of the isotherms in plot C would merge onto a single line and all of the isobars in plot D would be horizontal. The DOCIpµsp probe exhibits a strong temperature dependence and a weak pressure dependence (plots A and B for isotherms and isobars, respectively). The pressure response is more likely the cause of spectral leakage, i.e., short wavelength PtTFPP emission passing through the 550 nm filter, rather than being due to pressure dependence of the DOCIpµsp probe's fluorescence. Note that in general the emission temperature- and pressure-dependence of the two luminophores as detected by using the CCD camera is similar to that observed using a fluorescence spectrometer (Figure 8.5).

As shown in Figure 8.10 (plot D), when a temperature gradient exists, the PtDOCIpµsp/VPDMS coating shows a change in the emission intensity ratio. If the user were to assume that the coating was temperature independent, then changes in the intensity-ratio response would incorrectly be attributed to a change in pressure. Figure 8.11 shows the calculated pressure field across the coupon surface using the 310 K isotherm

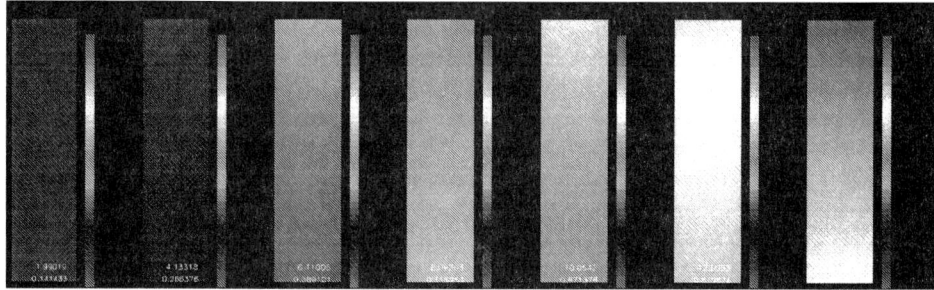

Figure 8. 11. (Please see Color Inserts Section) Pressure calculated images using the PtDOCIpµsp/VPDMS PSP and assuming a constant temperature of 310 K. Emission collected at 650 nm and excitation at 465 nm. Intensity scale bars: 0 – 18.0 psi.

pressure response as a calibration and assuming that the coating is temperature-independent. Ideally, each processed image should exhibit a single color representing a constant pressure field. Of course in reality it is known that the pressure field across the surface is constant. However, because the top of the coupon is warmer (> 310 K) the emission intensity in this region is lower and this leads to an increase in the emission intensity-ratio (I_{ref}/I). When the calibration is applied, this falsely appears as a region of higher pressure. The opposite is true for the region of lower temperature (< 310 K) near the bottom of the coupon. Table 8.1 (uncorrected) lists the average and standard deviation of the pressure over the surface of the coupon. For all pressure conditions, the calculated average pressure overestimates the pressure by 2 to 3% and the standard deviation is 6 to 8% of the average, indicating a large variance in the calculated pressure across the surface. For the atmospheric case, the standard deviation corresponds to +/- 1.2 psi, clearly illustrating the relatively large error induced by assuming the luminescence

A DUAL LUMINOPHORE PRESSURE SENSITIVE PAINT

intensity of PtTFPP is not temperature dependent.

The DOCIpµsp intensity image data acquired at 550 nm is used to correct for the temperature dependence of the PtTFPP emission intensity. Image data was analyzed using least squares regression macros written in MatLab. First, the calibration surfaces were determined as follows. The 550 nm emission image intensity data in Figures 8.10A and B was fitted by a least-squares analysis to a two-parameter (P,T) linear surface of the form $[I_{ref}/I]_{550} = a_1 + b_1P + c_1T$ in order to define the parameters $a_1 - c_1$. And similarly the 650 nm image data in Figures 8.10C and D was fitted with a two-parameter quadratic surface of the form $[I_{ref}/I]_{650} = a_2 + b_2P + c_2T + d_2P^2 + e_2T^2 + f_2PT$ to define the parameters $a_2 - f_2$. (It was necessary to use a higher order polynomial to properly define the surface for the 650 nm data. This equation was chosen empirically.)

Table 8.1. Pressure results for the PtDOCIpµsp/VPDMS coating assuming temperature independent response (uncorrected), and with temperature correction (corrected).[22]

P(psi)[a]	uncorrected		corrected	
	P_{avg}[b]	σ_{avg} %[c]	P_{avg}[b]	σ_{avg} %[c]
2.00	2.06	6.8	1.95	4.61
4.05	4.20	6.4	4.06	2.22
6.03	6.18	6.3	6.04	1.32
8.07	8.17	6.4	8.03	1.00
10.07	10.14	6.7	10.02	0.70
12.04	12.11	7.1	11.92	0.60
14.79	15.09	7.8	14.88	0.90

[a] Pressure of the calibration chamber measured using a pressure gauge.
[b] Average pressure determined using luminescence image data.
[c] σ_{avg} % = (σ_{avg}/P_{avg}) x 100%.

Next, the calibrations were used to determine the full-field, temperature corrected images. These calculations are performed for every point in the image matrices as follows. An iterative process was performed which uses the equations derived from the calibration procedure described above to convert the intensity ratio images to pressure images. First, by assuming that the temperature is at the reference condition ($T = T_{ref}$), the pressure, P_i, is calculated using $[I_{ref}/I]_{650}$, the 650 nm surface coefficients ($a_2 - f_2$), and T_{ref}. This pressure is then used to calculate the temperature, T_i, using $[I_{ref}/I]_{550}$, the 550 nm surface coefficients ($a_1 - c_1$), and P_i. Then, the new T_i is used to calculate P_{i+1} and so on until a specified tolerance is achieved for P and T (this process generally requires 2 – 3 iterations).

Figure 8.12 shows the calculated pressure images using both the 550 nm and 650 nm data image sets. Note that by including the 550 nm image data in the pressure analysis, the color is homogeneous in each image which indicates that the analysis corrects for the temperature-dependence of the PtTFPP luminescence. Table 8.1 (corrected columns) lists the average and standard deviation of the pressure over the surface of the coupon. For all conditions, the calculated surface pressure standard deviation is significantly lower than was obtained without the 550 nm data, indicating a more uniform pressure distribution (as expected). For the atmospheric case, the standard deviation corresponds to +/- 0.13 psi, a substantial improvement in the pressure error. In addition, note that the estimated pressures are also much closer to the true pressures.

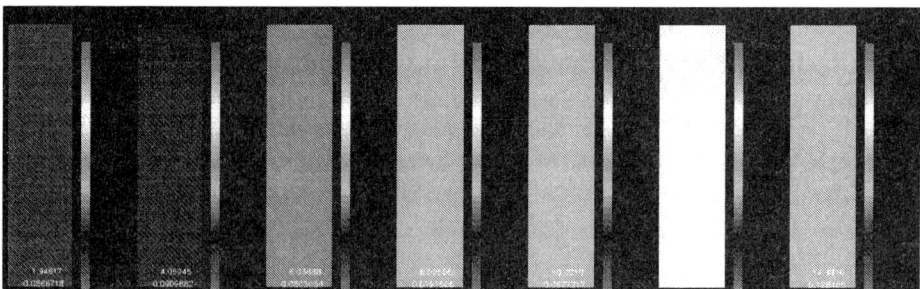

Figure 8.12. (Please see Color Inserts Section) Pressure image maps calculated using the PtDOClpμsp/VPDMS PSP and the 550 nm and 650 nm emission data sets. Emission imaged at 550 nm and 650 nm and excitation at 465 nm. Intensity scale bars: 0–18.0 psi.

8.7. SUMMARY AND RELATIONSHIP TO THE OTHER WORK IN THE FIELD

In this chapter we describe the preparation and characterization of a dual-luminophore pressure sensitive paint formulation that consists of a polystyrene microsphere-encapsulated temperature sensing luminophore (DOCI) and a pressure sensing luminophore (PtTFPP) distributed within a polydimethylsiloxane binder. Fluorescence microscopy and SEM imaging were employed to evaluate the molecular distribution of the luminophores in the coatings. The luminescence from the coating was imaged within a static pressure-temperature calibration cell to evaluate its overall performance to changes in pressure under non-isothermal conditions. The results of the calibration chamber test validate the coating, and demonstrate its ability to accurately determine pressure on a surface that is subjected to a temperature gradient.

In parallel with the research described above, several other research groups have been working towards the development of multi-luminophore pressure sensitive paint formulations. In particular, Gouterman, Khalil and co-workers have worked extensively to develop luminophores and polymer binders that can be used to correct for the temperature-dependence in the pressure calibration. In one report they describe the incorporation of an inorganic phosphor ($BaMg_2Al_{16}O_{27}:Eu^{3+}$) into a PtTFPP-based pressure sensitive paint formulation. The luminescence from the Eu phosphor varies with temperature, but it does not vary with pressure. However this coating displayed problems due to competitive absorption of the excitation light by PtTFPP, absorption of the phosphor's emission by PtTFPP, and an inhomogeneous distribution of the phosphor particles. Expanding on earlier work with platinum porphyrins, the same group explored the application of silicon octaethylporphine (SiOEP) as a pressure-insensitive, temperature-sensitive luminophore. The silicon porphyrin complex suffers from an increase in fluorescence intensity as the temperature rises due to repopulation of the singlet excited state via the triplet state. The rise in fluorescence intensity interferes with calibration runs at vacuum pressure.[21]

Gouterman and co-workers also explored the use of a polymer binder which features a low activation energy to oxygen diffusion. The objective of this work is that decreasing the activation barrier for oxygen diffusion will lead to a reduction in the temperature dependence of the luminescence intensity from the pressure probe. A temperature sensitivity of -0.6 %-°C^{-1} for PtTFPP dispersed in the fluoroacrylic polymer

FIB was been reported. This temperature gradient is significantly less than the temperature sensitivity of platinum octaethylporphine in the silicone polymer binder Genesee GP-197(-1.7 %-°C^{-1}).[24] Finally, in a recent report the same group described a dual luminophore PSP formulation which utilizes a tris(β-diketonate) phenanthroline europium complex as temperature sensor and PtTFPP as pressure sensor. They use the emission lifetime of the luminophores to measure the pressure and temperature.[26]

Oxygen sensors based on the quenching of luminescence of ruthenium(II) polypyridine complexes entrapped in sol-gel films have been studied extensively. The sol-gel process facilitates control over the film composition which in turn makes it possible to tune the pressure sensitivity of the sensor. Sol-gel derived silica films exhibit excellent sensitivity and reversibility.[27-29]

The Japanese group of Asai and co-workers have reported the synthesis of several fluoro-acrylate co-polymers that can be used as binders for pressure sensitive paints. Specifically, poly(isobutylmethacrylate-*co*-trifluoroethylmethacrylate),[30] poly(styrene-co-pentafluorostyrene)[31] and poly(styrene-co-trifluoroethylmethacrylate)[32] exhibit linear Stern-Volmer correlations when luminophores such as PtTFPP and [Ir(ppy)$_3$] are dispersed in the matrices. The same group reported a dual luminophore PSP formulation which uses poly(isobutylmethacrylate-*co*-trifluoroethylmethacrylate) as the binder and PtTFPP as the pressure probe. Temperature correction is achieved by using Rhodamine B as a temperature-dependent, pressure-independent luminophore. However, while the emission from both probes can be observed, its use is limited because of the relatively low temperature dependence of Rhodamine B between 15 °C and 30 °C.[33]

In summary, it is evident that the dual-luminophore PSP formulation that we describe has properties which are among the best of the systems that have been prepared to date. Nevertheless, due to the importance of the temperature problem in PSP application, it is evident that additional work in this field is needed to allow the development of robust multi-luminophore PSP formulations.

8.8. REFERENCES

1. M. Gouterman, Oxygen quenching of luminescence of pressure sensitive paint for wind tunnel research, *Journal of Chemical Education* **74**(6), 697-702 (1997).
2. M. J. Morris, J. F. Donovan, J. T. Kegelman, S. D. Schwab, R. L. Levy, and R. C. Crites, Aerodynamic applications of pressure sensitive paint, *AIAA Journal* **31**(3), 419-425 (1993).
3. R. H. Engler, C. Klein, and O. Trinks, Pressure sensitive paint systems for pressure distribution measurements in wind tunnels and turbomachines, *Measurement Science and Technology* **11**(7), 1077-1085 (2000).
4. P. Hartmann, and W. Trettnak, Effects of polymer matrixes on calibration functions of luminescent oxygen sensors based on porphyrin ketone complexes, *Analytical Chemistry* **68**(15), 2615-2620 (1996).
5. A. Mills, Controlling the sensitivity of optical oxygen sensors, *Sensors and Actuators, B: Chemical* **B51**(1-3), 60-68 (1998).
6. A. Mills, Effect of plasticizer viscosity on the sensitivity of an [Ru(bpy)32+(Ph4B-)2]-based optical oxygen sensor, *Analyst (Cambridge, United Kingdom)* **123**(5), 1135-1140 (1998).
7. A. Mills, and F. C. Williams, Chemical influences on the luminescence of ruthenium diimine complexes and its response to oxygen, *Thin Solid Films* **306**(1), 163-170 (1997).
8. P. Hartmann, Photochemically induced energy-transfer effects on the decay times of ruthenium complexes in polymers, *Analytical Chemistry* **72**(13), 2828-2834 (2000).
9. J. N. Demas, and B. A. DeGraff, Design and applications of highly luminescent transition metal complexes, *Analytical Chemistry* **63**(17), 829A-837A (1991).
10. E. R. Carraway, J. N. Demas, B. A. DeGraff, and J. R. Bacon, Photophysics and photochemistry of oxygen sensors based on luminescent transition-metal complexes, *Analytical Chemistry* **63**(4), 337-42 (1991).

11. J. N. Demas, and B. A. DeGraff, Applications of luminescent transition metal complexes to sensor technology and molecular probes, *Journal of Chemical Education* **74**(6), 690-695 (1997).
12. S.-K. Lee, and I. Okura, Photoluminescent oxygen sensing on a specific surface area using phosphorescence quenching of Pt-porphyrin, *Analytical Sciences* **13**(4), 535-540 (1997).
13. A. Mills, and A. Lepre, Controlling the response characteristics of luminescent porphyrin plastic film sensors for oxygen, *Analytical Chemistry* **69**(22), 4653-4659 (1997).
14. D. B. Papkovsky, G. V. Ponomarev, W. Trettnak, and P. O'Leary, Phosphorescent complexes of porphyrin ketones: Optical properties and application to oxygen sensing, *Analytical Chemistry* **67**(22), 4112-17 (1995).
15. J. H. Bell, E. T. Schairer, L. A. Hand, and R. D. Mehta, Surface pressure measurements using luminescent coatings, *Annual Review of Fluid Mechanics* **33**, 155-206 (2001).
16. O. Stern, and M. Volmer, The extinction period of fluorescence, *Physik. Z.* **20**, 183-8 (1919).
17. W. Xu, R. C. McDonough, III, B. Langsdorf, J. N. Demas, and B. A. DeGraff, Oxygen sensors based on luminescence quenching: Interactions of metal complexes with the polymer supports, *Analytical Chemistry* **66**(23), 4133-41 (1994).
18. K. A. Kneas, W. Xu, J. N. Demas, and B. A. DeGraff, Oxygen sensors based on luminescence quenching: interactions of tris(4,7-diphenyl-1,10-phenanthroline)ruthenium(II) chloride and pyrene with polymer supports, *Applied Spectroscopy* **51**(9), 1346-1351 (1997).
19. K. S. Schanze, B. F. Carroll, S. Korotkevitch, and M. J. Morris, Concerning the temperature dependence of pressure sensitive paints, *AIAA Journal* **35**, 306-310 (1997).
20. M. A. Woodmansee, and J. C. Dutton, Treating temperature-sensitivity effects of pressure sensitive paint measurements, *Experiments in Fluids* **24**(2), 163-174 (1998).
21. L. M. Coyle, D. Chapman, G. Khalil, E. Schibli, and M. Gouterman, Non-monotonic temperature dependence in molecular referenced pressure-sensitive paint (MR-PSP), *Journal of Luminescence* **82**(1), 33-39 (1999).
22. J. M. Bedlek-Anslow, *Development and characterization of luminescent oxygen sensing coatings* (University of Florida, Gainesville, 2000).
23. D. Eastwood, and M. Gouterman, Porphyrins. XVIII. Luminescence of cobalt, nickel, palladium, platinum complexes, *Journal of Molecular Spectroscopy* **35**(3), 359-75 (1970).
24. E. Puklin, B. Carlson, S. Gouin, C. Costin, E. Green, S. Ponomarev, H. Tanji, and M. Gouterman, Ideality of pressure-sensitive paint. I. Platinum tetra(pentafluorophenyl)porphine in fluoroacrylic polymer, *Journal of Applied Polymer Science* **77**(13), 2795-2804 (2000).
25. K. Li, and H. D. H. Stover, Synthesis of monodisperse poly(divinylbenzene) microspheres, *Journal of Polymer Science, Part A: Polymer Chemistry* **31**(13), 3257-63 (1993).
26. B. Zelelow, G. E. Khalil, G. Phelan, B. Carlson, M. Gouterman, J. B. Callis, and L. R. Dalton, Dual luminophor pressure sensitive paint II. Lifetime based measurement of pressure and temperature, *Sensors and Actuators, B: Chemical* **B96**(1-2), 304-314 (2003).
27. J. Hradil, C. Davis, K. Mongey, C. McDonagh, and B. D. MacCraith, Temperature-corrected pressure-sensitive paint measurements using a single camera and a dual-lifetime approach, *Measurement Science and Technology* **13**(10), 1552-1557 (2002).
28. F. Z. Jiang, R. Xu, D. Y. Wang, X. D. Dong, and G. C. Li, Preparation and properties of novel aerodynamic pressure-sensitive paint via the sol-gel method, *Journal of Materials Research* **17**(6), 1312-1319 (2002).
29. S. K. Lam, M. A. Chan, and D. Lo, Characterization of phosphorescence oxygen sensor based on erythrosin B in sol-gel silica in wide pressure and temperature ranges, *Sensors and Actuators, B: Chemical* **B73**(2-3), 135-141 (2001).
30. Y. Amao, K. Asai, T. Miyashita, and I. Okura, Photophysical and photochemical properties of optical oxygen pressure sensor of platinum porphyrin-isobutylmethacrylate-trifluoroethylmethacrylate copolymer film, *Polymer Journal (Tokyo)* **31**(12), 1267-1269 (1999).
31. Y. Amao, K. Asai, T. Miyashita, and I. Okura, Novel optical oxygen sensing material: platinum porphyrin-styrene-pentafluorostyrene copolymer film, *Analytical Communications* **36**(11/12), 367-369 (1999).
32. Y. Amao, K. Asai, T. Miyashita, and I. Okura, Novel optical oxygen pressure sensing materials: platinum porphyrin-styrene-trifluoroethyl methacrylate copolymer film, *Chemistry Letters*(10), 1031-1032 (1999).
33. K. Mitsuo, K. Asai, M. Hayasaka, and M. Kameda, Temperature correction of PSP measurement using dual-luminophor coating, *Journal of Visualization* **6**(3), 213-223 (2003).

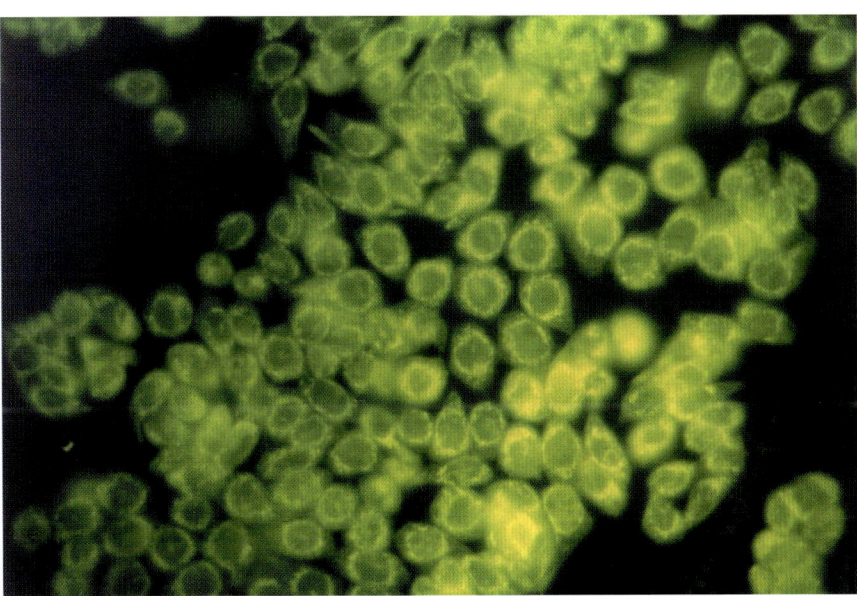

Figure 1.12. (Page 28, I. Prodi *et al.*) HT29 cells (human adenocarcinoma cell line), grown on glass coverslip and stained with 37 (50 uM).

Figure 7.6A. (Page 257, Verhoeven) Fluorescence colors shown at room temperature by Fluoroprobe (FP) under UV illumination in a series of solvents with increasing polarity. From left to right : cyclohexane, benzene, toluene, 1,4-dioxane, chloroform, and pyridine.

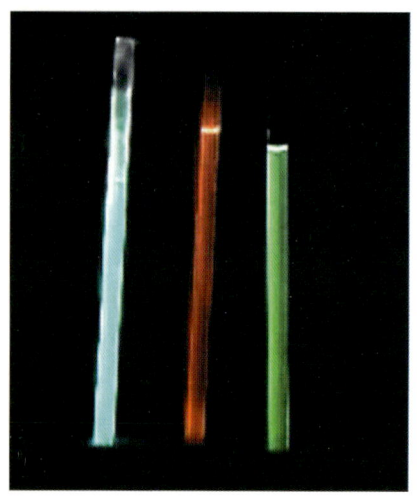

77 K   140 K   293 K

Figure 7.6B. (Page 257, Verhoeven) Fluorescence of FP in MTHF at three temperatures

Figure 7.6C. (Page 257, Verhoeven) Blue light emitted by a spin coated PLED based on 9 in PVK

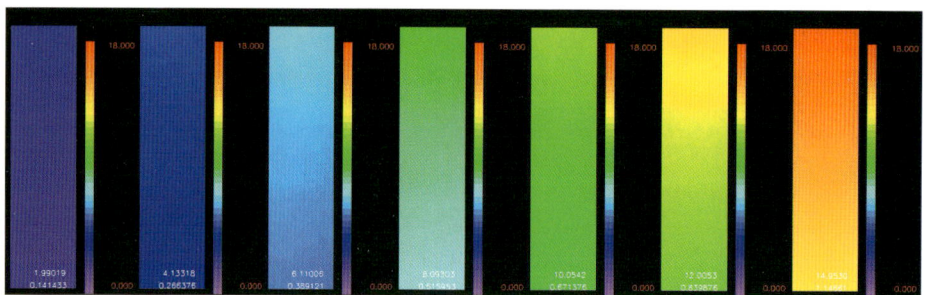

Figure 8.11. (Page 298, M.E. Kose *et al.*) Pressure calculated images using the PtDOCIp☐sp/VPDMS PSP and assuming a constant temperature of 310 K. Emission collected at 650 nm and excitation at 465 nm. Intensity scale bars: 0 – 18.0 psi.

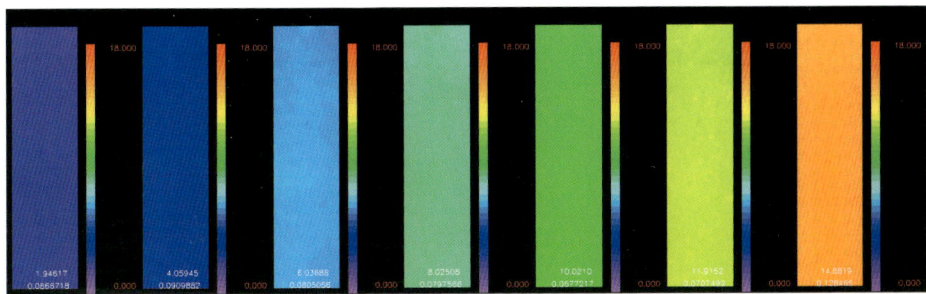

Figure 8.12. (Page 300, M.E. Kose *et al.*) Pressure image maps calculated using the PtDOCIp☐sp/VPDMS PSP and the 550 nm and 650 nm emission data sets. Emission imaged at 550 nm and 650 nm and excitation at 465 nm. Intensity scale bars: 0 –18.0 psi.

INDEX

Acetate, fluorescent sensors for, 70
  photoinduced electron transfer (PET) sensors, 234
  ruthenium (II)- and rhenium (I)-bipyridyl alix[4]diquinone sensors, 102, 103
Acetonitrite, sigma-coupled charge-transfer probe fluorescence in, 255, 256–257, 258
Acridine
  as anion sensor, 86
  as nitric oxide sensor, 169, 170
Acrylate polymer, fluorinated, as pressure sensitive paint component, 285
Acryloly fluorescein dye system, of carbon dioxide optical sensors, 124
Acyclic polyamines, as photoinduced electron transfer (PET) sensors, 231, 233
Acyclic receptor-based sensors
  for anions, 95–98
  for cations/metal ions
    for alkali or alkaline earth metal ions, 14–15, 24, 25–35
    amidic group-based, 12–13
    anthracene chromophore/polyamine chain-based, 7, 8–9
    anthryl group/-SCH2CH2CH2-based, 10–11
    for cadmium ions, 14, 18–19, 20, 21
    carboxylate-based, 7–8
    for copper ions, 7, 8, 11, 12–13, 18–19
    with dansyl groups, 17–20
    dioxo-tetraamines-based, 11, 12–13
    exciton-coupled circular dichroism (ECLD) spectra of, 17
    with isotropic and anisotropic properties, 17
    for nickel ions, 7, 8, 11, 12–13

Acyclic receptor-based sensors (*cont.*)
  for cations/metal ions (*cont.*)
    photoinduced electron transfer-based fluorescence quenching of, 6–7
    photoinduced electron transfer via conformational rearrangement of, 15–16
    polyoxyethylene chains-based, 14–16
    1,2,4-thiadiazole derviatives-based, 14
    use in aqueous medium, 7, 8
    for zinc, 7, 8, 9, 10, 14, 18–19, 20–21
Acyclic[Ru(bpy)$_3$]$^{2+}$-ferrocene, as anion sensor 96, 97
Acyclic water-soluble rhenium (II) bipyridyl polyaza, as anion receptor/sensor, 97
6-acyl-2-methoxynaphthalene (AMN), 40, 41
Adenosine diphosphate (ADP), fluorescent sensors for, 68
  azacrown ether acridine-based, 82
  4,9-diazapyrenium derivative-based, 87–88
Adenosine monophosphate (AMP), fluorescent sensors for
  4,9-diazapyrenium derivative-based, 87–88
  phenanthridinium cyclophanes-based, 87–88
Adenosine triphosphate (ATP)
  fluorescent sensors for, 67, 68
  azacrown ether acridine-based, 82
  4,9-diazapyrenium derivative-based, 87–88
  hydrolysis catalyst for, 82
Aerodynamic models, pressure sensitive paint-based surface air pressure distribution measurement on, 285, 286
Albumin, 45
Alchoxysilane derivatives, as silica nanoparticle modifiers, 47, 48–50
Alkali or alkaline earth metal ions
  acyclic receptor-based sensors for, 24, 25–35

307

Alkali or alkaline earth metal ions (*cont.*)
  acyclic receptor-based sensors for (*cont.*)
    with derivatized polyamine chains, 14–15
  crown ethers-based sensors for, 24, 25–35
    coumarin-derivatized, 25, 27
    dibenzo-, 26, 27
    8-hydroxyquinoline-derivatized, 27–28
    tribenzo-, 26, 27
  cyclic receptor-based sensors for, 25–37
    1-benzothiazol-3-(4-donor)-phenyl-substituted prop-en-1-ones, 35, 36
    bi-functional, 36
    calixarene derivatives, 38–42
    crown ethers as, 24, 25–35
  fluorescent redox-active device detection of, 200
Alkane diammonium ions, photoinduced electron transfer (PET) sensors for, 229
1,3-Alternate calix[4]arene-crowns, 38–39, 41–42
Alzheimer's disease, nitric oxide in, 163–164
Amides, as rhenium (I)-bipyridyl anion sensor components, 95, 96, 97
Amino acids, fluorescent sensors for, 68
  anion complexation agent-based
    anthracene-based, 75
    β-cyclodextrins-based, 65
  fluorophore-copper complex-based sensors, 94
    copper complex-modified β-cyclodextrin cyclic receptor-based, 42
  photoinduced electron transfer (PET) sensors, 229–230
γ-Aminobutyric acid zwitterions, photoinduced electron transfer (PET) sensors for, 229–230
5-Aminocoumarine, as anion sensor, 90, 91
8-Aminoquinoline appended diaza-18-crown-6 ligands, 29, 32
Aminotroponimine ligands, dansyl-containing, Co(II) complexes of, 176–181
Analyte detection, processes in, 3
Anion complexation agents, fluorescent, 59–118
  binding sites, 59
  design, 59
    binding unit-spacer-signalling unit structure, 91
  as dual electrochemical and photochemical sensors, 77–78

Anion complexation agents, fluorescent (*cont.*)
  electron energy transfer and, 60
  electrostatic binding by, 59–60
  highest occupied molecular orbit (HOMO) of, 60
  inorganic, 94–113
    lanthanide-based, 108–113
    rhenium(I)-bipyridyl-based, 94–104
    ruthenium(II)-based, 102, 103
  for interlocked structures, 73
  as logic gates, 72–73
  lowest unoccupied molecular orbit (LUMO) of, 60
  organic, 60–94
    with anthracene signaling units, 66–79
    with biarylthiourea signaling units, 82
    with condensed or conjugated heteroaromatic ring signaling units, 82–89
    with naphthalene signaling units, 60–66
    with pyrene signaling units, 79–81
  photoinduced electron transfer (PET) reactions and, 60, 61
  pH window, 60
  rigidity effect in, 60, 61
  signaling subunits, 59
  solvent effects on, 60
Anions: *see also specific anions*
  as environmental pollutants, 59
  fluorescence-based detection of, 219
  fluorescent redox-switchable device-based detection of, 202–203, 208–211
  functions of, 59
  photoinduced electron transfer (PET) sensors for, 234, 237
  redox-switchable sensors for, 202–203, 208–211
Anisotropy, fluorescence, for zinc ion detection, 46
Anthracene
  as anion sensor signaling unit, 66–79
    covalent bonds-based, 78–79
    electrostatic interactions-based, 66–73
    hydrogen bonding-based, 66–73
    metal coordinative interaction-based, 74–78
  as cation/metal ion sensor, 7, 8–11
    with dioxo tetraamines, 11, 12–13
    excimer formation with, 10
    with polyamine chains, 7, 8–9

Anthracene (cont.)
 fluorescence quantum yield of, 225
 as nitric oxide sensor, 169, 170
 as photoinduced electron transfer (PET) fluorophore, 234, 237
 as polyazamacrocycle receptor luminophore, 21–22
Anthracene-based redox-switchable devices, 194, 196
Anthrylpolyammonium
 disubstituted, as anion sensor, 67, 68
 monosubstituted, as anion sensor, 66–67
Anthrylthiourea, as anion sensor, 70–72
Apoprotein, 46
L-Arginine, nitric oxide synthesis from, 163, 164
Atmosphere, carbon dioxide content measurements in, 120
Azacrown ethers, as anion sensors
 with acridine, 82, 83
 with guanidinium, 69
 with rhenium bipyridyl and ruthenium, 101
Azamacrocycles, as photoinduced electron transfer (PET) sensors, 228–229

2,2-Baipyridine metallic complexes, as anion sensors, 85
BAPTA recognition moiety, 234
Barium ions
 carboxylate-based sensors for, 7
 crown ethers-based sensors for, as electronic tongues, 35
Benzene, sigma-coupled charge-transfer probe fluorescence in, 255, 256, 257, 258
Benzenesulfonamide, *para*-substituted, 46
Benzene sulfonamido groups, of crown ether-based cation sensors, 32, 33
1-Benzothiazol-3-(4-donor)-phenyl-substituted prop-en-1-ones, 35, 36
Benzoxadiazole sulfonamide derivatives, 46
Biarylthiourea, as anion sensor, 82
Bicarbonate, detection in mineral water, 93
Biimidazole diamides, as anion sensors, 84, 85
Binaphthylisothiouronium complex, as anion sensor, 64–65
Biochemical analytes, fluorescence-based detection of, 219
2,3-Biphosphoglycerate, lanthanide-based sensors for, 112
Biraphthyl, as calixarene-based metal ion sensor chromophore, 39

Bis(2-pyridyl-methyl)amine, as photoinduced electron transfer (PET) sensor, 233–234
Block-copolymers, charge-transfer fluorescence phase transition in, 277
BODIPY, as nitric oxide sensor, 169, 170
Boric acids, photoinduced electron transfer (PET) sensors for, 232, 233
Boron-dipyrromethane (DPE) dye, 35
Boronic acid-guanidinium ditopic anion receptor/sensor, 79, 80
Boronic acids
 as anion sensors, effect of KF titration on, 65–66
 as photoinduced electron transfer (PET) sensors, 230, 235–236, 237
 photoinduced electron transfer (PET) sensors for, 232, 233
Brewing industry, carbon dioxide measurement applications in, 119
Bromide, anthracene-based sensors for, 71
Bromothymol blue, as FRET acceptor dye, 131, 133
Butane-1, 4-diammonium ions, photoinduced electron transfer (PET) sensors for, 229
p-tert-Butylcalix[4]-arene, 39–40, 41
di-n-Butylether, sigma-coupled charge-transfer probe fluorescence in, 255, 258, 266, 269

Cadmium ions
 acyclic receptor-based sensors for, 21
  with dansyl groups, 18–19, 20
  with 1,2,4-thiadazole derivative receptor units, 14
 cyclic receptor-based sensors for
  anthracene-based, 21, 22
  5-chloro-8-methoxy quinoline appended diaza-18-crown-6, 29, 31
  coumarin-derivatized crown ethers, 25, 27
  crown ethers-based, 25, 27, 31
  with dansyl groups, 23
 differentiated from zinc ions, 22
Caesium ions: *see* Cesium ions
Caffeine, extraction from coffee beans, 120
Calcium Crimson dye, 235
Calcium Green dye, 235
Calcium ions
 carboxylate-based sensors for, 7

Calcium ions (*cont.*)
  crown ethers-based sensors for, as electronic tongues, 35
  cyclic receptor-based sensors for
    as bi-functional fluorionophores, 36
    dibenzo-crown ethers-based, 27
    tribenzo-crown ethers-based, 27
  photoinduced electron transfer (PET) sensors for, 234, 235
Calcium Orange dye, 235
Calixarene derivatives, as metal ion sensors, 38–42
Calixarenes
  as photoinduced electron transfer (PET)-based sensors, 238, 239
  as rhenium(I)-bipyridyl anion sensors, 102–103
Calixpyrroles, as anion sensors
  with anthracene signaling units, 72
  second-generation, 89, 90
Capnography, for carbon dioxide measurement, 119
  with dry carbon dioxide optical sensors, 139
Carbohydrates, boronic acid-guanidinium ditopic anion sensors for, 79
Carbonate, fluorescent redox-switchable device-based detection of, 209, 210
Carbon dioxide
  as "greenhouse gas," 120
  use in modified atmosphere packaging (MAP), 119–120
Carbon dioxide measurement
  applications of, 119–121
  with gas chromatography, 120–121
  with infra-red spectroscopy, 120
  with Severinghaus electrode, 120–121, 122, 123, 155
Carbon dioxide optical sensors, 119–161
  dry, 137–155
    for dissolved CO_2 detection, 139
    effect of acidic vapors on, 139
    effect of Hofmann β-hydrogen elimination reaction on, 139
    encapsulated ion-paired dye component, 138–140
    encapsulating medium for, 141–143
    equilibrium process in, 138–139
    fluorescein dye system, 141
    fluorescence resonance energy transfer (FRET)-based, 147–151

Carbon dioxide optical sensors (*cont.*)
  dry (*cont.*)
    1-hydroxypyrene-3,6,8-trisulfonic acid dye component, 141, 142, 143–145
    ion-pair technology of, 138–140
    with ruthenium (II) tris[4,4′-diphenyl-1,2,2′-bipyridyl] cation, 142, 143
    with ruthenium (II) tris(4,7-diphenyl-1,10-phenanthroline) cation, 143
    sulforhodamine 101 component, 141
    tetraoctyl ammonium hydroxide-based, 138
    tetra t-butyl ammonium hydroxide-based, 138
    with tributyl phosphate plasticizer, 138–139
    with tris-(thenoyltrifluoroacetonato europium (III)), 141
  dry luminescence intensity-based, 140, 141–142, 143–147
  europium (II) complex-based, 145–147
  FRET-based, 147–151
  dry luminescence lifetime-based, 140, 142–143
  FCO2-R system, 156
  fluorescence resonance energy transfer (FRET)-based
    acceptor-donor combinations of, 131, 133, 135
    applications of, 155–157
    bromothymol blue acceptor, 131, 133
    comparison with europium (II) complex-based sensors, 146
    cost of, 134–135, 150
    dry, 147–151
    dual-luminophore referencing (DLR), 151–155
    with eosin donor, 131, 133
    laser use in, 134
    with phase modulation, 131–133, 149–150
    with phenol red acceptor, 131
    practical systems of, 155–157
    with rhodamine donor, 131, 133
    with ruthenium (dph-bpy)$^{2+}$-(TMS)$_2$/MCPTMDA$^+$/EC film sensor, 150, 151
    with ruthenium (dpp)$_3$2+-(TMS)$_2$/Sudan (III)-TOAH/(silica sol-gel/EC) film sensor, 150, 151

INDEX

Carbon dioxide optical sensors (cont.)
  fluorescence resonance energy transfer
    (FRET)-based (cont.)
    with SR/TB-TOAH/EC film, 147–149, 150
    with Texas Red hydrazine donor, 131, 133
    wet, 146, 147
  luminescence intensity-based, limitations to, 155
  luminescence lifetime-based, limitations to, 155
  1-naphthylmethylamine-based, 137
  OxySense[99] system, 156–157
  pH-sensitive quenchers, 137
  temperature sensitivity of, 155
  wet, 121–137
    acryloyl fluorescein dye system, 124
    basic characteristics of, 121, 122
    c-SNAFLc dye, 125
    c-SNARF dye, 125
    dye deprotonation process in, 128
    effect of carbonic anhydrase on, 136
    effect of osmotic pressure on, 137–138
    effect of water vapor on, 137–138
    encapsulating medium for, 124–126
    eosin dye, 126
    fluorescein dye system, 124
    fluorescence resonance energy transfer
      (FRET)-based, 146
    gas-permeable membrane of, 121, 122
    7-hydroxycoumarin-4-acetic acid dye, 125
    1,hydroxypyrene-3,6,8-trisulfonic acid dye
      system, 123, 124, 125, 126, 127–128
    as hyperbolic response sensors, 129
    Lübbers and Opitz, 128–129
    β-methylumbelliferon dye system, 124, 128–129
    pH-sensitive dyes, 123
    principles of operation of, 122–123
    ptfe/1,hydroxypyrene-3,6,8-trisulfonic
      acid/agarose/sodium bicarbonate, 129, 130
    rhodamine 6 G dye, 126
    ruthenium (II) tris[2-(2-
      pyrazinyl)thiazole], 126, 135–137
    Texas red hydrazine dye, 126
  wet luminescence-intensity, 123–130
    humidity effects on, 128
    ionic strength effects on, 128
    sensor calibration in, 133
    temperature effects on, 128

Carbon dioxide optical sensors (cont.)
  wet luminescence-intensity (cont.)
    wavelength-ratiometric, 134
  wet luminescence lifetime, 126, 130–135
    comparison with luminescence-intensity, 133, 134, 135
    fluorescence resonance energy transfer
      (FRET)-based, 130–135
    phase-modulated, 131–133
  YSI-8500 system, 155–156
Carbonic anhydrase, effect on CO_2 optical sensors, 136
Carbonic anhydrase II, as zinc optical sensor, 46
4-Carboxy-2′,7′-difluorofluorescein
  succinylmidyl ester, 173
5-Carboxyfluoresceine, 92, 93
Carboxylates, as fluorescent sensors
  with anthracene-based signaling units, 79
  as metal ion cations sensors, 7–8
  with zinc(II) complex receptors, 75–76
Cation probes and sensors
  with acyclic receptors, 6–21
    amidic group-based, 12–13
    anthracene-based, 8–11
    anthracene chromophore/polyamine chain-based, 7, 8–9
    anthryl group/-SCH2CH2CH2-based, 10–11
    in aqueous medium, 8
    carboxylates-based, 7–8
    with dansyl groups, 17–20
    dioxo-tetraamines-based, 11, 12–13
    excimer formation with, 10
    exciton-coupled circular dichroism
      (ECLD) spectra of, 17
    with isotropic and anisotropic properties, 17
    luminescence intensity-modulating
      property of, 9–10
    photoinduced electron transfer-based
      fluorescence quenching of, 6–7
    photoinduced electron transfer via
      conformational rearrangement of, 15–16
    polyoxyethylene chains-based, 14–16
    1,2,4-thiadiazole derviatives-based, 14
    use in aqueous medium, 7, 8
  crown ethers-based, 21, 25–35
    with benzene sulfonamido groups, 32, 33

Cation probes and sensors (*cont.*)
   crown ethers-based (*cont.*)
      with 1,8-dioxyanthone residue, 34
      with 4,5-dioxyanthone residue, 34–35
      as electronic tongues, 35
   with cyclic receptors, 21–42
      calixarene derivatives, 38–42
      crown ethers, 21, 25–37
      exciplex emission from, 25
      intramolecular charge transfer absorption of, 26
      polyazamacrocyles, 21–25
   dendrimer-based, 42
   proteins- or polypeptides-based, 42–47
   silica nanoparticles-based, 47–51
Cations: *see also specific cations*
   chemosensors for, 5–6; *see also* Cation probes and sensors
   fluorescence-based detection of, 219
Cavitand structure
   of photoinduced electron transfer (PET) fluorescent sensors, 227, 238, 239
   of rhenium (I) bipyridyl fluorophore-based anion receptors, 104
Cesium ions, fluorescent sensors for, 33
   crown ethers-based sensors for, as electronic tongues, 35
   cyclic receptor-based sensors for
      1,3-alternate calix[4]arene-crown-6 ligands, 41–42
      calixcrowns, 39
Charge-transfer optical probes, sigma-coupled, 249–284
   applications of, 271–281
      as mobility change probes in (semi)crystalline material, 276–277
      penetration of solvents and vapors in polymers, 279–280
      as polymerization probes, 277–279
      as solvation probes in glass-forming solvents, 271–275, 278
      thin-film electroluminescence, 280–281
   donor/acceptor-π-coupled systems, 249, 254
   4-dimethylamine-4′-nitrosostilbene (DMANS), 250, 254
   D site metal ion complexation/protonation-type, 249
   electron donor/acceptor complexes of, 249
      conjugated pi-system-linked, 249
      directly-linked (D/A), 249

Charge-transfer optical probes, sigma-coupled (*cont.*)
   fluorogenic, 267–271
   MFP, 268–270, 277, 278, 279
   MFT, 268–270, 277
   Fluoroprobe (FP), 253–271
      absorption spectra of, 254, 255–256, 258, 267
      charge-transfer absorption improvement in, 263–267
      charge-transfer transfer maxima of, 269
      in PMMA, 278, 279
      comparison with
         fluorogenic derivatives, 269
         FluoroTrope, 265–267
      dark decay rates of, 258, 260–261
      electroluminesence of, 280
      fluorescence maxima of, 255–256, 266
      fluorescence quantum yields and lifetimes of, 258–263, 266
      fluorescence under UV illumination, 256–257
      fluorogenic derivative, with covalent attachments, 267–271
      gas phase of, 263
      in glass-forming solvents, 271–275
      Lippert-Mataga plots for, 256, 263
      maleimide derivatives-based, 268–271
      radiative rates of, 258, 260–261, 262, 264, 265, 266, 267
      radiative spectra of, 267
      solvatochromism of, 255, 256–257, 263
      structure of, 254, 255
      under supersonic jet-cooled conditions, 263
      three-state model, 261
      through-bond-interactions (TBI) in, 262, 264, 265
   Fluoroprobe-like, fluorescence quantum yields and lifetimes of, 258–263
   FluoroTrope (FT)
      charge-transfer absorption spectra of, 265, 267
      comparison with Fluoroprobe, 265–267
      electroluminesence of, 281
      fluorogenic derivative, with covalent attachments, 267–271
      maleimide derivatives-based, 268–271
      radiative rates of, 266
      solvatochromism of, 265
      structure of, 265

Charge-transfer optical probes, sigma-coupled (*cont.*)
  Lippert-Mataga equation for, 273, 274
  with saturated hydrocarbon bonds, 254
  solvatochormic, 249–250
    DMABN, 250
    DMANS, 250
    $E_T(30)$, 249, 250–251
    Z-probe, 249, 250–251
  solvatochromism maximization of, 250–253
    dipole moments in, 250, 251
    ellipsoidal solvent cavity models of, 251, 252
    equation, 251
    Lippert-Mataga plots of, 251
    Mulliken two-state model of, 253, 261
    non-zero electronic donor/receptor coupling (V_{DA}) and, 253
    spherical solvent cavity models of, 251, 252
    Van der Waals surface in, 251, 252
  through-sigma-bond interaction in, 254
Chelator-based photoinduced electron transfer (PET) fluorescent sensors, 227, 231–236
Chemical sensors
  advantages of, 1
  applications of, 1
  fluorescence-based, 2
Chemosensors
  definition of, 2
  fluorogenic ligands of, 5–6
  goal of, 2, 5
  luminescent, 3
    immobile, 4–5
    mobile, 4
  signal transduction mechanisms of, 3–4
  for metal ions (cations), 5–6
  with optical transduction, 5
  signal transduction mechanisms of, 5–6
Chloride ions, fluorescent sensors for, 70
  anthracene-based, 71
  naphthylurea-based sensors, 63
  photoinduced electron transfer (PET) sensors, 234
Chloroform, sigma-coupled charge-transfer probe fluorescence in, 255, 257, 258
5-Chloro-8-methoxy quinoline appended diaza-18-crown-6, as cadmium ion sensor, 21, 29

Chromium ions, luminescence intensity-enhancing property of, 6
Citrate
  as commercial beverage component, 92
  fluorescent sensors for, 67
    fluoresceine-based, 92
    phenanthroline-based, 86
L-Citrulline, nitric oxide synthesis from, 164
Cleft-type mixed metals, as anion receptors/sensors, 98, 99
Cobalt complexes, as nitric oxide sensors, 176–181
  with dansyl aminotroponimine ligand scaffold, 176–181
  FATI sensors, 180–181
  SATI sensors, 176–179
Cobalt ions
  dansyl group-based acyclic receptor-based sensors for, 18–19, 20
  fluorescence anistropic detection of, 46
  heteroditopic cryptand cyclic receptor-based sensors for, 37, 38
  luminescence intensity-enhancing property of, 6
  silica nanoparticle sensors for, 51
Coffee, decaffeinated, 120
Consensus peptide (CP), 45–46
Copolymers, charge-transfer fluorescence phase transition in, 277
Copper ions, fluorescent sensors for
  acyclic receptor-based
    anthracene-based, 12–13
    carboxylates-based, 7
    with dansyl groups, 18–19
    dioxo-tetraamines-based, 11, 12–13
    photoinduced electron transfer-based fluorescence quenching of, 6
    at sub-ppm levels, 7
    with 1,2,4-thiadazole derivative receptor units, 14
  cyclic receptor-based, 35
    anthracene-based with polyamine receptors, 8
    1-benzothiazol-3-(4-donor)-phenyl-substituted prop-en-1-ones-based sensors, 35, 36
    with dansyl groups, 23, 24–25
    dioxocyclam-based, 24
    heteroditopic cryptand, 37, 38
    triazacyclononane, 24–25

Copper ions, fluorescent sensors for (cont.)
  dendrimer-based sensors, 42
  fluorescence anistropic detection of, 46
  luminescence intensity-enhancing property of, 6
  photoinduced electron transfer (PET) sensors, 223
  polypeptides-based sensors, 42–44, 45
  silica nanoparticle sensors, 51
  tripodal ligands-based sensors, 17
Coronard (crown)-based photoinduced electron transfer (PET) fluorescent sensors, 227–230
Coumarin, 236
  as nitric oxide sensor, 169, 170
Coumarin 153, 25, 26–27
Coumarine, as anion receptor-based sensor, 90, 91, 93, 94
Crown ethers
  as alkali or alkaline earth metal ion sensors, 24, 25–35
    comparison with calixarene-based sensors, 39–40
    coumarin-derivatized, 25, 27
    dibenzo-crown ethers, 26, 27
    8-hydroxyquinoline-derivatized, 27–28
    tribenzo-crown ethers, 26, 27
  as anion sensors, with rhenium bipyridyl and ruthenium fluorophores, 101–102
  as cation/metal ion sensors, 21, 25–35
    for alkali or alkaline earth metal ions, 24, 25–35
    with 8-aminoquinoline appended diaza-18-crown-6 ligands, 29, 32
    anthracocryptand, 25
    with benzene sulphonamido group, 32, 33
    as bi-functional fluorionophores, 36
    coumarin 153 derivatized, 26–27
    with 1,8-dioxyanthone residue, 34
    with 4,5-dioxyanthone residue, 34–35
    as electronic tongues, 35
    exciplex formation with, 25
  metal ion selectivity of, 25
Cryptand structure
  heteroditopic, for transition metal ions, 37–38
  of photoinduced electron transfer (PET) sensors, 227, 236–237, 238
  of zinc-complex anion sensors, 74, 75
c-SNAFLc dye, 125

c-SNAFL dye, structure of, 127
c-SNARF dye, 125
Cyanoanthracene, 43, 241
Cyclen, acridine-appended, 86
Cyclen-based anion receptors, 110–111
  with hydrogen carbonate, 111
  with phenanthridium, 109
  terbium (III) complexes of, 110
Cyclen macrocycle, as cyclic receptor-based cation sensor, 22–23
Cyclic receptor-based cation probes and sensors, 21–42
  for alkali and alkaline metal ions, 24, 25–35
    1-benzothiazol-3-(-4-donor)phenyl-substituted prop-en-1-ones-based, 35, 36
    bifunctional, 36
    calixarene derivatives-based, 38–42
    crown ethers-based, 24, 25–35
  for cadmium ions
    anthracene-based, 21, 22
    5-chloro-8-methoxy quinoline-appended diaza-18-crown 6, 29, 31
    coumarin-derivatized crown ethers, 25, 27, 31
    with dansyl groups, 23
  for calcium ions
    as bifunctional fluorionophores, 36
    dibenzo-crown ethers-based, 27
    tribenzo-crown ethers-based, 27
  for copper ions, 35
    1-benzothaizol-3-(-4-donor)phenyl-substituted prop-en-1-ones-based, 35, 36
    with dansyl groups, 23, 24–25
    dioxocyclam-based, 24
    heteroditopic cryptand-based, 37, 38
    triazacyclononane-based, 24–25
  crown ethers as, 21, 24, 25–35
  intramolecular charge transfer absorption of, 26
  for nickel ions, 8
    coumarin-derivatized crown ethers-based, 25, 27
    crown ethers-based, 25, 27
    dioxocyclam-based, 24
    heteroditopic cryptand, 37, 38
    polyazamacrocyles, 21–25
  for potassium ions
    crown ethers as, 27, 33, 35

Cyclic receptor-based cation probes and
    sensors (cont.)
  for potassium ions (cont.)
    dibenzo-crown ethers as, 27
    as electronic tongues, 35
  for zinc ions
    anthracene-based, 21, 22
    carboxylates-based, 7
    coumarin-derivatized crown ethers-based,
      25, 27
    crown ethers-based, 25, 27, 32, 33
    with dansyl groups, 22–23
    heteroditopic, 37–38
β-Cyclodextrin, 242–243
  with copper complexes, 42
  dansyl-substituted, as amino acids sensor, 65
γ-Cyclodextrin-pyridinium, as anion sensor, 66
γ-Cyclodextrins, as anion sensors, 66
Cyclohexane
  absorption spectrum of, 265
  sigma-coupled charge-transfer probe
    fluorescence in, 255, 256–257, 258,
    266, 269
trans-1,4-Cyclohexanedicarboxylate, 97
Cyclophane-based anion receptors, 105, 106
Cytochrome c', 173–174, 183

Dacda receptors, 113
DAF-2, as nitric oxide sensor, 168, 169–171
DAF-2 diacetate, as nitric oxide sensor, 169–
    170
Dansyl
  as acyclic receptor-based sensor component,
    17–20
  as cyclic receptor-based sensor component, 22
  of polyaminic receptors, 24–25
  β-cyclodextrin-appended, 42
  definition of, 44
  luminescence bands of, 17–18
  as polypeptide-based metal ion sensor
    component, 44
  as silica nanoparticle modifier, 47–48, 50, 51
Dendrimer receptor-based sensors, 42, 43, 44
Deoxyadenosine, 86
Deoxyguanosine, 86
Deoxythimidine, 86
Deoxytidine, 86
α,ω-Diacetyl polyethylene glycols, charge-
    transfer fluorescence phase transition
    in, 276–277

o-Diamine-based compounds, as nitric oxide
    sensors, 166–171
  fluorophore systems for, 169, 170
  with diamine fluorescein platform
    systems, 167–169
5,6-Diamine-1,3-naphthalene disulfonic acid,
    169
1,2-Diaminoanthraquinone, 169
Diaminofluorescein-based nitric oxide sensors,
    167–169
2,3-Diaminonaphthalene (DAN), 166–167, 183
Diaminorhodiamine-based nitric oxide sensors,
    169
Diazacrown ether-based photoinduced electron
    transfer (PET) sensors, 227–228,
    239, 240
4,9-Diazapyrenium derivatives, as nucleotide
    sensors, 87–88
Diboronic acids, as photoinduced electron
    transfer (PET) sensors, 235–236, 237
Dicarboxylate, stereoselective receptors for, 97
Dichloromethane
  as polymer swelling agent, 279–280
  sigma-coupled charge-transfer probe
    fluorescence in, 255, 258, 266, 269
Dichlororfluorescin, as nitric oxide sensor, 171,
    174
Diethanolamine, as photoinduced electron
    transfer (PET) sensor, 232, 233
Diethylether, sigma-coupled charge-transfer
    probe fluorescence in, 255, 258, 266,
    269
3,3'-Diethyloxacarbocyanine iodide-loaded
    microspheres, 289, 290, 297, 298,
    299, 300
  axis length of, 294, 295
  emission spectra of, 290, 291
  fluorescence microscopy imaging of, 294–295
  pressure measurement accuracy assessment
    of, 296, 297–298, 299, 300
  scanning electron microscopy imaging of,
    292–296
  temperature-dependent photoluminescence
    of, 292
Dihydrogen phosphate
  fluorescent sensors for, 70
  photoinduced electron transfer (PET) sensors
    for, 234
1,2-Dihydroxybenzenes, fluorescent redox-
    active system detection of, 197

Diisopropylether, sigma-coupled charge-transfer probe fluorescence in, 255, 258
1,2-diMeO-ethane, sigma-coupled charge-transfer probe fluorescence in, 258
9,10-Dimethoxyanthracene, fluorescence quantum yield of, 225
L-3,4-Dimethoxy-DOPA, 43
1,2-Dimethoxyethane, sigma-coupled charge-transfer probe fluorescence in, 255
4-Dimethylamine-4'-nitrosostilbene π-coupled charge-transfer probe, 250, 254
Dimethylaminobenzoate-guanidinium, as anion receptor-based sensor, 91
p-N,N-Dimethylaminobenzoic acid, lanthanide-based sensor for, 110
N-(4-Dimethylaminophenyl)maleimide, 270, 271
N,N-Dimethylaniline, 75
9,10-Dimethylanthracene, fluorescence quantum yield of, 225
Dimethyl phosphate, fluorescent sensors for, 67
Diols, photoinduced electron transfer (PET) sensors for, 230, 232
1,4-Dioxane, sigma-coupled charge-transfer probe fluorescence in, 255, 256, 257, 258
Dioxcyclam, as cyclic metal ion receptor, 24
Dioxo-tetraamines, as nickel and copper ion receptors/sensors, 11, 12–13
1,8-Dioxyanthone residue, 34–35
  complexed with crown ethers, 34
4,5-Dioxyanthone residue, 34–35
9,10-Diphenylanthracene, fluorescence quantum yield of, 225
4,4-Diphenyl-2,2-dipyridyl, structure of, 140
4,7-Diphenyl-1,10-phenanthroline, structure of, 140
Dipicolylamine, as photoinduced electron transfer (PET) sensor, 232, 235
2,3-Dipyridine-2-ylquinoxaline, as anion receptor/sensor, 83
Dipyrrolylquinoxalines, as anion sensors, 83, 104
Dirhodium tetracarboxylate-based nitric oxide sensors, 181–184
Distyrylbenzene units, synthesized as Z-Z isomer, 33
DNA (deoxyribonucleic acid), 59
Dual-luminophore referencing (DLR), as carbon dioxide measurement method, 151–155

Electroluminescence, of charge-transfer probes, 280–281
Electronic tongues
  for alkali and alkaline earth metals, 35
  for transition metals, 23–24
Electron transfer fluorescent sensors: *see also* Photoinduced electron transfer (PET) fluorescent sensors
Endothelial cells, nitric oxide detection in, 170
Energy transfer, in photoinduced electron transfer (PET) sensing, 240
Environmental monitoring, carbon dioxide measurement applications in, 120
Eosin dye
  as carbon dioxide indicator dye, 126
  as fluorescence resonance energy transfer (FRET) donor dye, 131, 133
Eosine Y, selective binding with pyrophosphate, 94
Ethylacetate, sigma-coupled charge-transfer probe fluorescence in, 255, 258, 266, 269
Ethylstearate, charge-transfer fluorescence phase transition in, 276
Europium complexes
  of carbon dioxide dry optical sensors, 145–147
  of lanthanide-based anion sensors, 111–113
    with coordinated nitrate anions, 111
    cyclen-based hydrogencarbonate, 111
  of phenanthridium-based anion sensors, 109
Eutrophication, 59
Eutropium, photoinduced electron transfer (PET) sensors for, 240
Excimer formation
  with anthracene-based cation sensors, 10
  with photoinduced electron transfer (PET) sensors, 239–240
  with redox-switchable fluorescent devices, 199

FATI sensors, 180–181
Fermentation, carbon dioxide monitoring during, 119, 134
Ferrocene, as anion sensor, 77–78
Ferrocene-based redox-switchable devices, 190
Ferrocene/ferrocene redox couple-based redox-switchable devices, 195
Ferrocene-naphthalimide-based redox-switchable devices, 195

INDEX

Ferrocenium/ferrocene redox couple redox control units, 192
Fiber-optic sensors, for nitric oxide, 173–174
Fluorescein
  as photoinduced electron transfer (PET) sensor, 233–234
  as silica nanoparticle modifier, 49–50
  as zinc sensor platform, 16
Fluorescein dye system, of carbon dioxide optical sensors, 124
Fluorescence quantum yield
  of fluorescent molecules, 225, 226
  of naphthylalkyamines, 226
Fluorescence resonance energy transfer (FRET)-based carbon dioxide optical sensors, 130–135
  acceptor-donor combinations of, 131, 133, 135
  applications of, 155–157
  bromothymol blue acceptor, 131, 133
  comparison with europium (II) complex-based sensors, 146
  cost of, 134–135, 150
  dry, 147–151
    with phase modulation, 149–150
    with SR/TB-TOAH/EC film, 147–149, 150
  dual-luminophore referencing (DLR), 151–155
  with eosin donor, 131, 133
  laser use in, 134
  phase-modulation technique in, 131–133
  with phenol Red acceptor, 131
  practical systems of, 155–157
  with rhodamine donor, 131, 133
  with ruthenium (dph-bpy)$^{2+}$-(TMS)$_2$/MCPTMDA$^+$/EC film sensor, 150, 151
  with ruthenium (dpp)$_3$2+-(TMS)$_2$/Sudan (III)-TOAH/(silica sol-gel/EC) film sensor, 150, 151
  with Texas Red hydrazine donor, 131, 133
  wet, 146, 147
Fluorescent molecules, fluorescence quantum yield of, 225, 226
Fluorescent nitric oxide cheletropic traps (FNOCTs), 172–173, 183
"Fluorescent nose," 280
Fluoride, fluorescent sensors for
  with anthracene signaling units, 71
  naphthylurea-based, 63

Fluoroacryclic polymer binder (FIB), for pressure sensitive paint, 300–301
Fluoro-acrylate co-polymer, as pressure sensitive paint binder, 301
Fluorometry, redox, 190
Fluorophore-copper complex ensemble, 93, 94
Fluorophore-nitroxide (fluorescence-spin molecular)-based redox-switchable devices, 197–198
Fluorophore-nitroxide-type fluorosensor, 197, 198
Fluoroprobe (FP), 253–271
  absorption spectra of, 254, 255–256, 258, 267
  charge-transfer absorption improvement in, 263–267
  charge-transfer fluorescence maxima of, 269
  in PMMA, 278, 279
  comparison with
    fluorogenic derivatives, 269
    FluoroTrope, 265–267
  dark decay rates of, 258, 260–261
  electroluminesence of, 280
  fluorescence maxima of, 255–256, 266
  fluorescence quantum yields and lifetimes of, 258–263, 266
  fluorescence under UV illumination, 256–257
  fluorogenic derivative, with covalent attachments, 267–271
  gas phase of, 263
  Lippert-Mataga plots for, 256, 263
  maleimide derivatives-based, 268–271
  radiative rates of, 258, 260–261, 262, 264, 265, 266, 267
  radiative spectra of, 267
  solvatochromism of, 255, 256–257, 263
  structure of, 254, 255
  under supersonic jet-cooled conditions, 263
  three-state model, 261
  through-bond-interactions (TBI) in, 262, 264, 265
FluoroTrope (FT)
  charge-transfer absorption spectra of, 265, 267
  comparison with
    fluorogenic derivatives, 269
    Fluoroprobe, 265–267
  electroluminesence of, 281
  fluorogenic derivative, with covalent attachments, 267–271

FluoroTrope (FT) (*cont.*)
  maleimide derivatives-based, 268–271
  radiative rates of, 266, 267
  solvatochromism of, 265
  structure of, 265
Food packaging industry, carbon dioxide measurement applications in, 119–120
FRET: *see* Fluorescence resonance energy transfer (FRET)-based carbon dioxide optical sensors

D-Galaturonic acid, fluorescent sensors for, 87
Gas chromatography, carbon dioxide measurement applications of, 120–121
Gases, fluorescence-based detection of, 219
Global warming, 120
D-Glucarate, boronic acid-guanidinium ditopic anion sensor for, 79
D-Gluconate, boronic acid-guanidinium ditopic anion sensor for, 79
D-Glucose, boronic acid-guanidinium ditopic anion sensor for, 79
D-Glucuronic acid, boronic acid-guanidinium ditopic anion sensor for, 79
D-Glucuronic acid, fluorescent sensors for, 87
Glutarate, anthrylthiourea receptor/sensor for, 71
Gold colloid-coated optical fibers, 173
"Greenhouse effect," 120
Griess assay, 164–165
Guanidinium, 59
Guanidinium-carboxyfluoresceine, tripodal, 92, 93
Guanosine monophosphate, anthracene-based sensors for, 76
Guanylate cyclase, soluble, heme domain of, 173–174, 183

Heavy metals: *see also specific heavy metals*
  redox-switchable fluorosensor-based detection of, 199–208, 211
  Cd^{II} complex-based, 205–206
  Cu^{II} complex-based, 207, 208
  Hg^{III} complex-based, 205–207, 208
  $1,6,6\lambda^4$-trithia-3,4-diapentadene-based, 203–208
Henry's law, 286–287
Heparin, fluorescent sensors for, 67
Heparinase, fluorescent sensors for, 67
Heteroaromatic rings, as anion recognition signaling units, 82–89
Heterometallic cyclophane anion receptors, 105, 106
n-Hexane, sigma-coupled charge-transfer probe fluorescence in, 255, 258, 266, 269
Histidine, fluorophore-copper complex-based sensors for, 94
Holoprotein, 46
HPTS/hexadecyl trimethyl ammonium hydroxide/RTV-silicone rubber, 144–145
H^R DATI ligand-based nitric oxide sensors, 176–181
Huntington's disease, nitric oxide in, 163–164
Hydrocarbonated compounds, as anion sensors, 82
Hydrogen bonding, in anion receptor binding, 61, 63
Hydrogen carbonate, fluorescent redox-switchable device-based detection of, 209, 210
Hydrogen phosphate
  fluorescent redox-switchable device-based detection of, 209, 210
  photoinduced electron transfer (PET) sensors for, 234
Hydroquinone, as photoinduced electron transfer (PET) sensor, 230
Hydrosphere, carbon dioxide measurements in, 120
7-Hydroxycoumarin-4-acetic acid dye, 125
1-Hydroxypyrene-3,6,8-trisulfonate
  as carbon dioxide optical sensor dye, 123, 124, 125, 127–128
  photochemical and chemical characteristics of, 126
1-Hydroxypyrene-3,6,8-trisulfonic acid
  as carbon dioxide optical sensor dye, 123, 124, 125, 126, 127–128
  structure of, 127
1-Hydroxypyrene-3,6,8-trisulfonic acid/tetra octyl ammonium hydroxide/ethyl cellulose/tributyl phosphate sensor film, 143–144, 145

Imidazolate, as zinc receptor/sensor, 74
Iminoyl thiourea
  as anion sensor, 62
  as cation sensor, 14

INDEX

Inositol-1,4,5-triphosphate, 93
International Union of Pure and Applied Chemistry (IUPAC), sensor definition of, 1, 2
Iridium (III)-based anion receptors, 107–108
Iron complex-based nitric oxide probes, 174–176
Iron (II) dithiocarbamate complex, 175
Iron (II) methoxycoumarinomethyl cyclam scaffold, 175, 176
Iron (II) quinolone pendant cyclam complex, 174, 175
Iron ions, luminescence intensity-enhancing property of, 6
Isothiuronium groups, as anion sensors, 63
Isothiuronium salts, as photoinduced electron transfer (PET) sensor, 234, 237

Lanthanide-based anion sensors, 108–113
Lasers, use in fluorescence resonance energy transfer (FRET) systems, 134
Lead ions
  acyclic receptor-based sensors for, with 1,2,4-thiadazole derivative receptor units, 14
  cyclic receptor-based sensors for
    calixarene derivatives-based,
      calixarene derivatives-based, 40–42
    coumarin-derivatized crown ether-based, 25, 27
    crown ethers-based, 25, 27
    heteroditopic cryptand, 37
Light-emitting diodes (LEDs), 280–281
Lithium, photoinduced electron transfer (PET) sensor for, 228
Logic gates
  ditopic ion-pair receptors as, 72–73
  photoinduced electron transfer (PET) sensors as, 241–243
    AND gates, 241, 242
    INHIBIT gates, 242–243
    NOT gates, 242
    OR gates, 241, 242
Long-term potentiation, nitric oxide in, 164–165
Luminescence, as sensing and switching signaling feature, 189
Luminescence-intensity carbon dioxide optical sensors
  dry, 140, 141–142, 143–147
    europium (II) complex-based, 145–147

Luminescence-intensity carbon dioxide optical sensors (cont.)
  dry (cont.)
    fluorescence resonance energy transfer (FRET)-based, 147–151
  wet, 123–130
    humidity effects on, 128
    ionic strength effects on, 128
    sensor calibration of, 133
    wavelength-radiometric, 134
Luminescence-lifetime carbon dioxide optical sensors, 126, 130–135
  comparison with luminescence-intensity sensors, 133, 134, 135
  fluorescence resonance energy transfer (FRET)-based, 130–135
  phase-modulated, 131–133
Luminophores: *see also specific luminophores*
  photophysical modulation of, 2

Macma receptors, 113
Macroacyclic anion receptors, with rhenium (I)-bipyridyl subunit, 97, 99–102, 105
Magnesium ions
  acyclic receptor-based sensors for, 21
  crown ether-based cyclic receptor sensors for, 26, 28
  photoinduced electron transfer (PET) sensors for, 234, 235
Maleimide groups, as fluorogenic charge-transfer probe donor moiety, 267–271
  copolymerization with acrylic monomers, 268, 277–278
Malonate, anthrylthiourea-based sensor for, 71
Manganese ions, acyclic receptor-based sensors for, at 10-110 ppm levels, 7
Marcus law, 16
Mercury ions
  acyclic receptor-based sensors for, 21
    with 1,2,4-thiadazole derivative receptor units, 14
    1,2,4-thiadiazole, 14
  cyclic receptor-based sensors for, 29–30, 35
    anthracene-based, 21–22
    1-benzothiazol-3-(4-donor)-phenyl-substituted prop-en-1-ones, 35, 36
    with dansyl groups, 23
    with phenoxazene scaffold, 36–37
Metabolism, redox-active fluorescent system analysis of, 196

Metal ions: *see also pecific metal ions*
  fluorescence-based detection of, 219
Metal-ligand interactions, of anthracene-based anion sensors, 74–78
9-Methoxyanthracene, fluorescence quantum yield of, 225
6-Methoxy-(8-*p*-toluenesulfonamido)quinoline (TSQ), 32
9-Methylanthracene, fluorescence quantum yield of, 225
Methylcyclohexane, 273, 274, 278
Methylmethacrylate polymerization, charge-transfer probe fluorescence during, 277–278
1-Methylnaphthalene, fluorescence quantum yield of, 225
2-Methylnaphthalene, fluorescence quantum yield of, 225
2-Methyltetrahydrofuran, 273, 274, 275
β-Methylumbelliferon, structure of, 127
β-Methylumbelliferon dye system, of carbon dioxide optical sensors, 124, 128–129
MFP fluorogenic charge-transfer probe, 268–270, 277, 278, 279
MFT fluorogenic charge-transfer probes, 268–271, 277
Microscopy, mobile sensor molecules in, 4
Mineral water, bicarbonate detection in, 93
Modified atmosphere packaging (MAP), 119–120
Molecular recognition, supramolecular, 219
Monoazacrown ether, as photoinduced electron transfer (PET) sensor, 227, 228
Multiple sclerosis, nitric oxide in, 163–164

Naphthalene
  as anion sensor signaling unit, 60–66
    γ-cyclodextrin-pyridium complex-based, 66
    dative-covalent bond-based, 65–66
    electrostatic interaction and/or hydrogen binding-based, 59, 60–65
    metal coordinative interaction-based, 65
  as calixarene-based metal ion sensor chromophore, 39
  fluorescence quantum yield of, 225
  as photoinduced electron transfer fluorophore, 225
2,3-Naphthotriazole, 166
Naphthylalkylamines, fluorescence quantum yield of, 226

Naphthylthioureas, as anion sensors, 61
  hydroxymethyl-substituted, 62, 63
Naphthylurea/thiourea, polysubstituted, as anionic nucleotide sensor, 63, 64
Neurodegenerative diseases, nitric oxide in, 163–164
Neutral molecules
  fluorescence-based detection of, 219
  photoinduced electron transfer (PET) sensors for, 230
Nickel ions
  acyclic receptor-based sensors for, 7
    anthracene-based, 8, 12–13
    with anthracene chromophore and polyamine receptors, 8
    with dioxo-tetraamine ligands, 11, 12–13
    at 10-110 ppm levels, 7
  cyclic receptor-based sensors for
    coumarin-derivatized crown ethers-based, 25, 27
    crown ethers-based, 25, 27
    dioxocyclam-based, 24
    heteroditopic cryptand, 37, 38
  dendrimer-based sensors for, 42
  photoinduced electron transfer (PET) sensors for, 223, 238
  silica nanoparticle sensors for, 51
Nile Red family, 36–37
Nitrate, as environmental pollutant, 59
Nitric oxide
  definition of, 163
  fluorescence-based detection of, 163–188
  functions of, 164
  in long-term potentiation, 164–165
  non-fluorescent-based detection of, 164–165
  properties of, 163, 164
  as retrograde neurotransmitter, 164, 165
  synthesis of, 163, 164
Nitric oxide sensors
  calcium ion-based, 165
  cheletropic trap-based, 172–173, 183
  o-diamine-based, 167–169, 170, 183
  diaminofluorescein platform-based, 167–169
  2,3-diaminonaphthalene (DAN)-based, 166–167, 183
  diaminorhodiamine-based, 169
  dichlororfluorescein-based, 171, 174
  dirhodium tetracarboxylate-based, 181–184
  fiber-optic systems, 173–174
  fluorophore systems, 169, 170

INDEX

Nitric oxide sensors (cont.)
  non-reversible, 173
  reversible, 173–174, 181–183
  rhodamine B hydrazide-based, 171–172, 183
  transitional metals-based, 173
    cobalt complexes, 176–181, 183–184
    iron complexes, 174–176
  Znpyr Zn^{2-}-based, 165
Nitric oxide synthase
  endothelial, 163
  inducible, 163–164
  neuronal, 163, 164
Nitrogen species, redox-active fluorescent system detection of, 196
4-(N,N-dimethylamino)benzoate, 91
Nucleobases, imide-containing, fluorescent sensors for, 86
Nucleoside triphosphates, anthracene-based sensors for, 77
Nucleotides
  anionic, 63, 64
  anthracene-based sensors for, 76
  4,9-diazapyrenium derivatives-based sensors for, 87–88

Oceans, carbon dioxide content measurements of, 120
Octadecyl octadecanoate, charge-transfer fluorescence phase transition in, 276
OLEDs (organic light-emitting diodes), 280, 281
Optical fibers, 5
  gold-coated, 173
Organic light-emitting diodes (OLEDs), 280, 281
Orotic acid (vitamin B13), 76
Oxalate, 76
OxySense[99] system, 156–157

Packaging, modified atmosphere (MAP), 119–120
Palladium, 105, 106
Parkinson's disease, nitric oxide in, 163–164
Pentane-1,4-diammonium ions, 236
di-n-Pentylether, sigma-coupled charge-transfer probe fluorescence in, 266
Peptides, phosphorylated
  disubstituted anthryl-zinc(II) complex sensors for, 77
  photoinduced electron transfer (PET) sensors for, 232, 235

Perchlorate anion, 105
Pertechnetate, 59
PET sensors: see Photoinduced electron transfer (PET) fluorescent sensors
Phenanthridinium-based anion sensors, 109
Phenanthridinium cyclophanes, as nucleotide sensors, 88, 89
Phenanthridinium ions, N-alkylated, 109
Phenanthroline
  as anion sensors, 86, 87
  as rhenium (I)-bipyridyl ligand, 98
Phenol derivatives, as anion sensors, 89, 90
Phenoxazene scaffold, of cyclic receptor-based mercury sensor, 36–37
9-Phenylanthracene, fluorescence quantum yield of, 225
Phosphate
  anthracene derivative-based sensors for, 77
  as environmental pollutant, 59
  pyrene-based sensors for, 80–81
Phosphodiesters, fluorescent sensors for, 98
Photoinduced electron transfer (PET) fluorescent sensors, 3, 219–247
  anthracene-based acyclic cation, 8, 9
  basic concept of, 220
  classification of, 227–230
    cavitand-based, 227, 238, 239
    chelator-based, 227, 231–236
    coronard (crown)-based, 227–230
    cryptand-based, 227, 236–237, 238
    energy transfer-related, 240
    excimer formation-related, 239–240
    podand-based, 227, 231–236
    polymer-supported, 238, 239
  design of, 225–227
    criteria for, 225–226
    fluorophore unit, 225, 226, 227
    receptor unit, 225, 226–227
    spacer unit, 225, 226, 227
  fluorescence quenching and, 220–221
  as logic gates, 241–243
    AND gates, 241, 242
    INHIBIT gates, 242–243
    NOT gates, 242
    OR gates, 241, 242
  metal-triggered metal-centered emission with, 240
  with nonmacrocyclic receptors, 231, 232
  OFF-ON-OFF switches, 224–225, 232
  OFF-ON switches, 220–222, 226, 231, 238

Photoinduced electron transfer (PET)
  fluorescent sensors (cont.)
  ON-OFF switches
    chelation-enhanced quenching with, 222
  ON-OFF switches, 222–223, 228, 230, 234, 238
  pH-sensing, 224, 231, 232
  principles of, 220–225
  redox-active, 222, 223
  Zn^{2+}-induced folding, 223
Photosynthesis, 119
Phthalic acid, 112
Picolinic acid, 112–113
Platinum (II) *meso*-tetrakis(pentafluorophenyl)-porphine, as pressure sensitive paint luminophore, 289, 292–295, 296, 297, 298, 299, 300
  correlation with fluoro-acrylate co-polymer binders, 301
  emission spectra of, 289–290, 291
  with fluoroacryclic polymer binders (FIBs), 300–301
  with inorganic phosphor, 300
  oxygen-quenching efficiency of, 290
  pressure measurement accuracy assessment of, 296, 297–299, 300
  with silicon octaethyl/porphine (SIOEP) luminophore, 300, 301
  with sol-gel-derived silica films, 301
  temperature-dependent photoluminescence of, 291, 296, 298–299, 300
Platinum (II) porphyrin, as pressure sensitive paint luminophore, 285
Podand-based photoinduced electron transfer (PET) fluorescent sensors, 227, 231–236
Polyamide anion receptors, 80–81
Polyamines
  as anthracene-based cation sensor components, 7, 8
  anthryl linear, as anion sensor components, 67–68
  as cation receptors, 8
  as cyclic receptors, combinatorial approach to, 24–25
  as silica nanoparticle receptors, 50, 51
Polyammonium, cationic, 59
Polyanthracene, as amino acids sensor, 75
Polyazamacrocycles
  as cation sensors, 21–25

Polyazamacrocycles (*cont.*)
  as photoinduced electron transfer (PET) sensors, 228–229
Polybutylene terephthalate, charge-transfer fluorescence phase transition in, 277
Polydimethyl siloxane, vinyl-terminated, 289, 290, 291, 292, 294, 295, 296, 297, 298, 299, 300
Polyethylene glycol, charge-transfer fluorescence phase transition in, 277
Poly(isobutylmethacrylate)-*co*-trifluoroethylmethylate, as pressure sensitive paint binder, 301
Poly-L-glutamate, fluorescent sensors for, 67
Polymeric anion receptors, 78
Polymerization processes, charge-transfer probe fluorescence of, 277–279
Polymer light-emitting diodes (PLEDs), 280–281
Polymers, solvent and vapor penetration of, Fluoroprobe monitoring of, 279–280
Polymer-supported photoinduced electron transfer (PET) fluorescent sensors, 238, 239
Polymethylmethacrylate (PMMA), charge-transfer probe fluorescence during, 277, 278
Polynucleotides, sensors for, 88
Polyoxyethylene chains, derivatized, as cation sensors, 14–16
Polypeptides, as cation probes and sensors, 42–47
Polypyrenylthiourea, as anion sensor, 80
Polypyridine ruthenium (II) complex, as pressure sensitive paint luminophore, 285
Polystyrene latex, penetration by dichloromethane, 279–280
Polytetrafluoroethylene (PTFE)/1,hydroxypyrene-3,6,8-trisulfonic acid/sodium carbonate, 135, 138
Polyvinylcarbazole, as light-emitting diode polymer matrix, 281
Porphyrin, as photoinduced electron transfer (PET) sensor, 230
Porphyrin-based redox-switchable devices, 194–195
Potassium fluoride, titrated, effect on boronic acid fluorophores, 65–66

INDEX

Potassium ions, cyclic receptor-based sensors for
  crown ethers as, 27, 33, 35
  dibenzo-crown ethers as, 27
  as electronic tongues, 35
Pressure sensitive paint, dual-luminophore, 285–305, 288–301
  applications of, 285
  catalyst solution for, 289
  definition of, 285
  3,3'-diethyloxacarbocyanine iodide-absorbed microsphere component, 289, 291, 292–295, 296, 297, 298, 299, 300
  dyed microsphere component, 289, 290
  emission spectra of, 289–291
  fluorescence microscopy of, 294–295, 300
  fluoro-acrylate co-polymer binder of, 301
  global temperature mapping with, 288–289
  long-lifetime luminophore, 285, 286
  low-oxygen diffusion activation barrier polymer binder of, 300
  luminescence intensity of, 286
  nanometer-to-micrometer-sized capsules component, 289
  oxygen-based luminescence quenching mechanism of, 286–288
    Stern-Volmer plots for, 287, 288
  platinum (II) *meso*-tetrakis-(pentafluorophenyl)porphine luminophore, 289–292, 294, 296–301
    correlation with fluoro-acrylate co-polymer binder, 301
    with fluoroacryclic polymer binder (FIB), 300–301
    with inorganic phosphor, 300
    with poly(isobutylmethacrylate)-*co*-trifluoroethylmethylate binder, 301
    with poly(styrene-co-pentafluorostyrene) binders, 301
    with poly(styrene-co-trifluoroethyl methacrylate) binders, 301
    with silicon octaethyl/porphine (SIOEP) luminophore, 300, 301
    with sol-gel-derived silica films, 301
  polymer binders of, 285, 286, 289
  preparation and application of, 289, 290
  pressure measurement accuracy of, 296–300
    with full-field, temperature-corrected images, 299–300
    intensity-ratio results, 297–299

Pressure sensitive paint, dual-luminophore *(cont.)*
  primer layer of, 285, 286
  rationale for, 288
  with rhodamine B luminophore of, 301
  scanning electron microscopy of, 292–294, 300
  second layer of, 285, 286
  Stern-Volmer plots for, 291, 292
  temperature effects on, 288–289
  typical formation of, 285, 286
  vinyl-terminated polydimethylsiloxane polymer binder of, 289, 290, 291, 292, 294, 295, 296, 297, 298, 299, 300
Proline, cyclic receptor-based sensor for, 42
Pronase, fluorescent sensors for, 67
di-n-Propylether, sigma-coupled charge-transfer probe fluorescence in, 266
Proteins
  as cation probes and sensors, 42–47
  as divalent metal ion receptors, 42–47
Pseudorotaxane, anion-switched, 73
Pt TFPP: *see* Platinum (II) *meso*-tetrakis(pentafluorophenyl)porphine
Pyrene
  as anion sensor signaling unit, 79–81
    electrostatic interactions-based, 79–81
    hydrogen bonding-based, 79–81
    hydrophobic effect-based, 81
  fluorescence quantum yield of, 225
  as metal ion cation sensor, 7
  as photoinduced electron transfer (PET) fluorophore, 225
Pyrenium, as nucleotide sensor, 87–88
Pyrenophane, as anion sensor, 81
pyrenylguanidinium, as anion sensor, 79
Pyridine
  as photoinduced electron transfer (PET) sensor, 232, 235
  sigma-coupled charge-transfer probe fluorescence in, 255, 257, 258
Pyrophosphatase
  fluorescent sensors for, 68
  photoinduced electron transfer (PET) sensors for, 231
Pyrophosphate, fluorescent sensors for
  anthrylthiourea-based sensors, 71
  fluorophore-copper complex ensemble-based, 93, 94
  quinoxaline-pyrrole-based sensors, 84

Quinone/hydroqinone redox couple-centered redox control units, 192, 193, 194
Quinoxaline-pyridinium, as anion sensor, 83
Quinoxaline-pyrrole, as anion sensor, 84

Reactive nitrogen oxide species, 163–164
Reactive oxygen species, redox-active fluorescent system detection of, 196
Redox-switchable fluorescent devices, 189–218
  analyte recognition function of, 192, 194
  analyte-responsive, 199–211
    for anion detection, 202–203, 208–211
    with bifunctional receptors, 202–204
    charge transfer-based, 199, 202, 203–204
    definition of, 199
    electron transfer-based, 199–200, 202, 203–204
    exciplex/excimer forming-type, 199
    for heavy and transition metals detection, 199–208, 211
    for optical and electrochemical analyte sensing, 199, 200–202
    redox control of, 202–211
    signal generation in, 199–200
    1,2,4-thiadazole/N-iminoyl thiourea-based, 203–208
    1,6,6aλ^4trithia-3,4-diapentadene-based, 203, 204
  applications of, 190
  bistable, reversible OFF/ON, 194
  chromophore emission switching by, 191
  composite, 190–191, 194–196
    alkyl spacers, 190, 191
    anthracene-based, 194, 196
    bistable redox-active control units, 190, 191
    copper-based, 196
    ligand-stabilized redox-active transition metal ion-based, 196
    luminescent signaling modules, 190, 191
    nickel-based, 196
    porphyrin-based, 194–195
    ruthenium complex-based, 194
    tetrathiafulvalene-based, 194–195
    trimetallic complex-based, 196
  Cu^{II}/Cu^{I} redox couple for, 192
  electron transfer (ET) property of, 190–191, 192
    interactions with simple fluorescent redox switches, 194

Redox-switchable fluorescent devices (cont.)
  electron transfer (ET) property of (cont.)
    in redox-active control units, 192, 194
    in tetrathiafulvalene-based systems, 195
  emission properties control by, 190–191
  energy transfer (EnT) property of, 190–191
    interactions with simple fluorescent redox switches, 194
    in redox-active control units, 192
  ferrocene-based, 190
  ferrocene/ferrocene redox couple-based, 195
  ferrocene-naphthalimide-based, 195
  fluorescent signaling of redox-active analytes with, 196–198
    fluorophore-nitroxide (fluorescence-spin molecular)-based, 197–198
    zinc porphyrin-based, 197
  future developments in, 211
  multiple-receptor, 199
  OFF/OFF, 196
  OFF/ON, electron transfer-type, 196
  OFF/ON or ON/OFF, 191
  ON/OFF, 196
  redox-activated optical signal generation with, 194
  redox-activated signal generation by, 190–198
  redox-active control units of, 192–194
    electro-active tetrathiafulvane, 192
    ferrocenium/ferrocene redox couple, 192
    ligand-active stabilized transition metal ions, 192
    quinone/hydroqinone redox couple-centered, 192, 193, 194
    semihydroquinone-centered, 192, 193
  redox-active transition metal ion-based, 190
  signal transduction by, 189
  simple fluorescent redox switches, 193, 194
Rhenium (I)-bipyridyl, as anion sensor, 94–104
  acyclic, 95–98
  calixarene-based, 102–103
  crown ethers-based, 101–102
  with hydrogen-bonding groups, 95
  as ion-pair receptor, 101–102
  macrocyclic, 97, 99–102
Rhenium (I)-bipyridyl-boronic acid, as saccharides sensor, 98
Rhenium (I)-bipyridyl calix[4]arene, as anion sensor, 102, 103
Rhenium (I)-pallidium (II) macrocycle anion sensor, 105

Rhenium-pyridyl cleft anion receptors, 106
Rhodamine, as fluorescence resonance energy transfer (FRET) donor, 131, 133
Rhodamine B, as pressure sensitive paint luminophore, 301
Rhodamine B hydrazide, as nitric oxide sensor, 171–172, 183
Rhodamine 6 G dye, 126
  as fluorescence resonance energy transfer (FRET) donor dye, 131, 133
Ruthenium $(bpy)_{32}+$-dioxo-tetraaamines, as nickel and copper sensors, 11, 12–13
Ruthenium complex-based redox-switchable devices, 194
Ruthenium (II), as anion sensor, 97, 102, 103
Ruthenium (II) calix[4]arene, as anion sensor, 102, 103
Ruthenium (II) tris[2-(2-pyrazinyl)thiazole], structure of, 127
Ruthenium (II) tris[2-(2-pyrazinyl)thiazole] cation-based carbon dioxide optical sensor, 126, 135–137
Ruthenium terpyridine-based anion receptors, 107

Saccharides, fluorescent sensors for
  photoinduced electron transfer (PET) sensors, 230, 235–236, 237
  rhenium (I)-bipyridyl-boronic acid-based, 98
SATI sensors, 176–179
Sensors: see also Chemical sensors; Chemosensors
  definition of, 1, 2
  fluorescent molecular, classification of, 220
Severinghaus electrode, 120–121, 122, 123, 155
Sialic acid, fluorescent sensors for, 87
Sigma Scan Pro (SPSS Inc.), 294
Signal amplification
  of dendrimer-based sensors, 42, 47
  in silica nanoparticle systems, 51
Silica nanoparticles, 5
  as cation sensors, 47–51
    dansylated, 47–48, 50, 51
    fluorescein-modified, 49–50
    signal amplification with, 51
    triethoxysilane-modified, 48–49
Silicone, as pressure sensitive paint component, 285
Silicon octaethyl/porphine (SIOEP), as pressure sensitive paint luminophore, 300, 301

Silver ions
  acyclic receptor-based sensors for, 10, 11
  cyclic receptor-based sensors for, 25, 35, 36
    1-benzothiazol-3-(4-donor)-phenyl-substituted prop-en-1-ones-based, 35, 36
    heteroditopic cryptand, 37, 38
    exciplex formation, 25
cSNAFL, structure of, 127
Sodium
  crown ethers-based sensors for, as electronic tongues, 35
  cyclic receptor-based sensors for
    calixarene derivatives-based, 40, 41
    dibenzo-crown ethers-based, 27
  photoinduced electron transfer (PET) sensors for, 238, 239
Sol-gel-derived silica films, 301
D-Sorbitol, boronic acid-guanidinium ditopic sensor for, 79
Spectroscopy
  infra-red, carbon dioxide measurement applications of, 120
  luminescence, 2
Squaramide-fluoresceine ensemble, as sulfate sensors, 93
Stearylstearate, charge-transfer fluorescence phase transition in, 276
Strontium ions
  carboxylate-based acyclic sensors for, 7
  crown ethers-based sensors for, as electronic tongues, 35
Sucrose octaacetate, 275
Sulfate, dimethylaminobenzoate-guanidinium sensor for, 91
Sulforhodamine, structure of, 140

Terbium complexes
  of cyclen-based anion sensors, 110
  of lanthanide-based anion sensors, 111, 112–113
  of phenanthridium-based anion sensors, 110
Terpyridine, 107
Tetrahydrofuran, sigma-coupled charge-transfer probe fluorescence in, 255, 258, 266
Tetra-N-oxide bipyridine-europium complex, of lanthanide-based anion receptors, 112
Tetraoctyl ammonium hydroxide, as CO_2 optical sensor encapsulating medium, 141, 142, 143, 144, 146

Tetra t-butyl ammonium hydroxide, as dry carbon dioxide optical sensor, 138
Tetrathiafulvalene-based redox-switchable devices, 194–195
Texas Red hydrazine dye, 126
  as fluorescence resonance energy transfer (FRET) donor dye, 131, 133
Thia-anthracene receptors, 10
1,2,4-Thiadiazole, as anion sensor, 62
1,2,4-Thiadiazole-N-iminoyl thiourea redox-switchable receptor system
  for anion detection, 208–211
  for heavy and transition metals detection, 203–208
1,2,4-Thiazole derivatives, as metal ion sensors, 14
Thiourea-based anion sensors, with anthracene signaling units, 70–71
Thiourea-based photoinduced electron transfer (PET) sensors, 237
Thymidine 5′-monophosphate (TMP), anthracene signaling unit-based sensors for, 76
Toluene, Fluoroprobe fluorescence and absorption maxima of, 257
Transitional metal-based nitric oxide sensors, 173
Transition metal ions: *see also specific transition metal ions*
  heteroditopic cryptand-based sensors for, 37–38
  photoinduced electron transfer (PET) sensors for, 6
Transition metals: *see also specific transition metals*
  chemosensors for, 5
  photoinduced electron transfer (PET) sensors for, 232, 233
2,4,6-triamino-1,3,5-trimethoxycyclohexane, cation-binding affinity of, 76
Triazacyclononane, as copper ion sensor, 24–25
1,2-Triazole, 166, 167
Triethoxysilane, as silica nanoparticle modifier, 48–49
Trimetallic complex-based redox-switchable devices, 196
Triphenylacetate, 75, 76
Tripodal ligands, as cation receptors, 16–17
1,6,6aλ⁴Trithia-3,4-diapentadene-based redox-switchable devices, 203, 204

TSQ, 32
TSQ (6-methoxy-(8-p-toluenesulfonamido)-quinoline), 32

Urea-based photoinduced electron transfer sensor, 237
5′-Uridine diphosphate (UDP), fluorescent sensors for, 78
Uridine monophosphate (UMP), anthracene signaling unit-based sensors for, 76
5′-Uridine monophosphate (UMP), fluorescent sensors for, 78
5′-Uridine phosphates (UTP), fluorescent sensors for, 78
Uronic acids, fluorescent sensors for, 87

Vitamin B13 (orotic acid), 76

Wavelength-ratiometric method, for luminescence intensity measurement, 134
Wind tunnel studies, of pressure sensitive paint, 285, 286, 287

Yellow Springs Instrument, 155–156

Zinc, as anion complexation agent
  with anthracene signaling units, 74–78
  cryptand, 74, 75
Zinc finger consensus peptide (CP), 45–46
Zinc finger domains, 43–44
Zinc ions
  acyclic receptor-based sensors for, 20–21
  anthracene-based, 8, 10
  anthracene-based with polyamine chains, 8, 9
  with anthryl group/SCH2CH2CH28 spacers, 10
  aqueous ratiometric, 9, 10
  carboxylates-based, 7
  with dansyl group chromophore, 18–19, 20
  with excimer formation, 10
  1,2,4-thiazole derivative receptor units, 14
  carbonic anhydrase II-based optical sensors for, 46
  consensus peptide-based sensor for, 45–46
  cyclic receptor-based sensors for
    anthracene-based, 21, 22
    carboxylates-based, 7
    coumarin-derivatized crown ethers, 25, 27
    crown ethers-based, 25, 27, 32, 33

Zinc ions (*cont.*)
    cyclic receptor-based sensors for (*cont.*)
        with dansyl groups, 22–23
        heteroditopic cryptand, 37–38
    differentiated from cadmium ions, 22
    fluoresceine platform-based sensors for, 16
    fluorescence anistropy detection of, 46
    luminescence intensity-enhancing property of, 6

Zinc ions (*cont.*)
    photoinduced electron transfer (PET) sensors for, 228–229, 232, 233, 234, 238
        cryptand-based, 236
    polypeptides-based sensors for, 42–46
    tripodal ligands-based sensors for, 16, 17
Zinc porphyrin-based redox-switchable devices, 197
Zinquin, 32